T0271915

Sintering of
Ceramics

Sintering of Ceramics

Mohamed N. Rahaman
University of Missouri-Rolla, USA

CRC Press
Taylor & Francis Group
Boca Raton London New York

CRC Press is an imprint of the
Taylor & Francis Group, an **informa** business

CRC Press
Taylor & Francis Group
6000 Broken Sound Parkway NW, Suite 300
Boca Raton, FL 33487-2742

© 2008 by Taylor & Francis Group, LLC
CRC Press is an imprint of Taylor & Francis Group, an Informa business

International Standard Book Number-13: 978-0-8493-7286-5 (Hardcover)

Library of Congress Cataloging-in-Publication Data

Rahaman, M. N., 1950-
   Sintering of ceramics / Mohamed N. Rahaman.
     p. cm.
   Includes bibliographical references and index.
   ISBN-13: 978-0-8493-7286-5 (alk. paper)
   ISBN-10: 0-8493-7286-0 (alk. paper)

   1. Ceramics--Heat treatment. 2. Firing (Ceramics) 3. Ceramics--Analysis. 4. Sintering. I. Title.
TP814.R34 2007   666'.04--dc22   2007002280

Visit the Taylor & Francis Web site at
http://www.taylorandfrancis.com

and the CRC Press Web site at
http://www.crcpress.com

Contents

Preface

Owing to the generally high melting points of the raw materials, the fabrication of ceramics commonly includes a heat treatment step in which a powder or porous material, already formed into the required shape, is converted to a useful solid. This heat treatment step is referred to as *sintering* or, in the more traditional ceramics industry, as *firing*. Sintering has its origins in the ancient civilizations of thousands of years ago. With the achievement of controlled sintering of metals and ceramics in the early twentieth century, the technological background was established for developing the modern theory and practice of sintering, which experienced rapid growth after the mid-1940s.

The widespread use of sintering for the production of ceramics (and some metals) has led to a variety of approaches to the subject. In practice, the ceramist, wishing to prepare a material with a particular set of properties, identifies the required final microstructure and tries to design processing conditions that will produce this required microstructure. A key objective of sintering studies is, therefore, to understand how the processing variables influence the microstructural evolution, in order to develop a useful framework for designing processing conditions to achieve the required microstructure.

This book emphasizes the principles of sintering and how they are applied to the practical effort of producing materials with the required microstructure. Although the focus is on the sintering behavior of ceramics, the book can also be used for developing a background in the sintering of metals and other materials because of the basic treatment of the principles. My intention has been to prepare a text that is suitable for an introductory graduate-level course in sintering, but the book will also be useful to scientists and engineers in industrial research and development who are involved in the production of materials by sintering or who may wish to develop a background in sintering. Parts of the book will also be useful for introductory graduate courses in kinetic processes in materials and for upper-level undergraduate courses in thermal processes in ceramics.

The arrangement of the book provides a logical development of the principles and practice of sintering. Following a treatment of sintering fundamentals, such as driving forces for sintering, defect chemistry, and diffusion in Chapter 1, sintering and microstructural development of solid-state materials and viscous sintering of amorphous materials are discussed in Chapters 2 and 3. Sintering of the basic material systems is continued in Chapter 4 with a treatment of sintering in the presence of a liquid phase. Some special topics and difficult issues in sintering, such as constrained sintering, solid solution additives, morphological stability of continuous phases, and sintering with concurrent reaction or crystallization, are considered in Chapter 5. The treatment of sintering is concluded in Chapter 6 with a discussion of practical methods used in sintering and the effects of process variables on sintering behavior. The topics are covered in sufficient depth to enable the reader to understand publications in the sintering literature without difficulty, and numerous references to key publications are given at the end of each chapter to assist the reader with further reading.

Several people have been very helpful to me in the writing of this book. While limited space does not allow me the opportunity to name them all, I am greatly indebted to them for their help. I am particularly indebted to Frank L. Riley and Richard J. Brook at the University of Leeds and Lutgard C. De Jonghe at the Lawrence Berkeley National Laboratory, who provided me with the opportunity to learn sintering in their laboratories. I wish to thank the many authors and publishers who have allowed me permission to reproduce their figures in this book. Last but not least, I would like to thank my wife, Vashanti, for her patience and understanding.

Author

Mohamed N. Rahaman is Professor of Ceramics in the Department of Materials Science and Engineering, University of Missouri–Rolla. He received B.A. (Hons) and M.A. degrees from the University of Cambridge, England, and a Ph.D. degree from the University of Sheffield, England. Prior to joining the University of Missouri in 1986, Dr. Rahaman held positions at the University of Leeds, England; the University of the West Indies, Trinidad; and the Lawrence Berkeley National Laboratory, Berkeley, California. Dr. Rahaman is the author of three books and the author or coauthor of more than 135 publications, most of them in the area of processing and sintering of ceramics.

1 Sintering of Ceramics: Fundamentals

1.1 INTRODUCTION

The heat treatment process in which a powder or porous material, already formed into a required shape, is converted to a useful solid is referred to as *sintering*. The sintering process has its origins in the ancient civilizations of thousands of years ago [1]. With the achievement of controlled sintering of metals and ceramics in the early twentieth century, the technological background was established for the development of modern theory and practice of sintering, which experienced rapid growth after the mid-1940s.

This chapter examines some fundamental concepts in sintering. For sintering to occur, there must be a decrease in the free energy of the system. The *curvature* of the free surfaces and, when used, the applied pressure, provide the main motivation or *driving force* for sintering to occur. However, to accomplish the process within a reasonable time, we must also consider the *kinetics* of matter transport. In crystalline ceramics, matter transport occurs predominantly by *diffusion* of atoms, ions, or other charged species. Solid-state diffusion can occur by several paths that define the *mechanisms of diffusion* and, hence, the mechanisms of sintering. The rate of diffusion depends on the type and concentration of *defects* in the solid, so an understanding of the defect structure and the changes in the defect concentration (the defect chemistry) is important. We must also understand how key variables in the sintering process, such as temperature, gaseous atmosphere, and solutes (dopants), control the defect chemistry.

To predict how the rate of sintering depends on the primary processing variables, equations for the flux of matter must be formulated and solved subject to the appropriate boundary conditions. Matter transport can be viewed in terms of the flux of atoms (ions) or, equivalently, in terms of the counterflow of vacancies. Following Fick's laws of diffusion, the flux can be analyzed in terms of the concentration gradient of the diffusing species, but the equations take a more generalized form when expressed in terms of the *chemical potential* (the molar Gibbs free energy). In this view, matter transport occurs from regions of higher chemical potential to regions of lower chemical potential. In inorganic solids, the different ions or charged species diffuse at different rates, but matter transport must take place in such a way that the stoichiometry and electroneutrality of the solid are preserved. The diffusion of the ions is therefore coupled, and this coupled diffusion is referred to as *ambipolar diffusion*.

1.2 THE SINTERING PROCESS

Sintering processes are commonly divided into several categories, depending on the type of system. One category is that for a pure, single-phase, polycrystalline material, such as α-Al_2O_3. Sintering is achieved by heating the consolidated mass of particles, referred to as the green body or powder compact, to a temperature that is in the range between approximately 50% and 80% of the melting temperature. For Al_2O_3 with a melting temperature of 2073°C, the sintering temperature is commonly between 1400°C and 1650°C. The powder does not melt; instead, the joining together of the particles and the reduction in the porosity (i.e., densification) of the body, as required in the fabrication process, occur by atomic diffusion in the solid state. This type of sintering is referred to as *solid-state sintering*. For many polycrystalline ceramics, the required density or microstructure of the final article is difficult to achieve by solid-state sintering, or the solid-state sintering temperature

TABLE 1.1
Sintering Processes for Some Ceramic Compositions

Composition	Sintering Process	Application
Al_2O_3	Solid-state sintering with MgO additive	Sodium vapor arc lamp tubes
	Liquid-phase sintering with a silicate glass	Furnace tubes; refractories
MgO	Liquid-phase sintering with a silicate glass	Refractories
Si_3N_4	Liquid-phase sintering with oxide additives (e.g., Al_2O_3 and Y_2O_3) under nitrogen gas pressure or under an externally applied pressure	High-temperature structural ceramics
SiC	Solid-state sintering with B and C additives; liquid-phase sintering with Al, B and C, or oxide additives	High-temperature structural ceramics
ZnO	Liquid-phase sintering with Bi_2O_3 and other oxide additives	Electrical varistors
$BaTiO_3$	Liquid-phase sintering with TiO_2-rich liquid	Capacitor dielectrics; thermistors
Pb (Zr,Ti)O_3 (PZT) (Pb,La)(Zr,Ti)O_3 (PLZT)	Sintering with a lead-rich liquid phase; hot pressing	Piezoelectric actuators and electro-optic devices
ZrO_2/(3–10 mol% Y_2O_3)	Solid-state sintering	Electrical-conducting oxide for fuel cells
Mn-Zn and Ni-Zn ferrites	Solid-state sintering under a controlled oxygen atmosphere	Soft ferrites for magnetic applications
Porcelain	Vitrification	Electrical insulators; tableware
SiO_2 gel	Viscous sintering	Optical devices

is too high for the intended fabrication process. One solution is the use of an additive that forms a small amount of liquid phase between the particles or grains at the sintering temperature. This method is referred to as *liquid-phase sintering*. The liquid phase provides a high-diffusivity path for transport of matter into the pores to produce densification, but it is insufficient, by itself, to fill up the porosity. Another solution to the difficulty of inadequate densification is the application of an external pressure to the body during heating in either solid-state or liquid-phase sintering. This method is referred to as *pressure-assisted sintering* or *pressure sintering*; hot pressing and hot isostatic pressing are well-known examples. *Viscous sintering* refers to the sintering process in amorphous materials, such as glasses, in which matter transport occurs predominantly by viscous flow. In clay-based ceramics, such as porcelains, sintering is often achieved by the formation of a large-volume fraction of a liquid phase that fills up the pores. This type of sintering is referred to as *vitrification*. The liquid phases are molten silicates, which remain as glass after cooling. This gives the ceramic ware a glassy appearance, and such ceramics are referred to as vitrified. Table 1.1 provides a list of some ceramic compositions and the methods by which they are sintered.

The heating schedule in the sintering of industrial ceramics can have several temperature–time stages, but in laboratory-scale studies, a simple schedule involving *isothermal sintering* or *constant heating rate sintering* is often used. In isothermal sintering, the temperature is increased rapidly to a fixed (isothermal) sintering temperature, maintained at this temperature for the required time, and finally lowered to room temperature. Sintering models commonly assume idealized isothermal conditions. In constant heating rate sintering, the sample is heated at a constant rate to a specified temperature and then cooled to room temperature.

1.2.1 CHARACTERIZATION OF SINTERING

While several measurements can be used to characterize the sintering and microstructural evolution of a compacted mass of particles, by far the most widely used measurements are the density (or shrinkage) and the grain size. Methods for measuring density, shrinkage, and grain size are described in Chapter 6. The progress of sintering is often determined from the density or the linear shrinkage of the powder compact as a function of time or temperature during the heat treatment. The *bulk density* is defined as the mass divided by the external volume of the body. A better parameter is the *relative density* ρ, defined as the bulk density divided by the theoretical density of the solid. Relative density and porosity P of a body are related by

$$\rho = 1 - P \tag{1.1}$$

The linear shrinkage is defined as $\Delta L/L_o$, where L_o is the original length, L is the length at a given time or temperature, and $\Delta L = L - L_o$. If the shrinkage is isotropic, the relationship between the relative density and the shrinkage is

$$\rho = \frac{\rho_o}{(1 + \Delta L/L_o)^3} \tag{1.2}$$

where ρ_o is the initial relative density and $\Delta L/L_o$ is negative as defined. Measurements of ρ or $\Delta L/L_o$ are easy to perform and provide substantial information about the rate of sintering. It is often required to determine the *densification rate*, defined as $(1/\rho)(d\rho/dt)$, where t is time, which is equivalent to a volumetric strain rate. The average grain size and its standard deviation are often the microstructural parameters of most interest, but a distribution of the grain sizes can also be shown, if necessary, as a histogram or a continuous function. The mean grain intercept length determined from two-dimensional sections is conventionally taken as a measure of the average grain size. Measurement of the average pore size and the pore size distribution of the partially sintered samples using mercury porosimetry provides additional information about the character-istics of the open pores and the homogeneity of the microstructure.

1.2.2 APPROACH TO SINTERING

The widespread use of sintering for the production of ceramics and some metals has led to a variety of approaches to the subject. In practice, the ceramist, wishing to prepare a material with a particular set of properties, identifies the required final microstructure and tries to design processing conditions that will produce this required microstructure. The key objective of sintering studies is therefore to understand how the processing variables influence the microstructural evolution. In this way, useful information can be provided for the practical effort of designing processing conditions for producing the required microstructure.

   One approach to developing an understanding of sintering is to connect the behavior or changes in behavior during sintering to controllable variables and processes. This can be achieved, on one hand, empirically by measuring the sintering behavior under a set of controlled conditions, and, on the other hand, theoretically by modeling the process. The theoretical analyses and experimental studies performed over the last 50 years or so have produced an excellent qualitative understanding of sintering in terms of the driving forces, the mechanisms, and the influence of the principal processing variables such as particle size, temperature, and applied pressure. However, the data base and models are far less successful at providing a quantitative description of sintering for most systems of interest. For this shortcoming, the sintering models have received some criticism.

   Table 1.2 lists some of the important parameters in sintering that may serve to illustrate the scope of the problem [2]. In general, the processing and material parameters provide a useful set

TABLE 1.2
Some Important Parameters in the Sintering of Ceramics

Behavior	**Processing and Material Parameters**
General morphology	Powder preparation: particle size, shape, and size distribution
Pore evolution: size, shape, interpore distance	Distribution of dopants or second phases
Grain evolution: size and shape	Powder consolidation: density and pore size distribution
Density: function of time and temperature	Sintering: heating rate and temperature
Grain size: function of time and temperature	Applied pressure
Dopant effects on densification and grain growth	Gaseous atmosphere

Models	**Characterization Measurements**
Neck growth	Neck growth
Surface area change	Shrinkage, density, and densification rate
Shrinkage	Surface area change
Densification in the later stages	Grain size, pore size, and interpore distance
Grain growth: porous and dense systems, solute drag, pore drag, pore breakaway	Dopant distribution
Concurrent densification and grain growth	Strength, conductivity, and other microstructure-dependent properties

Data Base

Diffusion coefficients: anion and cation, lattice,
  grain boundary and surface
Surface and interfacial energies
Vapor pressure of components
Gas solubilities and diffusivities
Solute diffusivities
Phase equilibria

Source: Coble, R.L., and Cannon, R.M., Current paradigms in powder processing, in *Processing of Crystalline Ceramics,* Mater. Sci. Res., Vol. 11, Palmour, H. III, Davis, R.F., and Hare, T.M., Eds., Plenum Press, New York, 1978, p. 151.

of variables for model experimental and theoretical studies. Some parameters, such as the sintering temperature, applied pressure, average particle size, and gaseous atmosphere, can be controlled with sufficient accuracy. Others, such as the powder characteristics and particle packing, are more difficult to control but have a significant effect on sintering. While partial information exists in the other areas of behavior, characterization measurements, and the data base, much critically needed information is severely lacking. This is especially serious in the data base for the fundamental parameters, such as the surface and grain boundary energies and the diffusion coefficients. This lack of information, coupled with the complexity of practical ceramic systems, makes quantitative predictions of sintering behavior very difficult, even for the simplest systems.

1.3 DRIVING FORCE FOR SINTERING

As with all other irreversible processes, sintering is accompanied by a lowering of the free energy of the system. The sources that give rise to this lowering of the free energy are commonly referred to as the *driving forces for sintering*. Three possible driving forces are: the curvature of the particle surfaces, an externally applied pressure, and a chemical reaction.

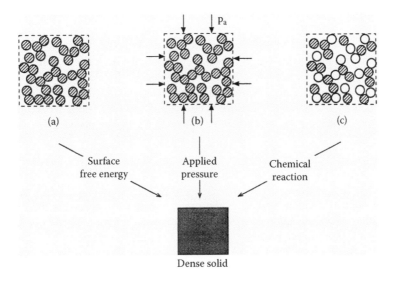

FIGURE 1.1 Schematic diagram illustrating three main driving forces for sintering: surface free energy, applied pressure, and chemical reaction.

1.3.1 SURFACE CURVATURE

In the absence of an external stress and a chemical reaction, surface curvature provides the driving force for sintering (Figure 1.1a). To see why this is so, let us consider, for example, one mole of powder consisting of spherical particles with a radius a. The number of particles is

$$N = \frac{3M}{4\pi a^3 \rho} = \frac{3V_m}{4\pi a^3} \tag{1.3}$$

where ρ is the density of the particles, which are assumed to contain no internal porosity, M is the molecular weight, and V_m is the molar volume. The surface area of the system of particles is

$$S_A = 4\pi a^2 N = 3V_m/a \tag{1.4}$$

If γ_{SV} is the specific surface energy (i.e., the surface energy per unit area) of the particles, then the surface free energy associated with the system of particles is

$$E_S = 3\gamma_{SV} V_m/a \tag{1.5}$$

E_S represents the decrease in surface free energy of the system if a fully dense body were to be formed from the mole of particles and provides a motivation for sintering. For Al_2O_3 particles, taking $\gamma_{SV} \approx 1$ J/m^2, $a \approx 1$ μm, and $V_m = 25.6 \times 10^{-6}$ m^3, then $E_S \approx 75$ J/mol. The decrease in free energy given by Equation 1.5 assumes that the dense solid contains no grain boundaries, which would be the case for a glass. As discussed in Chapter 2, for polycrystalline materials, the grain boundaries play a role that must be considered in determining the magnitude of the driving force. The decrease in surface free energy is accompanied by an increase in energy associated with the boundaries.

1.3.2 APPLIED PRESSURE

In the absence of a chemical reaction, an externally applied pressure normally provides the major contribution to the driving force when the pressure is applied over a significant part of the heating process as in hot pressing and hot isostatic pressing (Figure 1.1b). Surface curvature also provides a contribution to the driving force, but for most practical situations it is normally much smaller than that provided by the external pressure. The external pressure does work on the system of particles and, for 1 mole of particles, the work done can be approximated by

$$W = p_a V_m \tag{1.6}$$

where p_a is the applied pressure and V_m is the molar volume. W represents the driving force for densification due to the application of an external pressure. For $p_a = 30$ MPa, which is a typical value of the stress applied in hot pressing, and for Al_2O_3 with $V_m = 25.6 \times 10^{-6}$ m^3, then $W = 750$ J.

1.3.3 CHEMICAL REACTION

A chemical reaction can, in principle, provide a driving force for sintering if it can be used to aid the densification process (Figure 1.1c). The change in free energy accompanying a chemical reaction is

$$\Delta G^o = - RT \ln K_{eq} \tag{1.7}$$

where R is the gas constant (8.314 J K^{-1} mol^{-1}), T is the absolute temperature, and K_{eq} is the equilibrium constant for the reaction. Taking $T = 1000$ K and $K_{eq} = 10$, then $\Delta G^o \approx 20,000$ J/mol. This decrease in energy is significantly greater than the driving force due to an applied stress. In practice, a chemical reaction is hardly ever used deliberately to drive the densification process in advanced ceramics because microstructure control is difficult when a chemical reaction occurs concurrently with sintering (see Chapter 5).

1.4 DEFECTS IN CRYSTALLINE SOLIDS

The driving forces provide a motivation for sintering, but the actual occurrence of sintering requires transport of matter, which in crystalline solids occurs by a process of diffusion involving atoms, ions, or molecules. Crystalline solids are not ideal in structure. At any temperature they contain various imperfections called *defects*. It is the presence of these defects that allows diffusional mass transport to take place. Because defects control the rate at which matter is transported, they control the rates of processes such as sintering, grain growth, and creep.

Defects in crystalline solids occur for structural reasons, because the atoms (or ions) are not arranged ideally in the crystal when all the lattice sites are occupied, and for chemical reasons, because inorganic compounds may deviate from the fixed composition determined by the valence of the atoms. There are different types of structural defects in a crystalline solid, which are normally classified into three groups: *point* defects, *line* defects, and *planar* defects. Point defects are associated with one lattice point and its immediate vicinity. They include missing atoms or *vacancies*, *interstitial* atoms occupying the interstices between atoms, and *substitutional* atoms sitting on sites that would normally be occupied by another type of atom. These point defects are illustrated in Figure 1.2 for an elemental solid (e.g., a pure metal). The point defects that are formed in pure crystals (i.e., vacancies and interstitials) are sometimes referred to as intrinsic or native defects.

Most of our discussion will be confined to point defects in ionic solids (ceramics). Line defects, commonly referred to as *dislocations*, are characterized by displacements in the periodic structure

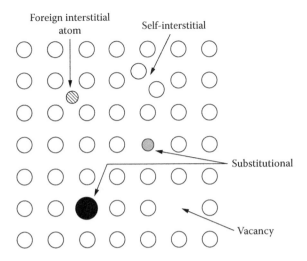

FIGURE 1.2 Point defects in an elemental solid.

of the lattice in certain directions. They play their most important role in the plastic deformation of metals. Planar defects include free surfaces, grain boundaries, stacking faults, and crystallographic shear planes.

1.4.1 POINT DEFECTS AND DEFECT CHEMISTRY

A key distinction between point defects in ceramics and those in metals is that in ceramics, the defects can be electrically charged. For a pure ionic compound with the stoichiometric formula MO consisting of a metal M (with a valence of +2) and oxygen O (valence –2), the types of point defects that may occur are vacancies and interstitials of both M and O. These *ionic* defects may be either charged or neutral. In addition to the single defects, it is possible for two or more defects to associate with one another, leading to the formation of defect clusters. There may also be *electronic* (or valence) defects consisting of quasi-free electrons or holes (missing electrons). If the compound contains a small amount of solute (or impurity) atoms M_f, substitutional or interstitial defects of M_f will also occur, and these may be either charged or neutral.

Another important distinction between ceramic and metallic systems is that the composition of the ceramic may become nonstoichiometric by annealing in a suitable gaseous atmosphere (e.g., a controlled oxygen partial pressure). The compound seeks to equilibrate itself with the partial pressure of one of its components in the surrounding atmosphere. This equilibration leads to a change in the composition and a change in the type and concentration of the defects. For example, annealing the compound MO in an atmosphere with a low oxygen partial pressure may lead to an oxygen-deficient oxide, MO_{1-x}, in which oxygen vacancies predominate. On the other hand, annealing in an atmosphere with a higher oxygen partial pressure may lead to a metal-deficient oxide, $M_{1-y}O$, in which metal vacancies predominate.

Charged defects in solids can interact with one another in a way analogous to the interactions between ions (or between ions and electrons) in a solution. In the solid-state situation, the crystal may be viewed as a neutral medium into which the charged defects are dissolved. This similarity between solution chemical interactions and defect interactions in the solid state has resulted in the field of defect chemistry, which provides basic methods for studying the effects of point defects in solids. The methods are normally applicable to low defect concentrations. Generally, a broad distinction is made between *intrinsic defects,* those that are thermally generated in pure compounds, and *extrinsic defects,* those produced by external influences, such as impurities and gaseous atmospheres. Kroger [3] and Kofstad [4] provide a detailed discussion of point defects and defect chemistry in metal oxides.

1.4.2 KROGER–VINK NOTATION

A standard notation, referred to as the *Kroger–Vink notation*, is used for the description of point defects in ionic solids. In this notation, the defect is defined with respect to the perfect lattice and is described by three parts: the main symbol, a subscript, and a superscript. For example, in the notation M_L^C, the main symbol M denotes the particular type of atom; in the case of a vacancy, the main symbol is V. The subscript L denotes the site in the perfect lattice where the defect is located. The superscript C gives the *effective charge* (or relative charge) of the defect, equal to the difference in valence between the species on the L site and the valence of the atom that occupies the L site in the perfect lattice. The effective charge is represented as follows:

Positive effective charge: C = •
Negative effective charge: C = /
Neutral effective charge: C = ×

Electronic defects are specified as follows: a quasifree electron is represented as e', while a missing electron or hole is represented as $h^•$. Table 1.3 illustrates the use of the Kroger–Vink notation for some possible defects in Al_2O_3. The *concentration* of defects is denoted by square brackets, for example $[V_O^{••}]$ and $[Ti_{Al}^•]$. The concentration of electrons or holes, $[e']$ or $[h^•]$, is commonly written n or p, respectively.

1.4.3 DEFECT REACTIONS

The formation of defects can be viewed as a chemical reaction and, by this analogy, we can write defect reactions in a way similar to chemical reactions once the following three conservation rules are observed:

1. *Conservation of mass*: A mass balance must be maintained so that mass is neither created nor destroyed in the defect reaction. Vacancies have zero mass, whereas electronic defects are considered to have no effect on the mass balance.
2. *Electroneutrality*: The crystal must remain electrically neutral. This means that for the overall reaction, the sum of the positive effective charges must be equal to the sum of the negative effective charges.
3. *Site ratio conservation*: The ratio of the number of regular cation sites to the number of regular anion sites in the crystal remains constant. For example, in the compound MO, the ratio of the regular M and O sites must remain 1:1. Sites may be created or destroyed in the defect reaction, but they must occur in such a way that the site ratio in the regular lattice is maintained.

To see how these rules apply, let us consider the incorporation of MgO into Al_2O_3. Based on the similarity of the Mg^{2+} and Al^{3+} ionic radii, the Mg ions may enter the solid solution *substitutionally*.

TABLE 1.3
Kroger–Vink Notation for Some Possible Defects in Al_2O_3

$Al_i^{•••}$	Aluminum ion in the interstitial lattice site
$V_O^{••}$	Oxygen vacancy
Mg_{Al}'	Magnesium dopant on the normal Al lattice site
$Ti_{Al}^•$	Ti dopant on the normal Al lattice site
e'	Quasifree electron
$h^•$	Missing electron or hole

In the corundum structure, one-third of the octahedral sites between the close-packed O ions are vacant, so it is possible that the Mg ions could also enter the solid solution *interstitially*. It is not clear which incorporation reaction has the lower energy. In Al_2O_3, there are two cation sites to every three anion sites. Considering the substitutional process, if we incorporate two Mg atoms on cation sites, we must use two Al sites as well as two O sites. Since we have only two O sites, we can tentatively assume that the third O site for site conservation may be vacant. At this stage, on the basis of mass and site balance, we may write

$$2MgO \xrightarrow{\;Al_2O_3\;} 2Mg_{Al} + 2O_O + V_O \tag{1.8}$$

Assuming that the defects are fully ionized, which is believed to be a more realistic solid solution process, conservation of electroneutrality gives

$$2MgO \xrightarrow{\;Al_2O_3\;} 2Mg'_{Al} + 2O_O^x + V_O^{\bullet\bullet} \tag{1.9}$$

and we have mass, charge, and site ratio balance. Another possibility is the incorporation of the Mg ions interstitially, for which we may write

$$3MgO \xrightarrow{\;Al_2O_3\;} 3Mg_i^{\bullet\bullet} + 3O_O^x + 2V_{Al}''' \tag{1.10}$$

A third possibility is that the Mg is self-compensating and forms both substitutional and interstitial defects:

$$3MgO \xrightarrow{\;Al_2O_3\;} 2Mg'_{Al} + Mg_i^{\bullet\bullet} + 3O_O^x \tag{1.11}$$

The reader will verify the conservation of mass, electroneutrality, and site ratio in Equation 1.9 to Equation 1.11.

1.4.4 DEFECT CONCENTRATION

The concentration of defects can be determined through a statistical thermodynamics approach. We shall first consider the concentration of defects in a pure monatomic solid, such as a metal, which can contain vacancies or interstitials in only a single electronic state. For a crystal with N atoms and n vacancies (or interstitials), if the vacancies can move through the crystal, then the Gibbs free energy function can be written

$$G(T,p) = G_o(T,p) + n\Delta g_f - T\Delta S_c \tag{1.12}$$

where $G_o(T,p)$ is the Gibbs free energy function of a perfect crystal of N atoms subject to a uniform pressure p at a temperature T, Δg_f is the Gibbs free energy of formation of a vacancy, and ΔS_c is the increase in the configurational entropy of the crystal, given by the Boltzmann equation

$$\Delta S_c = k \ln \Omega = k \ln\left[\frac{(N+n)!}{N!n!}\right] \tag{1.13}$$

where k is the Boltzmann constant and Ω is the number of configurations in which the vacancies can be arranged. Substituting for ΔS_c from Equation 1.13 into Equation 1.12 gives

$$G(T,p) = G_o(T,p) + n\Delta g_f - kT \ln\left[\frac{(N+n)!}{N!n!}\right] \qquad (1.14)$$

The concentration of vacancies present in a crystal in thermal equilibrium at a given T and p will be such that G is a minimum, i.e.,

$$\left(\frac{\partial G}{\partial n}\right)_{T,p,N} = 0 \qquad (1.15)$$

For dilute concentration of vacancies, such that $(N + n) \approx N$, Equation 1.14 then gives

$$\Delta g_f + kT \ln\left(\frac{n}{N+n}\right) = 0 \qquad (1.16)$$

The concentration of vacancies, C_v, expressed as a fraction of the total number of lattice sites is given by

$$C_v = \frac{n}{N+n} = \exp\left(-\frac{\Delta g_f}{kT}\right) \qquad (1.17)$$

According to Equation 1.17, the defect concentration is exponentially dependent on Δg_f and temperature. Equation 1.17 can also be written as

$$C_v = \exp\left(-\frac{\Delta g_f}{kT}\right) = \exp\left(\frac{\Delta s}{k}\right)\exp\left(-\frac{\Delta h}{kT}\right) \qquad (1.18)$$

where Δs and Δh are the entropy and enthalpy of formation, respectively, of a defect. The term Δs is not the configurational entropy, which was already taken into account in the derivation of Equation 1.17, but is the nonconfigurational entropy associated with vibration frequencies accompanying the defect. It is often assumed that Δs is much smaller than the configurational entropy, such that $\exp(\Delta s/k) \approx 1$, so

$$C_v \approx \exp\left(-\frac{\Delta h}{kT}\right) \qquad (1.19)$$

An equivalent way is to view the formation of defects as a chemical reaction in which there is an equilibrium constant governed by the law of mass action. For a general reaction in which the reactants A and B lead to products C and D:

$$a\mathrm{A} + b\mathrm{B} \rightarrow c\mathrm{C} + d\mathrm{D} \qquad (1.20)$$

where a, b, c, and d are the number of atoms of the reactants and products. At equilibrium at a fixed temperature, the law of mass action applies, and assuming that the activities of the reactants and products are equal to their concentrations, the law gives:

$$K = \frac{[C]^c [D]^d}{[A]^a [B]^b} \tag{1.21}$$

where the brackets denote the concentrations and K, the *equilibrium constant*, is given by the Arrhenius equation

$$K = \exp\left(\frac{-\Delta g}{kT}\right) \tag{1.22}$$

In this equation, Δg is the free energy change for the reaction, k is the Boltzmann constant, and T is the absolute temperature. For defect reactions, the use of Equation 1.21 and Equation 1.22 requires that the concentration of the defects be expressed as site *fractions* (the number of defects divided by the total number of lattice sites). For low defect concentrations, the concentration of atoms or ions in their regular lattice sites is assigned a value of unity. For reactions involving gaseous species, the concentration of a gas is taken as its partial pressure.

1.4.5 INTRINSIC DEFECTS

Two of the more common types of intrinsic defects in ionic crystals are the *Schottky defect* and the *Frenkel defect*.

1.4.5.1 Schottky Defect

The formation of a Schottky defect can be viewed as the transfer of a cation and an anion from their regular lattice sites to an external surface, thereby creating extra perfect crystal and leaving behind vacancies (Figure 1.3a). To preserve electroneutrality of the crystal, vacancies must be formed in the stoichiometric ratio. For the compound MO and assuming that the defects are fully ionized, the defect formation reaction can be written

$$M_M^x + O_O^x \rightarrow V_M'' + V_O^{\bullet\bullet} + M_M^x + O_O^x \tag{1.23}$$

In this equation, M_M^x and O_O^x on both sides of the equation can be cancelled, so the net reaction is written

$$\text{null} \rightarrow V_M'' + V_O^{\bullet\bullet} \tag{1.24}$$

where null (also written nil or zero) denotes the creation of defects from a perfect lattice. Assuming that the reaction has reached equilibrium, applying the law of mass action gives

$$K_S = \left[V_M''\right]\left[V_O^{\bullet\bullet}\right] = \exp\left(\frac{-\Delta g_S}{kT}\right) \tag{1.25}$$

(a)

(b)

FIGURE 1.3 Schematic diagrams for the formation of (a) a Schottky defect and (b) a Frenkel defect.

where K_S is the equilibrium constant and Δg_S is the free energy change for the creation of Schottky defects as defined by Equation 1.24. For electroneutrality we must have $[M_M^{//}]=[V_O^{\bullet\bullet}]$; therefore,

$$\left[V_M^{//}\right]=\left[V_O^{//}\right]=\exp\left(\frac{-\Delta g_S}{2kT}\right) \tag{1.26}$$

Equation 1.26 is the same as Equation 1.17, except for the factor of 2 in Equation 1.26, which results from the formation of charged defects in pairs.

1.4.5.2 Frenkel Defect

A Frenkel defect is formed when an ion leaves its regular lattice site and occupies an interstitial site, leaving behind a vacant site (Figure 1.3b). For the compound MO and assuming fully ionized defects, the creation of a Frenkel defect at the cation site can be written

$$M_M^x \rightarrow M_i^{\bullet\bullet}+V_M^{//} \tag{1.27}$$

The equilibrium constant for the reaction is

$$K_F = [M_i^{\bullet\bullet}][V_M^{//}] = \exp\left(\frac{-\Delta g_F}{kT}\right) \qquad (1.28)$$

By invoking the electroneutrality condition $[M_i^{\bullet\bullet}] = [V_M^{//}]$ we obtain

$$\left[M_i^{\bullet\bullet}\right] = \left[V_M^{//}\right] = \exp\left(\frac{-\Delta g_F}{2kT}\right) \qquad (1.29)$$

Corresponding equations can be written for the formation of Frenkel defects on the anion sites. It should be noted that in their formation cation and anion Frenkel defects are not linked through an electroneutrality condition, so the cation interstitial concentration need not be equal to the anion interstitial concentration.

Table 1.4 lists the enthalpy of formation Δh for the most prevalent (lowest energy) intrinsic defects in several ionic systems.

1.4.6 Extrinsic Defects

Extrinsic defects are caused by external influences, such as the gaseous atmosphere (which may lead to nonstoichiometry) and solutes (or dopants). In this section we also discuss the Brouwer diagram, a way of describing changes in defect concentration.

TABLE 1.4
Values of the Energy of Formation,
Δh, **of Defects in Some Ceramics**

Compound	Defect Reaction	Δh (eV)
LiF	null $\to V_{Li}^{/} + V_F^{\bullet}$	2.4–2.7
LiCl	null $\to V_{Li}^{/} + V_{Cl}^{\bullet}$	2.1
NaCl	null $\to V_{Na}^{/} + V_{Cl}^{\bullet}$	2.2–2.4
KCl	null $\to V_K^{/} + V_{Cl}^{\bullet}$	2.6
AgCl	$Ag_{Ag}^x \to Ag_i^{\bullet} + V_{Ag}^{/}$	1.1
CaF$_2$	$F_F^x \to V_F^{\bullet} + F_i^{/}$	2.3–2.8
	$Ca_{Ca}^x \to V_{Ca}^{//} + Ca_i^{\bullet\bullet}$	~7
	null $\to V_{Ca}^{//} + 2V_F^{\bullet}$	~5.5
Li$_2$O	$Li_{Li}^x \to Li_i^{\bullet} + V_{Li}^{/}$	2.3
MgO	null $\to V_{Mg}^{//} + V_O^{\bullet\bullet}$	7.7
α-Al$_2$O$_3$	null $\to 2V_{Al}^{///} + 3V_O^{\bullet\bullet}$	20.1–25.7
	$Al_{Al}^x \to Al_i^{\bullet\bullet\bullet} + V_{Al}^{///}$	10.4–14.2
	$O_O^x \to O_i^{//} + V_O^{\bullet\bullet}$	7.6–14.5
TiO$_2$ (rutile)	null $\to V_{Ti}^{////} + 2V_O^{\bullet\bullet}$	5.2
	$O_O^x \to O_i^{//} + V_O^{\bullet\bullet}$	8.7
	$Ti_{Ti}^x \to Ti_i^{\bullet\bullet\bullet\bullet} + V_{Ti}^{////}$	12

1.4.6.1 Nonstoichiometry

Equilibration of ionic solids (e.g., an oxide MO) with an ambient gas that is also a constituent of the solid (e.g., O_2) can have significant effects on the defect structure. Oxides of elements tending to show fixed valency (e.g., MgO, Al_2O_3, and ZrO_2) have negligible deviation from stoichiometry, whereas those elements with variable valence (e.g., NiO, CoO, FeO, and TiO_2) can display a large departure from stoichiometry.

At any fixed temperature and composition, the oxide must be in equilibrium with a specific oxygen partial pressure. If it is not, the crystal will give up or take up oxygen until equilibrium has been reached. When the oxide gives up oxygen, in a *reduction* reaction, it will do so by creating *oxygen vacancies* in the lattice, with the electrons that were associated with the O^{2-} ions in the reaction being liberated within the solid. The overall reaction can be written as

$$O_O^x \rightarrow \frac{1}{2}O_2(g) + V_O^{\bullet\bullet} + 2e' \tag{1.30}$$

The creation of the vacant oxygen sites leads to a change in the cation to anion ratio, i.e., to *nonstoichiometry*. The equilibrium constant for the reaction is

$$K_R = n^2 \left[V_O^{\bullet\bullet}\right] p_{O_2}^{1/2} = K_R^o \exp\left(\frac{-\Delta g_R}{kT}\right) \tag{1.31}$$

where the concentration of the oxygen gas is taken as its partial pressure, p_{O_2}, K_R^o is a constant, and Δg_R is the free energy change for the reduction. For electroneutrality, $n = 2[V_O^{\bullet\bullet}]$ and substituting in Equation 1.31 gives

$$\left[V_O^{\bullet\bullet}\right] = (K_R/4)^{1/3} p_{O_2}^{-1/6} \tag{1.32}$$

When the oxide takes up oxygen, the *oxidation* reaction can be written as the consumption of oxygen vacancies, with the charge neutralized by combining with an electron, thus creating a missing electron or *hole* in the valence band:

$$\frac{1}{2}O_2(g) + V_O^{\bullet\bullet} \rightarrow O_O^x + 2h^{\bullet} \tag{1.33}$$

The equilibrium constant is

$$K_O = \frac{p^2}{[V_O^{\bullet\bullet}]p_{O_2}^{1/2}} = K_O' \exp\left(\frac{-\Delta g_O}{kT}\right) \tag{1.34}$$

where K_O' is a constant, and Δg_O is the free energy change for the oxidation.

It is important to note that since oxidation and reduction are the same thermodynamic process simply reversed, the reactions written to describe them are not independent. For example, taking the reduction reaction (Equation 1.30) in combination with the intrinsic electronic defect equilibrium

$$null \rightarrow e' + h^{\bullet} \tag{1.35}$$

yields the oxidation reaction (Equation 1.33). In fact, there are a number of ways of writing the oxidation and reduction reactions, which we may choose for convenience to show the formation or removal of particular defects. For example, oxidation can also be viewed as the buildup of the oxygen lattice and the creation of cation vacancies that take their charge by combining with an electron, thus creating a hole in the valence band. For an oxide MO, the reaction can be written as

$$\frac{1}{2}O_2(g) \rightarrow O_O^x + V_M^{//} + 2h^\bullet \tag{1.36}$$

Equation 1.36 can be obtained by adding the Schottky reaction to Equation 1.33.

1.4.6.2 Influence of Solutes

The term *solute* or *dopant* refers to a compound that is incorporated into a solid solution to modify the sintering rate, microstructure, or properties of the host ceramic. Although solutes may be present in concentrations as low as a fraction of a mole percent, they often have a significant influence on sintering and microstructure development. An *aliovalent* dopant is a solute in which the cation valence is different from that of the host cation, whereas a solute in which the cation valence is the same as that of the host is sometimes called an *isovalent* dopant. For aliovalent dopants, when the valence of the solute cation is *greater* than that of the host cation, the solute is referred to as a *donor* dopant. On the other hand, an *acceptor* dopant refers to a solute in which the cation valence is smaller than that of the host cation. For example, TiO_2 and MgO are donor and acceptor dopants, respectively, for Al_2O_3 (the host).

Let us now consider the incorporation of Al_2O_3 solute into MgO. Based on the similarity in ionic radii, we may assume that the Al will substitute for Mg, with charge compensation achieved by the simultaneous creation of a vacant Mg site:

$$Al_2O_3 \xrightarrow{\ \ MgO\ \ } 2Al_{Mg}^\bullet + V_{Mg}^{//} + 3O_O^x \tag{1.37}$$

As the incorporation reaction proceeds, the other defect equilibria, such as the creation of Schottky or Frenkel defects, are still present. Supposing that the intrinsic defects in MgO consist of Schottky disorder, then, following Equation 1.25, we can write

$$\left[V_O^{\bullet\bullet}\right]\left[V_{Mg}^{//}\right] = K_S \tag{1.38}$$

The electroneutrality condition is

$$\left[Al_{Mg}^\bullet\right] + 2\left[V_O^{\bullet\bullet}\right] = 2\left[V_{Mg}^{//}\right] \tag{1.39}$$

For very low Al_2O_3 concentration $[Al_{Mg}^\bullet] \ll [V_{Mg}^{//}]$, so

$$\left[V_O^{\bullet\bullet}\right] = \left[V_{Mg}^{//}\right] = K_S^{1/2} \tag{1.40}$$

indicating that the concentration of the intrinsic defects is independent of the Al_2O_3 concentration. As the incorporation reaction described by Equation 1.37 proceeds, the concentration of the Al in solid solution increases and the extrinsic defects begin to dominate. The electroneutrality condition now becomes

$$\left[Al_{Mg}^{\bullet}\right] = 2\left[V_{Mg}^{//}\right] \qquad (1.41)$$

Assuming that the Al_2O_3 has been completely incorporated in the reaction, the concentration of Al in solid solution is equal to the total atomic concentration of Al, so

$$\left[Al_{Mg}^{\bullet}\right] = [Al] \qquad (1.42)$$

Since Equation 1.38 must apply for the cation and anion vacancies, combination of Equation 1.38 with Equation 1.41 and Equation 1.42 yields

$$\left[V_{O}^{\bullet\bullet}\right] = \frac{2K_S}{[Al]} \qquad (1.43)$$

Equation 1.41 and Equation 1.42 also give

$$\left[V_{Mg}^{//}\right] = \frac{[Al]}{2} \qquad (1.44)$$

1.4.6.3 Brouwer Diagram

The changes in defect concentration as a function of temperature, oxygen partial pressure, or dopant concentration are sometimes described semiquantitatively in terms of a double logarithmic plot known as a *Brouwer diagram*. Figure 1.4 shows a Brouwer diagram for the effect of Al_2O_3 dopant on the defect chemistry of MgO that was considered above. Similar principles will apply for the variation of the defect concentration with temperature or oxygen partial pressure.

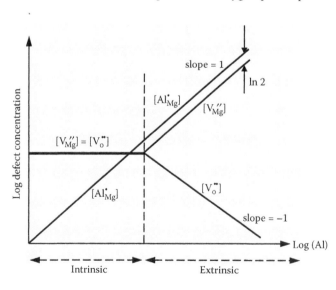

FIGURE 1.4 Brouwer diagram for MgO doped with Al_2O_3, showing the defect concentrations as a function of the concentration of Al.

As illustrated for the Al$_2$O$_3$-doped MgO system, the Brouwer diagram gives the net effect when the different defect reactions are considered simultaneously. In principle, the number of defect reactions that can be written for a given system is nearly limitless, but in practice the number of defects that must be included to describe a given defect-related property (e.g., electrical conductivity or sintering) is small. The majority defects are always important, but, in addition, certain minority defects may be relevant to a property of interest. The defect formation reactions will typically include those for:

1. Predominant intrinsic ionic defects (Schottky or Frenkel)
2. Intrinsic electronic defects
3. Oxidation and reduction
4. Solute incorporation

In addition, an equation for the electroneutrality condition for the bulk crystal is always necessary.

As an example, let us consider the construction of a Brouwer diagram for a pure oxide MO that forms predominantly Schottky defects at the stoichiometric composition, is oxygen deficient under reducing conditions, and is metal deficient under oxidizing conditions. Assuming doubly charged metal and oxygen defects, then the defect reactions can be written as follows:

For the oxygen-deficient case (Equation 1.30 and Equation 1.31):

$$O_O^x \rightarrow V_O^{\bullet\bullet} + 2e' + \frac{1}{2}O_2 \qquad K_R = \left[V_O^{\bullet\bullet} \right] n^2 p_{O_2}^{1/2} \tag{1.45}$$

where K_R is the equilibrium constant for the reduction reaction.

For the metal-deficient case (Equation 1.33 and Equation 1.34):

$$\frac{1}{2}O_2 \rightarrow V_M^{//} + 2h^{\bullet} + O_O^x \qquad K_O = \left[V_M^{//} \right] p^2 p_{O_2}^{-1/2} \tag{1.46}$$

where K_O is the equilibrium constant for the oxidation reaction.

The electronic defect condition is

$$\text{null} \rightarrow e' + h^{\bullet} \qquad K_i = np \tag{1.47}$$

where K_i is the equilibrium constant for the intrinsic ionization of electrons from the valence to the conduction band.

The Schottky defect reaction (Equation 1.24 and Equation 1.25) is

$$\text{null} \rightarrow V_M^{//} + V_O^{\bullet\bullet} \qquad K_S = \left[V_M^{//} \right]\left[V_O^{\bullet\bullet} \right] \tag{1.48}$$

Case A

In the oxygen-deficient region, oxygen vacancies dominate, so $[V_O^{\bullet\bullet}] = n/2 \gg [V_M^{//}]$. Then from Equation 1.45, we have

$$n = (2K_R)^{1/3} p_{O_2}^{-1/6} \qquad [V_O^{\bullet\bullet}] = \left(\frac{K_R}{4} \right)^{1/3} p_{O_2}^{-1/6} \tag{1.49}$$

From the Schottky and electronic defect conditions (Equation 1.47 and Equation 1.48), we have

$$\left[V_M^{//}\right] = \frac{K_S}{[V_O^{\bullet\bullet}]} = K_S \left(\frac{K_R}{4}\right)^{1/3} p_{O_2}^{1/6} \tag{1.50}$$

$$p = \frac{K_i}{n} = K_i (2K_R)^{-1/3} p_{O_2}^{1/6} \tag{1.51}$$

Case B

In the metal-deficient region, metal vacancies dominate, so $[V_M^{//}] = \dfrac{p}{2} \gg [V_O^{\bullet\bullet}]$. Then from Equation 1.46, we have

$$p = (2K_O)^{1/3} p_{O_2}^{1/6} \qquad \left[V_M^{//}\right] = \left(\frac{K_O}{4}\right)^{1/3} p_{O_2}^{1/6} \tag{1.52}$$

From the Schottky and electronic defect conditions (Equation 1.47 and Equation 1.48), we have

$$\left[V_O^{\bullet\bullet}\right] = \frac{K_S}{[V_M^{//}]} = K_S \left(\frac{K_O}{4}\right) p_{O_2}^{-1/6} \tag{1.53}$$

$$n = \frac{K_i}{p} = K_i (2K_O)^{-1/3} p_{O_2}^{-1/6} \tag{1.54}$$

Case C

In the stoichiometric region, $[V_O^{\bullet\bullet}] = [V_M^{//}] = K_S^{1/2} \gg p$ and n. Then from Equation 1.45, we have

$$n = K_R^{1/2} K_S^{1/4} p_{O_2}^{-1/4} \tag{1.55}$$

From the electronic defect condition (Equation 1.47), we have

$$p = \frac{K_i}{n} = K_i K_R^{-1/2} K_S^{-1/4} p_{O_2}^{1/4} \tag{1.56}$$

The Brouwer diagram is shown in Figure 1.5.

1.4.7 DEFECT CHEMISTRY AND SINTERING

To summarize at this stage, the point defect concentration and hence the rate of matter transport through the crystal lattice can be altered by manipulating three accessible variables: the temperature, the oxygen partial pressure (the gaseous atmosphere), and the concentration of dopants. Taking the Al_2O_3-doped MgO system as an example, if the sintering rate of MgO is controlled by the diffusion of oxygen vacancies, then Equation 1.43 indicates that the addition of Al_2O_3 acts to inhibit the sintering rate. On the other hand, if the diffusion of magnesium vacancies is the rate-controlling mechanism, then the addition of Al_2O_3 will increase the sintering rate (Equation 1.44). In practice, control of the

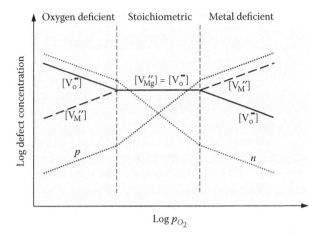

FIGURE 1.5 Brouwer diagram showing the defect concentrations as a function of oxygen pressure for a pure oxide MO that forms predominantly Schottky defects at the stoichiometric composition.

sintering rate through manipulation of the defect structure is not as straightforward as this example may indicate. A major problem is that the rate-controlling mechanism for sintering in most systems is not clear. Another problem is that more than one mechanism may operate during sintering, and the relative rates of transport by the different mechanisms may change with the conditions. Other factors, such as the particle packing homogeneity, have a considerable influence on the sintering rate and may often overwhelm the effects produced by changes in the defect structure.

1.5 DIFFUSION IN CRYSTALLINE SOLIDS

Diffusion is the process by which matter is transported from one part of a system to another as a result of atomic (or ionic or molecular) motions. The motion of an atom can be described in terms of the familiar *random walk* process, in which no atom has a preferred direction of motion. However, transfer of atoms occurs from the region of higher concentration to that of lower concentration. Detailed discussions of diffusion in crystalline solids appear in many texts, including a treatment of the concepts by Shewmon [5], a mathematical treatment by Crank [6], and a treatment of solid-state reactions by Schmalzried [7]. Review articles by Howard and Lidiard [8] and Atkinson [9] provide condensed treatments of key concepts.

1.5.1 DIFFUSION EQUATIONS

In an elementary view of diffusion, the movement of the diffusing species is considered to be driven by gradients in the concentration, and the atomic nature of crystal structure and atomic defects is not taken into account. This *continuum approach* to diffusion is quite similar to heat transfer [10]. The concentration can vary as a function of distance and time. When the concentration is independent of time, the mathematics of the diffusion process is described by *Fick's first law*, which states that the flux of the diffusing species J (number crossing unit area, normal to the direction of flux, per second) is proportional to the concentration gradient of the substance and occurs in the direction of decreasing concentration. For an isotropic medium, Fick's first law (in one dimension) is

$$J = -D\frac{dC}{dx} \tag{1.57}$$

where C is the concentration and x is the space coordinate measured normal to the section, and the negative sign arises because diffusion occurs in the direction opposite to that of increasing concentration. The constant of proportionality D is called the *diffusion coefficient* or *diffusivity* and has SI units of m²/s (but is more commonly expressed in cm²/s). The diffusion coefficient is a material property and is the most useful parameter for characterizing the rate of diffusive mass transport. It is usually a strong function of temperature, and it is also a function of composition, although in certain limiting cases, such as a dilute concentration of the diffusing species, it can be taken to be independent of the composition.

A concentration that is independent of time is often experimentally difficult to establish in a solid. It is more often convenient to measure the change in concentration as a function of time, t. This function is given by *Fick's second law*. For the one-dimensional case, when there is a concentration gradient only along the x-axis, Fick's second law is

$$\frac{dC}{dt} = D \frac{d^2C}{dx^2} \tag{1.58}$$

Fick's second law can be derived from his first law and an application of the principle of conservation of matter. For the one-dimensional case, consider the region between the two planes $[x_1, (x_1 + dx)]$ shown in Figure 1.6. The concentration C of the diffusing species is shown schematically as a function of distance x in Figure 1.6a. Since dC/dx at x_1 is $> dC/dx$ at $(x_1 + dx)$, the flux $J(x_1)$ will be $> J(x_1 + dx)$, as shown in Figure 1.6b. If $J(x_1) > J(x_1 + dx)$ and matter is conserved, the solute

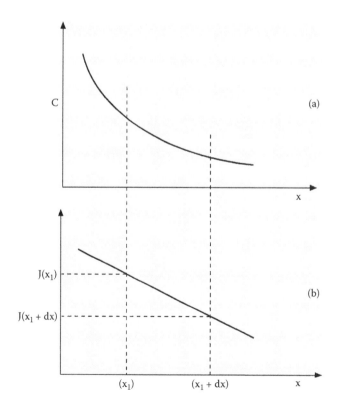

FIGURE 1.6 (a) Concentration C vs. distance x and (b) the resulting flux J vs. distance.

concentration in the region between x_1 and $x_1 + dx$ must increase. Considering a volume element of unit area normal to the x-axis and dx in thickness, the rate of change of concentration is

$$\left(\frac{dC}{dt}\right)_{x_1} dx = J(x_1) - J(x_1 + dx) \tag{1.59}$$

If dx is small, $J(x_1 + dx)$ can be related to $J(x)$ by

$$J(x_1 + dx) = J(x_1) + \left(\frac{dJ}{dx}\right)_{x_1} dx \tag{1.60}$$

Substituting Equation 1.60 into Equation 1.59 and using Equation 1.57 for J gives

$$\frac{dC}{dt} = -\frac{dJ}{dx} = -\frac{d}{dx}\left(-D\frac{dC}{dx}\right) = D\frac{d^2C}{dx^2} \tag{1.61}$$

1.5.1.1 Solutions of the Diffusion Equation

General solutions of Equation 1.58 can be obtained for a variety of initial and boundary conditions provided that the diffusion coefficient D is constant [6]. It is easy to see by differentiation that

$$C = \frac{A}{t^{1/2}} \exp\left(-\frac{x^2}{4Dt}\right) \tag{1.62}$$

where A is a constant, is a solution of Equation 1.58. The expression is symmetrical with respect to $x = 0$, tends to zero as x approaches infinity positively or negatively for $t > 0$, and for $t = 0$ vanishes everywhere except at $x = 0$, where it becomes infinite. For an amount of substance C_o deposited at $t = 0$ in the plane $x = 0$, the solution describing the spreading by diffusion in a cylinder of infinite length and unit cross section is [6]:

$$C = \frac{C_o}{2(\pi Dt)^{1/2}} \exp\left(-\frac{x^2}{4Dt}\right) \tag{1.63}$$

Figure 1.7 shows typical concentration profiles at three successive times. For a semi-infinite cylinder extending over the region $x > 0$, such that all the diffusion occurs in the direction of positive x, the solution is

$$C = \frac{C_o}{(\pi Dt)^{1/2}} \exp\left(-\frac{x^2}{4Dt}\right) \tag{1.64}$$

For a semi-infinite source, with a constant concentration C_o for $x < 0$, in contact with a semi-infinite solid, such that $C = 0$ for $x > 0$ at $t = 0$, the solution is

$$C = \frac{C_o}{2} erfc\left[\frac{x}{2(Dt)^{1/2}}\right] \tag{1.65}$$

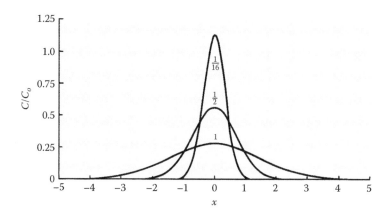

FIGURE 1.7 Graphical representation of the solution to Fick's second law for a thin planar source with a concentration C_o deposited at time $t = 0$ in the plane $x = 0$ of an infinite solid. The curves show concentration vs. distance for three values of Dt.

where *erfc* is the error-function complement, such that $erfc\ z = 1 - erf\ z$. Solutions to Equation 1.58 for several geometries and boundary conditions can be found in Crank [6], as well as in Atkinson [9] (by converting the solutions for heat transfer).

Equation 1.64 and Equation 1.65 describe the most common configurations that can be approximated experimentally for the determination of D. For example, a common technique is to deposit a very thin film of a mass isotope (or radioactive isotope), such as ^{18}O, on a plane surface of a thick sample and, after annealing for fixed times, determine the concentration of the diffusing species as a function of distance from the surface (so-called *depth profiles*). The diffusion coefficient is obtained from Equation 1.64 by plotting ln C vs. x^2 and calculating D from the slope. When Equation 1.65 applies, the argument of the complementary error function, *arg erfc* (C/C_o), is plotted against x to determine D. The diffusion coefficient of the isotope, obtained by these methods, is referred to as the *tracer diffusion coefficient*.

1.5.2 ATOMISTIC DIFFUSION PROCESSES

At the atomic level, diffusion may be viewed as the periodic jumping of atoms from one lattice site to another via an intermediate stage of higher energy that separates one site from another (Figure 1.8). The energy barrier that must be surmounted in the intermediate state before the jump can occur is called the *activation energy*. This periodic jumping of the atoms, in which the atom diffuses by a kind of Brownian motion or random walk, is sometimes referred to as *random diffusion*. In the simplest situations, we can express the diffusion coefficient D in terms of the mean square displacement $\langle x^2 \rangle$ of the atom after time t by using the Einstein formula:

$$D = \frac{\langle x^2 \rangle}{6t} \tag{1.66}$$

If the movements of the diffusing atoms are quite random, $\langle x^2 \rangle$ can be replaced by the number of jumps multiplied by the square of the displacement distance of each jump:

$$\langle x^2 \rangle = ns^2 \tag{1.67}$$

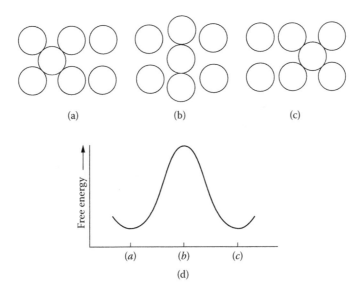

(a)   (b)   (c)

(d)

FIGURE 1.8 Schematic diagram showing the sequence of configurations when an atom jumps from one lattice site to another (a, b, and c) and the corresponding change in the free energy of the lattice (d).

From Equation 1.66 and Equation 1.67, a relation between D and the jump rate $\Gamma = n/t$ is obtained:

$$D = \frac{1}{6}\Gamma s^2 \tag{1.68}$$

An approximate equation relating D to the jump frequency and the jump distance can be derived without going through a rigorous treatment of the random-walk problem. Let us consider two adjacent planes, denoted A and B, a distance λ apart, in a crystalline solid that has a concentration gradient along the x-axis (Figure 1.9). Let there be n_A diffusing atoms per unit area in plane A and n_B in plane B. We consider only jumps to the left and right, that is, those giving a change in position

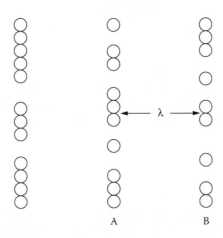

A   B

FIGURE 1.9 Planes of atoms with a gradient in concentration.

along the x-axis. The probability P that an atom will have, at any instant, sufficient energy to surmount the energy barrier q is

$$P = \exp\left(\frac{-q}{kT}\right) \tag{1.69}$$

where k is the Boltzmann constant and T is the absolute temperature. The number of atoms (per unit area of plane A) that have sufficient energy to surmount the energy barrier from A to B at any instant is $n_A P = n_A \exp(-q/kT)$. Assuming that the atoms vibrate (or oscillate) about their mean positions with a frequency v and that the vibration frequency is the same in all six orthogonal directions, then the *flux* of atoms (that is, the number jumping across unit area per second) from A to B is $(v/6)n_A \exp(-q/kT)$. Similarly, the flux of atoms from B to A is $(v/6)n_B \exp(-q/kT)$. The net flux of atoms from A to B is therefore

$$J = \frac{v}{6}(n_A - n_B)\exp\left(\frac{-q}{kT}\right) \tag{1.70}$$

We can relate the term $(n_A - n_B)$ to the concentration (that is, the number of atoms per unit volume) by observing that $n_A/\lambda = C_A$, and $n_B/\lambda = C_B$, giving

$$J = \frac{v\lambda}{6}(C_A - C_B)\exp\left(\frac{-q}{kT}\right) \tag{1.71}$$

Assuming that C changes slowly enough that $(C_A - C_B) = -\lambda(dC/dx)$, Equation 1.71 becomes

$$J = -\frac{v\lambda^2}{6}\exp\left(\frac{-q}{kT}\right)\frac{dC}{dx} \tag{1.72}$$

This equation is identical to Fick's first law (Equation 1.57), with the diffusion coefficient D given by

$$D = \frac{v\lambda^2}{6}\exp\left(\frac{-q}{kT}\right) \tag{1.73}$$

Equation 1.73 differs from Equation 1.68 in that the jump distance in three dimensions s may not be equal to the distance between the lattice planes λ.

For most diffusing atoms, q is an inconveniently small quantity and it is better to use the larger quantities $Q = N_A q$ and $R = N_A k$, where Q is the activation energy per mole, N_A is the Avogadro number, and R is the gas constant. In addition, the term $v\lambda^2/6$ is usually written as D_o, giving

$$D = D_o \exp\left(\frac{-Q}{RT}\right) \tag{1.74}$$

This method of writing D emphasizes the exponential dependence of D on temperature and gives a conveniently sized activation energy (in units of J/mol). Data for the diffusion coefficients of some common ceramics are given in Figure 1.10 [11].

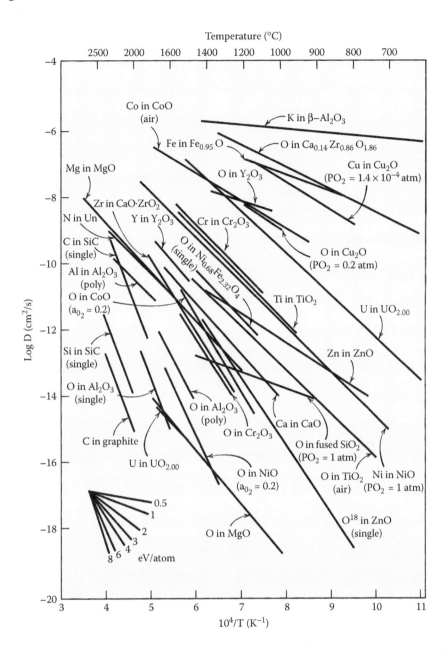

FIGURE 1.10 Diffusion coefficients in some common ceramics. Reported values vary over some 10 orders of magnitude. The activation energy can be estimated from the slope and the insert. (From Chiang, Y.M., Birnie, D., III, and Kingery, W.D., *Physical Ceramics*, Wiley, New York, 1997, p. 187. With permission.)

Taking the activation energy q as equal to $\Delta h - T\Delta s$, where Δh is the enthalpy and Δs is the entropy for atomic diffusion, then Equation 1.73 becomes

$$D = \frac{\nu\lambda^2}{6}\exp\left(\frac{\Delta s}{k}\right)\exp\left(\frac{-\Delta h}{kT}\right) \qquad (1.75)$$

A comparison of Equation 1.75 with Equation 1.74 indicates that

$$D_o = \frac{\nu\lambda^2}{6}\exp\left(\frac{\Delta s}{k}\right) \tag{1.76}$$

The vibration frequency of atoms about their mean positions (the Debye frequency) is between 10^{13} s^{-1} and 10^{14} s^{-1}, and $\lambda \approx 0.2$ nm, so $\nu\lambda^2/6 \approx 10^{-7}$ m^2/s. Data indicate that D_o for diffusion by the interstitial and vacancy mechanisms is in the range of 10^{-7} m^2/s to 10^{-3} m^2/s. It has been shown that for a given class of materials, the data for D_o and Q/RT_M (where T_M is the melting temperature) are nearly constant [12]. For the alkali halides (NaCl, LiF, etc.), the data give $D_o \approx 2.5 \times 10^{-3}$ m^2/s and $Q/RT_M \approx 22.5$, whereas for oxides (MgO, Al$_2$O$_3$, etc.), $D_o \approx 3.8 \times 10^{-4}$ m^2/s and $Q/RT_M \approx 23.4$.

1.5.3 Mechanisms of Diffusion

The different types of defects determine the path of matter transport, and diffusion along the major paths gives rise to the major mechanisms of matter transport: lattice diffusion (also referred to as volume or bulk diffusion), grain boundary diffusion, and surface diffusion.

1.5.3.1 Lattice Diffusion

Lattice diffusion takes place by the movement of point defects through the bulk of the lattice. Depending on the type of defects (vacancy or interstitial), lattice diffusion can occur by the *vacancy* mechanism or the *interstitial* mechanism. Vacancy and interstitial diffusion are by far the most important lattice diffusion mechanisms.

1.5.3.1.1 Vacancy Mechanism

Some lattice sites may be vacant, and a neighboring atom on a normal lattice site may then exchange its place with a vacant site (Figure 1.11a). As successive atoms move in this way, the vacancy migrates through the crystal. The movement of the atom is opposite to that of the vacancy, so we can track either the movement of the atom (i.e., atom diffusion) or, equivalently, the motion of the vacancy (i.e., vacancy diffusion). The diffusion coefficients of the atoms and the vacancies are related, but they are not equal. We can see this as follows: An atom can jump only if a vacancy is located on an adjacent lattice site, but a vacancy can jump to any of the occupied nearest neighbor sites. The number of atomic jumps will therefore be proportional to the fraction of sites occupied by vacancies, C_v. The atomic (or self) diffusion coefficient D_a and the vacancy diffusion coefficient D_v are related by

$$D_a = C_v D_v \tag{1.77}$$

Since C_v is usually much less than 1, it follows from Equation 1.77 that D_v is much greater than D_a. The vacancy concentration, as outlined earlier, is determined by the temperature, solute, and atmosphere. It should be noted that in vacancy diffusion, the flux of vacancies must be compensated by an equal and opposite flux of atoms; otherwise, vacancies would accumulate in the crystal and produce pores. Pore formation can actually occur during the interdiffusion of two atoms that have very different diffusion coefficients.

1.5.3.1.2 Interstitial Mechanism

If they are small enough, some atoms may occupy positions in the interstices of the lattice. An interstitial atom may move to another interstitial position (Figure 1.11b). A relationship analogous to Equation 1.77 will hold:

$$D_a = C_i D_i \tag{1.78}$$

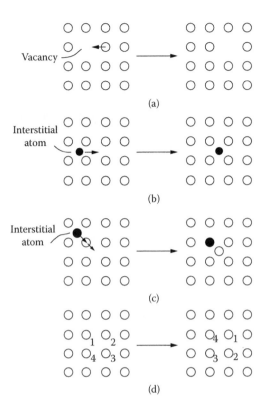

FIGURE 1.11 Lattice diffusion by (a) the vacancy mechanism, (b) the interstitial mechanism, (c) the interstitialcy mechanism, and (d) the ring mechanism.

where D_i is now the interstitial diffusion coefficient and C_i is the concentration of the interstitial atoms. In this case also, the defect concentration C_i is much less than 1, so D_i is much greater than D_a.

1.5.3.1.3 Interstitialcy Mechanism

If the distortion of the lattice becomes too large for interstitial diffusion to be favorable, then movement of the interstitial atoms may occur by the interstitialcy mechanism. The interstitial atom moves by pushing one of the neighboring normal lattice atoms into an interstitial site and itself taking up the normal lattice site (Figure 1.11c). The two need not be the same type of atom.

1.5.3.1.4 Exchange and Ring Mechanisms

A pair of neighboring atoms may simply exchange places, or a ring of neighboring atoms may execute a rotation in which each atom jumps and takes the place of the atom before it without the participation of a defect (Figure 1.11d). Several atoms can participate in a simultaneous exchange. The significant momentary distortion, coupled with the large energy changes arising from electrostatic repulsion, makes this mechanism improbable in ionic solids.

1.5.3.2 Grain Boundary Diffusion

In polycrystalline materials, the crystals or grains are separated from one another by regions of lattice mismatch and disorder called *grain boundaries*. Because of the highly defective nature of the grain boundary, we would expect grain boundary diffusion to be more rapid than lattice diffusion in the adjacent grains. Although not known accurately, the width of the grain boundary region is often observed to be 0.5–1 nm by high-resolution transmission electron microscopy. For a constant grain boundary width, the fraction of the solid that is occupied by the grain boundary increases with decreasing grain size, so the rate of grain boundary diffusion averaged over several grains is dependent on grain size.

1.5.3.3 Surface Diffusion

The free surface of a crystalline solid is not perfectly flat. It contains some vacancies (as in the bulk of the crystal), terraces, kinks, edges, and adatoms [13]. The migration of vacancies and the movement of adatoms provide the main mechanisms of surface diffusion. The diffusion process is assumed to be confined to a thin surface layer having a thickness of 0.5–1 nm.

1.5.3.4 Trends in Diffusion Coefficients

Since the atoms on the surface have fewer neighbors than those in the bulk of the lattice and are less tightly bound, we would expect the activation energy for surface diffusion to be smaller than that for lattice diffusion. This appears to be the case, to judge from the limited data available. Because of the lower activation energy, the relative importance of surface diffusion increases with decreasing temperature. As we shall see in Chapter 2, this has important consequences for sintering.

It has often been stated that the diffusion coefficients for lattice diffusion, D_l, grain boundary diffusion, D_{gb}, and surface diffusion, D_s, increase in the order $D_l < D_{gb} < D_s$, and that the corresponding activation energies vary as $Q_l > Q_{gb} > Q_s$. However, these relations may not always hold.

1.5.4 Types of Diffusion Coefficients

Several terminologies are used in the literature for specifying diffusion coefficients. They may describe the diffusion of a particular species (e.g., atom, interstitial, or vacancy), a particular diffusion path (e.g., lattice, grain boundary, or surface diffusion), or a particular process (e.g., chemical diffusion or ambipolar diffusion). Here we define terms as they are used in this text.

Self-diffusion coefficient: The self-diffusion coefficient D_{self} refers to the diffusion coefficient for the host atoms (or ions) in random diffusion and is the important diffusion coefficient in sintering. For lattice diffusion by the vacancy or interstitial mechanism D_{self} is given by Equation 1.77 or Equation 1.78.

Tracer diffusion coefficient: Direct measurement of D_{self} is often difficult. As outlined earlier, it more common to measure the diffusion coefficient for a mass isotope (or radioactive isotope). The tracer diffusion coefficient D^* measured in this type of experiment is close to but not identical to D_{self} because the motion of the tracer atom is not completely random. Successive jumps are, to a certain extent, correlated and dependent on previous jumps. D_{self} and D^* are related by

$$D^* = fD_{self} \tag{1.79}$$

where f is a correlation factor that depends on the crystal structure and the diffusion mechanism [5]. The calculated values for f are in the range of 0.6–1.0. For example, in the face-centered cubic lattice, $f = 0.78$ for the vacancy mechanism.

Lattice diffusion coefficient: The lattice diffusion coefficient D_l refers to any diffusion process in the lattice of the solid.

Grain boundary diffusion coefficient: The grain boundary diffusion coefficient D_{gb} (or D_b) refers to diffusion along the grain boundary.

Surface diffusion coefficient: The surface diffusion coefficient D_s refers to diffusion along a free surface.

Diffusion coefficients for defects: The diffusion coefficients for defects refer to a particular type of defect. We can, for example, specify the diffusion coefficient for a vacancy D_v or for an interstitial D_i. The diffusion coefficient of a defect is independent of its concentration and is given by Equation 1.73.

Chemical, effective, or interdiffusion coefficient: The chemical, effective, or interdiffusion coefficient $\tilde{D}$ refers to diffusion in a chemical composition (or chemical potential) gradient. This system is, by definition, not at chemical equilibrium, and it is less straightforward than tracer diffusion. However, it is important in many practical situations. A chemical composition gradient can, for example, be established experimentally by a gradient in oxygen activity, leading to migration of ions in response to the chemical gradient. Interdiffusion occurs when two different elements mix from opposite directions on the same sublattice of the crystal. The diffusion of the two elements is coupled to satisfy the requirements of conservation of lattice sites and the attainment of local chemical equilibrium. In ceramics, the diffusing species are ions, so electric fields develop in the interdiffusion zone, thereby generating an electric force to couple the fluxes of the interdiffusing ions.

1.6 THE CHEMICAL POTENTIAL

As outlined earlier, we can analyze matter transport in terms of Fick's laws by considering the flux of atoms or vacancies driven by gradients in the concentration. This is, however, a special case of mass transport. We should allow for the possibility that an atom is acted on by other types of driving forces, such as gradients in pressure, electric potential, and so on. One way to do this is to regard diffusion as driven by gradients in the *chemical potential* rather than gradients in concentration. The gradient in chemical potential could arise from gradients in pressure or electric potential, for example, as well as from concentration gradients.

Consider a phase with fixed mass and composition but under conditions of variable temperature, T, and pressure, p. For an infinitesimal reversible process, the change in the Gibbs free energy is

$$dG = \left(\frac{\partial G}{\partial T}\right)_p dT + \left(\frac{\partial G}{\partial p}\right)_T dp = -SdT + Vdp \tag{1.80}$$

where S is the entropy and V the volume of the system. For a pure substance in which the number of moles m of the substance is variable, the change in the Gibbs free energy is now given by

$$dG = \left(\frac{\partial G}{\partial T}\right)_{p,m} dT + \left(\frac{\partial G}{\partial p}\right)_{T,m} dp + \left(\frac{\partial G}{\partial m}\right)_{p,T} dm \tag{1.81}$$

Since the first two terms on the right-hand side of this equation are at constant mass and composition, we can use Equation 1.80. The third term represents the effect on the Gibbs free energy of changing the number of moles of the substance and can be expressed as

$$\mu = \left(\frac{\partial G}{\partial m}\right)_{T,p} \tag{1.82}$$

where μ is called the *chemical potential* of the substance. Equation 1.81 can now be written as

$$dG = -SdT + Vdp + \mu dm \tag{1.83}$$

Suppose we increase the number of moles of the substance while keeping T, p, and the composition constant. Then Equation 1.83 becomes

$$dG_{T,p} = \mu \, dm \tag{1.84}$$

Since μ depends only on T, p, and composition, it must remain constant, so Equation 1.84 can be integrated to give

$$G = \mu m \tag{1.85}$$

Equation 1.85 shows that the chemical potential of a pure substance is the *Gibbs free energy per mole* at the temperature and pressure in question.

1.6.1 CHEMICAL POTENTIAL OF A MIXTURE OF GASES

For 1 mole of an ideal gas at constant temperature T,

$$\left(\frac{\partial G}{\partial p} \right)_T = V = \frac{RT}{p} \tag{1.86}$$

where R is the gas constant and p is the pressure. On integration we obtain

$$G(T, p) = G_o(T) + RT \ln p \tag{1.87}$$

where G_o is a reference value. We can therefore write

$$\mu(T, p) = \mu_o(T) + RT \ln p \tag{1.88}$$

For a mixture of ideal gases at constant T and for a constant total pressure, following Equation 1.88, we can write for each component:

$$\mu_i(T, p) = \mu_{oi}(T) + RT \ln p_i \tag{1.89}$$

where p_i is the pressure of the ith component. It is more useful to relate the Gibbs free energy or the chemical potential of a particular component to its concentration defined by

$$C_i = \frac{n_i}{\sum n_i} = \frac{p_i}{p} \tag{1.90}$$

where n_i is the number of moles of each component in the mixture and p is the total gas pressure. Equation 1.89 can now be written

$$\mu_i(T, p, C_i) = \mu_o(T, p) + RT \ln C_i \tag{1.91}$$

Real gases do not show ideal behavior. The deviation from ideal behavior is, however, not too large over a fairly wide range of pressures, so Equation 1.89 to Equation 1.91 provide a good approximation to the chemical potential of real gases.

1.6.2 Chemical Potential of Solids and Liquids

For solid and liquid solutions, the chemical potential is defined by an expression similar to Equation 1.89 in which the p_i is replaced by a quantity a_i, called the *activity*:

$$\mu_i = \mu_{oi} + RT \ln a_i \tag{1.92}$$

The activity of pure liquids and solids under some specified standard conditions of temperature and pressure is taken as unity. For other systems, the activity is written

$$a_i = \alpha_i C_i \tag{1.93}$$

where α_i is called the activity coefficient and C_i is the concentration, usually expressed as a mole fraction. The chemical potential of the ith species in liquid or solid solution can now be written as

$$\mu_i = \mu_{oi} + RT \ln(\alpha_i C_i) \tag{1.94}$$

In the case of ideal solutions, $\alpha_i = 1$.

1.6.3 Chemical Potential of Atoms and Vacancies in a Crystal

Consider a crystal of a pure element that is perfect except for the presence of vacancies. If there are N atoms and n vacancies, the total number of lattice sites in the crystal is $N + n$. Following Equation 1.12, the Gibbs free energy of the crystal can be written as

$$G = U + n\Delta g_f + pV - TS \tag{1.95}$$

where U is the internal energy of the crystal, Δg_f is the energy of formation of a vacancy, p is the pressure, V is the volume of the crystal, T is the temperature, and S is the configurational entropy of the crystal, given by the Boltzmann equation (Equation 1.13). The chemical potential of the atoms is defined by

$$\mu_a = (\partial G / \partial N)_{T,p,n} \tag{1.96}$$

Application of Equation 1.96 to Equation 1.95 gives

$$\mu_a = \mu_{oa} + p\Omega_a + kT \ln\left(\frac{N}{N+n}\right) \tag{1.97}$$

where μ_{oa} is a reference value and Ω_a is the volume of an atom defined by

$$\left(\frac{\partial V}{\partial N}\right)_{T,p,n} = \Omega_a \tag{1.98}$$

If C_a is the fraction of lattice sites occupied by the atoms in the crystal, then Equation 1.97 can be written

$$\mu_a = \mu_{oa} + p\Omega_a + kT \ln C_a \tag{1.99}$$

Equation 1.98 shows that the chemical potential of atoms in a crystal depends on the pressure and on the atomic concentration. It is also worth noting that for small vacancy concentrations, the last term on the right-hand side of Equation 1.97 and Equation 1.99 is negligible.

The chemical potential of the vacancies is defined by

$$\mu_v = \left(\frac{\partial G}{\partial n}\right)_{T,p,N} \tag{1.100}$$

Following a procedure similar to that used for μ_a, we find that

$$\mu_v = \mu_{0v} + p\Omega_v + kT \ln C_v \tag{1.101}$$

where C_v is the vacancy concentration (i.e., the fraction of lattice sites occupied by the vacancies). The volume of a vacancy Ω_v may be greater or smaller than Ω_a, but from now on we will use the rigid lattice approximation and assume that $\Omega_a = \Omega_v = \Omega$.

1.6.4 Chemical Potential of Atoms and Vacancies beneath a Curved Surface

The atoms and vacancies beneath a curved surface will have their chemical potentials altered by the curvature of the surface, and this difference in chemical potential drives the diffusional flux of atoms to reduce the free energy of the system. Surfaces and interfaces play a dominant role in sintering and other diffusion-controlled processes, so the dependence of the chemical potential on curvature is vital to the analysis and understanding of sintering.

Consider a solid consisting of a pure element with adjoining convex and concave surfaces (Figure 1.12), and let us assume that vacancies are the only type of point defects present. The compressive pressure under a convex surface squeezes out vacancies, so we expect that the vacancy concentration beneath a convex surface will be below normal (e.g., relative to a flat surface). Following a similar type of argument, we expect that the vacancy concentration will be above normal beneath a concave surface. The difference in vacancy concentration leads to a diffusional flux of vacancies from the concave to the convex region or, equivalently, a diffusional flux of atoms from the convex to the concave region.

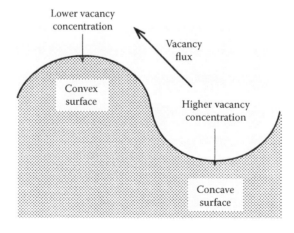

FIGURE 1.12 Schematic diagram showing the direction of flux for vacancies in a curved surface. The flux of atoms is equal and opposite to that of the vacancies.

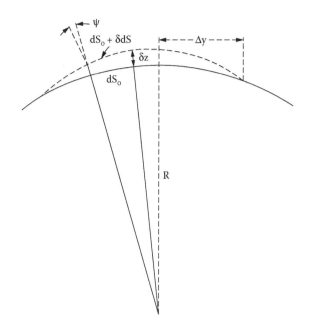

FIGURE 1.13 Infinitesimal hump formed by building up a curved surface. Full curve: original surface; dashed curve: built-up surface.

A method for determining the dependence of the chemical potential on the surface curvature was described by Herring [14]. Consider a smoothly curved surface as shown in Figure 1.13. Suppose we build an infinitesimal hump on the surface by taking atoms from beneath the surface and adding them to the surface. The change in surface-free energy, to a good approximation, is given by

$$\delta \left(\int \gamma_{SV} dS \right) = \int \delta(\gamma_{SV}) dS_o + \int \gamma_{SV} \delta(dS) \tag{1.102}$$

where δ represents a small change in a quantity, γ_{SV} is the specific surface energy, and dS is the change in surface area. We will assume that the surface is uniform and isotropic so the first term on the right-hand side of Equation 1.102 is zero. From Figure 1.13, the term $\delta(dS)$ is given by

$$\delta(dS) = (\sec \psi - 1) dS_o + \delta z \left(\frac{1}{R_1} + \frac{1}{R_2} \right) dS_o \tag{1.103}$$

where R_1 and R_2 are the two principal radii of curvature at dS_o. If Δy is small, then R_1 and R_2 may be taken as constant over the hump. The first term on the right-hand side of Equation 1.103 can be neglected, and substituting in Equation 1.102 gives

$$\delta \left(\int \gamma_{SV} dS \right) = \gamma_{SV} \left(\frac{1}{R_1} + \frac{1}{R_2} \right) \delta v \tag{1.104}$$

where δv is the volume of the hump.

The change in the volume term of the free energy due to the creation of the small hump is

$$\delta G_v = -p\delta v + \mu_v \frac{\delta v}{\Omega} \tag{1.105}$$

where p is the mean hydrostatic pressure in the crystal just beneath the surface, μ_v is the chemical potential of a vacancy, and $\delta v/\Omega$ represents the number of vacancies created by the transfer of atoms to the hump. In equilibrium, the sum of the energy changes defined by Equation 1.104 and Equation 1.105 must vanish, so

$$\mu_v = p\Omega + \gamma_{SV}\left(\frac{1}{R_1} + \frac{1}{R_2}\right)\Omega \tag{1.106}$$

The curvature K of the surface is defined by

$$K = \frac{1}{R_1} + \frac{1}{R_2} \tag{1.107}$$

where $K > 0$ for a convex surface. Substituting for K in Equation 1.106 gives

$$\mu_v = (p + \gamma_{SV}K)\Omega \tag{1.108}$$

In general, the chemical potential is measured relative to some reference value μ_{vo}, and, as found earlier, μ_v contains a vacancy concentration term. An expression for μ_v that incorporates not only curvature but also pressure and concentration effects can therefore be written as

$$\mu_v = \mu_{ov} + (p + \gamma_{SV}K)\Omega + kT\ln C_v \tag{1.109}$$

Using a similar procedure, the chemical potential of the atoms can be shown to be

$$\mu_a = \mu_{ov} + (p + \gamma_{SV}K)\Omega + kT\ln C_a \tag{1.110}$$

Since the terms containing C_a and C_v are normally very small, Equation 1.109 and Equation 1.110 show that μ_a and μ_v depend primarily on the hydrostatic pressure in the solid and on the curvature of the surface. The curvature term $\gamma_{sv}K$ has the units of pressure or stress, so that curvature and applied pressure effects in the sintering models can be treated by the same formulation.

1.6.5 EQUILIBRIUM VACANCY CONCENTRATION BENEATH A CURVED SURFACE

The equilibrium vacancy concentration in a crystal in thermal equilibrium at a given T and p will be such that G is a minimum, i.e.,

$$\mu_v = \left(\frac{\partial G}{\partial n}\right)_{T,p,N} = 0 \tag{1.111}$$

When externally applied pressure effects are absent, Equation 1.109 gives

$$C_v = C_{ov} \exp\left(-\frac{\gamma_{SV} K\Omega}{kT}\right) \tag{1.112}$$

where C_{ov} is a reference value, normally taken as the equilibrium vacancy concentration beneath a flat surface. For $\gamma_{SV} K\Omega \ll kT$, Equation 1.112 becomes

$$C_v = C_{ov}\left(1 - \frac{\gamma_{SV} K\Omega}{kT}\right) \tag{1.113}$$

1.6.6 VAPOR PRESSURE OVER A CURVED SURFACE

In sintering, matter transport by evaporation and condensation is normally treated alongside the solid-state diffusion mechanisms. The rate of transport is taken as proportional to the equilibrium vapor pressure over the surface, which can be related to the value of $(\mu_a - \mu_v)$ beneath the surface. Suppose a number dN of atoms is removed from the vapor and added to the surface with an accompanying decrease in the number of vacancies beneath the surface. The free energy change for this virtual operation must be zero, so that

$$\mu_{vap} = \mu_a - \mu_v \tag{1.114}$$

where μ_{vap} is the chemical potential of the atoms in the vapor phase. The vapor pressure is proportional to $\exp(\mu_{vap}/kT)$, so we can write

$$p_{vap} = p_o \exp\left(\frac{\mu_a - \mu_v - \mu_o}{kT}\right) \tag{1.115}$$

where p_o is a reference value of the vapor pressure corresponding to some standard value of the chemical potential, μ_o, and is normally taken as the value over a flat surface. Equation 1.109, Equation 1.110, and Equation 1.112 give

$$\mu_a - \mu_v = \mu_o + \gamma_{SV} K\Omega \tag{1.116}$$

Substituting into Equation 1.115 gives

$$p_{vap} = p_o \exp\left(\frac{\gamma_{SV} K\Omega}{kT}\right) \tag{1.117}$$

Equation 1.117 is commonly known as the Kelvin equation. For $\gamma_{SV} K\Omega \ll kT$, it becomes

$$p_{vap} = p_o\left(1 + \frac{\gamma_{SV} K\Omega}{kT}\right) \tag{1.118}$$

According to Equation 1.117 and Equation 1.118, for a given system under isothermal conditions, the vapor pressure increases with the curvature of the surface.

1.7 DIFFUSIONAL FLUX EQUATIONS

The theoretical analysis of sintering requires the formulation of equations for diffusional mass transport and their solutions, subject to the appropriate boundary conditions. There are two equivalent formulations: sintering can be viewed in terms of the diffusion of atoms or the diffusion of vacancies.

For an elemental solid, if the influence of the flux of the neutral atoms upon one another is neglected, then the flux of the atoms (in the x direction) can be expressed in a general form as [7]

$$J = - L_{ii} \frac{d\mu}{dx} \tag{1.119}$$

where the coefficients L_{ii} are called *transport coefficients* and μ is the chemical potential. By comparing Equation 1.119 with Fick's first law (Equation 1.57), we find that

$$L_{ii} = D \frac{dC}{d\mu} \tag{1.120}$$

Using the relation between chemical potential and concentration (Equation 1.94) to determine $dC/d\mu$, and substituting in Equation 1.120, we find

$$L_{ii} = \frac{D_i C}{kT} \tag{1.121}$$

where D_i is given by

$$D_i = D \left(1 + \frac{d \ln \alpha}{d \ln C} \right)^{-1} \tag{1.122}$$

For an ideal system, α is independent of the concentration, so $\ln \alpha$ does not vary with $\ln C$, and the second term in the brackets in Equation 1.122 is zero. By substituting Equation 1.121 into Equation 1.119, the atomic flux equation can be written

$$J = - \frac{D_i C}{kT} \frac{d\mu}{dx} \tag{1.123}$$

where, in the general case, D_i is given by Equation 1.122.

1.7.1 FLUX OF ATOMS

For a pure elemental solid in which the point defects consist of vacancies, diffusion of atoms or vacancies from one region to another does not produce a change in the total number of lattice sites. In a given region, the number of atoms and the number of vacancies change by equal and opposite amounts. The diffusional flux is determined by gradients in $(\mu_a - \mu_v)$, and following Equation 1.123, the flux of atoms is given by

$$J_a = -\frac{D_a C_a}{\Omega kT} \frac{d(\mu_a - \mu_v)}{dx} \tag{1.124}$$

where D_a is the atomic self-diffusion coefficient, and it must be remembered that C_a is the *fraction* of lattice sites occupied by the atoms. The diffusional flux of atoms can therefore be found from Equation 1.124, subject to the appropriate boundary conditions.

1.7.2 FLUX OF VACANCIES

When the flux of vacancies is considered, the atomic flux in sintering is taken as equal and opposite to the vacancy flux; that is,

$$J_a = -J_v \tag{1.125}$$

where J_v is given by

$$J_v = -\frac{D_v C_v}{\Omega kT} \frac{d\mu_v}{dx} = -\frac{D_v}{\Omega} \frac{dC_v}{dx} \tag{1.126}$$

where D_v is the vacancy diffusion coefficient. Determination of J_v requires an expression for C_v, which is normally taken as the *equilibrium* vacancy concentration (Equation 1.112 or Equation 1.113). For the equilibrium condition, we therefore have

$$J_a = \frac{D_v}{\Omega} \frac{dC_v}{dx} \tag{1.127}$$

Here again it should be remembered that C_v is the *fraction* of lattice sites occupied by vacancies.

1.8 DIFFUSION IN IONIC CRYSTALS: AMBIPOLAR DIFFUSION

In much of our discussion so far in this chapter, we have purposely considered the diffusing species to be uncharged atoms (or vacancies) so that electrostatic effects on the diffusing species have been avoided. However, the reader will know that in most crystalline ceramics, matter transport occurs by the motion of charged species (e.g., ions). For compounds containing more than one type of ion, we would expect that the different ions would have different diffusion rates. Matter transport from a given source to a given sink must occur in such a way that the stoichiometry and electroneutrality of the solid are preserved in the different regions of the solid (Figure 1.14). Other effects must also be taken into account. For example, if an external electric field is applied to the system, in addition to diffusion down a concentration gradient, the ions will migrate in response to the field. Even in the absence of an external field, the ions themselves can generate an internal field that will influence the motion.

Consider a diffusing species with a charge z_i (i.e., $z_i = +1$ for a sodium ion or $z_i = -2$ for a doubly charged oxygen ion). In a region where the electric potential is ϕ, the chemical potential of an ion is increased by an amount $z_i e \phi$, where e is the magnitude of the electron charge. In the

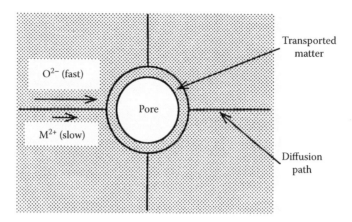

FIGURE 1.14 Schematic diagram illustrating that the diffusion of ions in an ionic solid must be coupled to preserve the stoichiometry and electroneutrality of the solid.

absence of an externally applied pressure, following Equation 1.99, we can write the chemical potential of an ion as

$$\mu_i = \mu_{oi} + kT \ln C_i + z_i e \phi \tag{1.128}$$

where C_i is the fraction of sites occupied by the ions in the crystal. Differentiating with respect to x and substituting into Equation 1.123 for the flux gives

$$J_i = -D_i \frac{dC_i}{dx} - B_i C_i z_i e E \tag{1.129}$$

where D_i is the diffusion coefficient for the ions, E is the electric field strength (equal to $d\phi/dx$), and $B_i = D_i/kT$ is the ionic mobility. The first term on the right-hand side of Equation 1.129 is the familiar diffusion term given by Fick's first law, whereas the second term arises from ion migration in the electric field.

Let us now apply Equation 1.129 to a system consisting of two different types of diffusing ions. One type has a charge z_+ (a positive number) and the other type has a charge z_- (a negative number), so this system may correspond to diffusion of metal and oxygen ions in a metal oxide. If no net current passes through the system, the electric current density must be zero, so

$$z_+ J_+ = -z_- J_- \tag{1.130}$$

Substituting for J_+ and J_- from Equation 1.129 gives

$$-z_+ D_+ \frac{dC_+}{dx} - B_+ C_+ z_+^2 e E = z_- D_- \frac{dC_-}{dx} + B_- C_- z_-^2 e E \tag{1.131}$$

After rearrangement, we can find an equation for E:

$$E = -\frac{1}{e\left(B_+ C_+ z_+^2 + B_- C_- z_-^2\right)}\left(z_+ D_+ \frac{dC_+}{dx} + z_- D_- \frac{dC_-}{dx}\right) \tag{1.132}$$

For electroneutrality we must have that

$$C_+ z_+ = -C_- z_- \tag{1.133}$$

Differentiating Equation 1.133 and multiplying by D_+ give

$$z_+ D_+ \frac{dC_+}{dx} = -z_- D_+ \frac{dC_-}{dx} \tag{1.134}$$

Substituting into Equation 1.132, we obtain

$$E = -\frac{z_-}{e(-B_+ z_+ C_- z_- + B_- z_- C_- z_-)} (D_- - D_+) \frac{dC_-}{dx} \tag{1.135}$$

Substituting for E in Equation 1.129 gives for the flux of the negative ions

$$J_- = -\frac{B_- z_- D_+ - B_+ z_+ D_-}{B_- z_- - B_+ z_+} \frac{dC_-}{dx} \tag{1.136}$$

Replacing the ionic mobility B_i by D_i/kT, Equation 1.136 becomes

$$J_- = -\frac{D_+ D_- (z_+ - z_-)}{D_+ z_+ - D_- z_-} \frac{dC_-}{dx} \tag{1.137}$$

Finally, applying Equation 1.129 and Equation 1.132 to Equation 1.137 gives

$$J_+ = -\frac{D_+ D_- (z_+ - z_-)}{D_+ z_+ - D_- z_-} \frac{dC_+}{dx} \tag{1.138}$$

By analogy with Fick's first law we can define an effective diffusion coefficient given by

$$\tilde{D} = \frac{D_+ D_- (z_+ - z_-)}{D_+ z_+ - D_- z_-} \tag{1.139}$$

To illustrate the use of Equation 1.139, let us consider the case of Al_2O_3. Assuming that the ions are fully ionized, the diffusing species are Al^{3+} and O^{2-} ions, and the effective diffusion coefficient is

$$\tilde{D} = \frac{5D_{Al^{3+}} D_{O^{2-}}}{3D_{Al^{3+}} + 2D_{O^{2-}}} \tag{1.140}$$

It is instructive to consider two limiting cases:

$$\text{Case 1: If } D_{O^{2-}} \gg D_{Al^{3+}}, \text{ then } \tilde{D} = \frac{5}{2} D_{Al^{3+}} \tag{1.141}$$

$$\text{Case 2: If } D_{Al^{3+}} \gg D_{O^{2-}}, \text{ then } \tilde{D} = \frac{5}{3} D_{O^{2-}} \tag{1.142}$$

We see from Equation 1.141 and Equation 1.142 that the more slowly diffusing ion determines the rate of matter transport, but the effect of the faster-diffusing species is to accelerate the motion of the slower ion. Physically, the faster-diffusing ion reduces its concentration gradient more rapidly than the slower-diffusing ion. However, only a small amount of such diffusion is necessary before a large potential gradient is set up. The potential gradient has the same sign as the concentration gradient, so it retards the transport of the faster ions and enhances the transport of the slower ions. The buildup of the potential occurs to the point where the fluxes are related by Equation 1.130.

The coupled diffusion of charged species is referred to as *ambipolar diffusion*. It has important consequences not only for sintering but also for other mass transport processes such as creep and the formation of oxide layers on materials. The slower (or slowest) diffusing species, as observed above, will control the rate of matter transport. A further complication arises when each ion may have more than one diffusion path (e.g., lattice diffusion and grain boundary diffusion). It is expected that matter transport will occur predominantly along the fastest path, so the rate-controlling mechanism becomes the slowest-diffusing species along its fastest path. Differences in the effective area and path length for mass transport must also be taken into account when considering the rate-controlling mechanism of a process.

As an example, Figure 1.15 shows a schematic plot of the log (creep rate) versus log (grain size) at constant temperature and applied stress for Al_2O_3, a system for which the relative values of the grain boundary and lattice diffusion coefficients are known: $D_l^{Al} > D_l^O$ and $D_b^O > D_b^{Al}$. The two lines with slope of –2 show the creep rate by lattice diffusion if Al and O are the rate-limiting ions, whereas the two lines with slope of –3 give the corresponding creep rate by grain boundary diffusion. The controlling species and path are given by the dotted line. We notice that each combination of species and path is controlling over a specific grain size range and that, in each of these regimes, it is not the

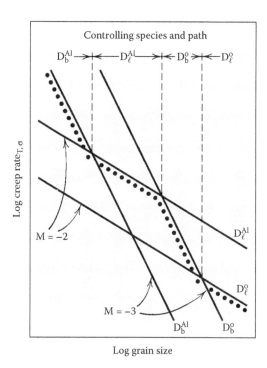

FIGURE 1.15 Diffusional creep in Al_2O_3 (schematic). The solid lines show the creep rate dependence on grain size for each assumed species and path. The dotted line traces the rate-limiting mechanism, given by the slower-diffusing species along its fastest path.

topmost line, giving the highest creep rate, that is rate controlling. The rate-controlling species and path are given by the second line, representing the other ion along its fastest path.

If the lattice and grain boundary coefficients are additive, the effective diffusion coefficient for an elemental solid, taking into account both diffusion paths, is [15]

$$\tilde{D} = D_l + \frac{\pi \delta_{gb} D_{gb}}{G} \tag{1.143}$$

where δ_{gb} is the width of the grain boundary region and G is the grain size. For a pure binary compound M_xO_y, the effective (or ambipolar) diffusion coefficient is given by [16]

$$\tilde{D} = \frac{(x+y)\left[D_l^M + \pi\delta_{gb}D_{gb}^M/G\right]\left[D_l^O + \pi\delta_{gb}D_{gb}^O/G\right]}{y\left[D_l^M + \pi\delta_{gb}D_{gb}^M/G\right] + x\left[D_l^O + \pi\delta_{gb}D_{gb}^O/G\right]} \tag{1.144}$$

In Equation 1.144, each of the terms in square brackets represents the effective diffusion coefficient for one of the species when both transport paths are taken into account. Examination of Equation 1.144 indicates that the slowest effective diffusion coefficient is rate controlling but, within each term, it is the faster of the two paths that dominates, confirming that the rate-controlling mechanism is the slowest-diffusing species along its fastest path.

1.9 CONCLUDING REMARKS

In this chapter we considered several topics that form the basis for understanding the sintering process. The concepts introduced will be used extensively in the next chapter, dealing with the sintering models, and in subsequent chapters. Matter transport during the sintering of polycrystalline ceramics occurs by diffusion, a thermally activated process, which can occur along different paths in the solid, giving rise to the different mechanisms of diffusion: lattice, grain boundary, and surface diffusion. The rate of atomic (or ionic) diffusion depends on the temperature and on the concentration of defects in the solid. The defect concentration can be manipulated by changing the temperature, the oxygen partial pressure (or atmosphere), and the concentration of dopants or impurities. Matter transport during sintering can be viewed in terms of the flux of atoms (ions) or, equivalently, in terms of the counterflow of vacancies. The flux of the diffusing species is driven by gradients in the concentration or, in the more general case, by gradients in the chemical potential. The diffusion of charged species (ions) in ceramics is coupled in such a way that stoichiometry and electroneutrality of the solid are maintained and, as a result of this coupling, the rate of sintering and other mass transport processes is controlled by the slowest-diffusing species along its fastest path.

PROBLEMS

1.1 Assuming that a powder has a surface energy of 1 J/m², estimate the maximum amount of energy that is available for densification for spherical particles with a diameter of 1 μm compacted to a green density of 60% of the theoretical density. Assume that there is no grain growth and the grain boundary energy is 0.5 J/m².

1.2 Assuming a constant grain boundary width of 0.5 nm, plot the fractional volume of a polycrystalline solid occupied by the grain boundaries as a function of the grain size for grain sizes in the range of 10 nm to 10 μm. To simplify the calculation, the grains may be assumed to have a cubic shape.

1.3 NaCl and MgO both have the rocksalt structure, and the dominant intrinsic defect is the Schottky defect. However, Table 1.4 shows that the enthalpy of formation of a Schottky defect for MgO (~7.7 eV) is much higher than that for NaCl (~2.3 eV). Explain.

Determine the concentration of Schottky defects in NaCl and in MgO at (a) 700°C and (b) $T/T_M = 0.95$, where T_M is the melting temperature, and comment on the values.

1.4   Write formally correct defect reactions and the corresponding mass action equilibrium constants for the following cases. If more than one reaction is possible, give the different possibilities and briefly discuss why these possibilities exist.

(a) Solid solution of $CaCl_2$ in NaCl
(b) Solid solution of Y_2O_3 in ZrO_2
(c) Reduction of CeO_2
(d) Solid solution of Nb_2O_5 in TiO_2
(e) Solid solution of Al_2O_3 in TiO_2
(f) Solid solution of $SrTiO_3$ in $BaTiO_3$
(g) Solid solution of Nb_2O_5 in $BaTiO_3$
(h) Solid solution of Al_2O_3 in $BaTiO_3$
(i) Solid solution of Y_2O_3 in $BaTiO_3$

1.5   The densification of a metal oxide MO is controlled by the diffusion of oxygen vacancies. The oxide has a native defect structure of oxygen vacancies. Develop equations for the defect equilibria and discuss how the densification rate can be enhanced.

1.6   Nickel oxide NiO is metal deficient with cation vacancies predominating. If it is doped with Li, which goes into solid solution on the regular Ni lattice sites, develop equations for the defect equilibria and construct the Brouwer diagram.

1.7   A pure oxide MO forms predominantly Schottky defects at the stoichiometric composition. It is oxygen deficient under reducing conditions and metal deficient under oxidizing conditions. If the oxide is doped with acceptor or donor dopants, develop the new equations for the defect equilibria and construct the Brouwer diagram.

1.8   The data in Figure 1.10 show that in MgO, Al_2O_3, and CoO, the cation self-diffusion coefficient is orders of magnitude greater than the oxygen self-diffusion coefficient. Explain.

1.9   List all possible mechanisms of diffusion in a polycrystalline ceramic.

(a) Which mechanism is expected to have the highest activation energy? The lowest activation energy? Explain.
(b) Which mechanism is expected to be the fastest? The slowest?
(c) Which mechanism is expected to dominate at lower temperatures? At higher temperatures?

1.10 The measured diffusion coefficient in ZnO is 5.0×10^{-5} cm²/s at 600°C and 2.0×10^{-5} cm²/s at 500°C. What would the diffusion coefficient be at 700°C?

1.11 The self-diffusion coefficient of a metal is 10^{-10} cm²/s at 800°C. A thin film of a mass isotope of the metal is evaporated onto a plane surface of the metal. Assuming that the concentration of the isotope is described by Equation 1.64:

(a) Plot the surface concentration as a function of time.
(b) Show that the average diffusion distance after an annealing time t is $(Dt)^{1/2}$.
(c) Calculate the average diffusion distance after 10 hours annealing at 800°C.

1.12 Estimate the difference in equilibrium vapor pressure for two spherical particles, one with a diameter of 0.1 μm and the other with a diameter of 10 μm, assuming a temperature of 500°C, an atomic radius of 10^{-10} m, and a surface energy of 1 J/m². Discuss what would happen if the two particles were placed in the same closed box at 500°C.

1.13 Determine the diffusional creep rate for a dense, fine-grained Al_2O_3 (average grain size of 1 μm) at 1400°C and under an applied pressure of 150 MPa. At this temperature, the values of the diffusion coefficients (in units of cm²/s) for Al_2O_3 have been measured as follows: $D_{gb}^{Al} = 5.6 \times 10^{-17}$; $D_{gb}^{O} = 7.0 \times 10^{-14}$; $D_{l}^{Al} = 4.0 \times 10^{-14}$; $D_{l}^{O} = 1.0 \times 10^{-17}$. (Relations for the creep rates by lattice diffusion [Nabarro–Herring creep] and by grain boundary diffusion [Coble creep] are given in Chapter 2.)

REFERENCES

1. Kingery, W.D., Sintering from prehistoric times to the present, in *Sintering '91*: Proc. Fifth Int. Symp. Science and Technology of Sintering, Chacklader, A.C.D., and Lund, J.A., Eds., Trans Tech, Brookfield, VT, 1992, p. 1.
2. Coble, R.L., and Cannon, R.M., Current paradigms in powder processing, in *Processing of Crystalline Ceramics,* Mater. Sci. Res., Vol. 11, Palmour, H. III, Davis, R.F., and Hare, T.M., Eds., Plenum Press, New York, 1978, p. 151.
3. Kroger, F.A., *The Chemistry of Imperfect Crystals*, North Holland Publishing Company, Amsterdam, 1964.
4. Kofstad, P., *Nonstoichiometry, Diffusion and Electrical Conductivity in Binary Metal Oxides*, Krieger Publishing Company, Malabar, FL, 1983.
5. Shewmon, P., *Diffusion in Solids*, The Metals Society, Warrendale, PA, 1989.
6. Crank, J., *The Mathematics of Diffusion*, 2nd ed., Clarendon Press, Oxford, 1975.
7. Schmalzried, H., *Solid State Reactions*, 2nd ed., Verlag Chemie, Weinheim, Germany, 1981.
8. Howard, R.E., and Lidiard, A.B., Matter transport in solids, *Rept. Prog. Phys.*, 27, 161, 1964.
9. Atkinson, A., Diffusion in ceramics, in *Materials Science and Technology, Vol. 11: Structure and Properties of Ceramics*, Cahn, R.W., Haasen, P., and Kramer, E.J., Eds., VCH, New York, 1994, p. 295.
10. Carlsaw, H.S., and Jaeger, J.C., *Conduction of Heat in Solids*, 2nd ed., Clarendon Press, Oxford, 1959.
11. Chiang, Y.M., Birnie, D. III, and Kingery, W.D., *Physical Ceramics*, Wiley, New York, 1997, p. 187.
12. Brown, A.M., and Ashby, M.F., Correlations for diffusion constants, *Acta Metall.*, 28, 1085, 1980.
13. Rahaman, M.N., *Ceramic Processing*, CRC Press, Boca Raton, FL, 2006, p. 129.
14. Herring, C., Surface tension as a motivation for sintering, in *The Physics of Powder Metallurgy*, Kingston, W., Ed., McGraw Hill, New York, 1951, p. 143.
15. Raj, R., and Ashby, M.F., On grain boundary sliding and diffusional creep, *Metall. Trans.*, 2, 1113, 1971.
16. Gordon, R.S., Mass transport in the diffusional creep of ionic solids, *J. Am. Ceram. Soc.*, 56, 147, 1973.

2 Solid-State and Viscous Sintering

2.1 INTRODUCTION

The reader will recall from Chapter 1 the basic types of sintering processes: solid-state sintering, liquid-phase sintering, viscous sintering, and vitrification. This chapter is devoted to an analysis of densification in solid-state and viscous sintering. The analysis of the densification process, by itself, is not very useful. We must also understand how the microstructure of the powder system evolves during sintering, which forms the subject of Chapter 3. To complete the picture of the basic theory and principles of sintering, liquid-phase sintering and vitrification are considered in Chapter 4.

The sintering phenomena in polycrystalline materials are considerably more complex than those in viscous sintering of amorphous materials because of the availability of several matter transport paths and the presence of grain boundaries. Matter transport in solid-state sintering can occur by at least six different paths, which define the mechanisms of sintering. In practice, more than one mechanism may operate during any given regime of sintering, and the occurrence of multiple mechanisms makes analysis of sintering rates and determination of sintering mechanisms difficult. Perhaps the most important consequence of grain boundaries is the occurrence of grain growth and pore growth during sintering, a process normally referred to as *coarsening*. The coarsening process provides an alternative route by which the free energy of the powder system can be reduced, so it reduces the driving force for densification. The interplay between the two processes, sometimes referred to as a competition between sintering (densification) and coarsening (grain growth), is discussed in Chapter 3.

In contrast to solid-state sintering, the analysis of viscous sintering appears relatively simple in principle. Matter transport occurs by a viscous flow mechanism, but the path along which matter flows is not specified. Instead, the equations for matter transport are derived on the basis of an *energy balance* concept, in which the rate of energy dissipation by viscous flow is put equal to the rate of energy gained by reduction in surface area. The models based on this energy balance concept are generally successful in describing the sintering kinetics of glasses and other amorphous materials.

A comprehensive theory of sintering should be capable of describing the entire sintering process as well as the evolution of the microstructure (i.e., grain size, pore size, and the distribution of the grain and pore sizes). However, in view of the complex nature of the process, it is unlikely that such a theory will be developed. A more realistic approach, which is adopted in this book, is first to develop an understanding of the densification and coarsening phenomena separately and then to explore the consequences of their interaction.

Several theoretical approaches have been used to analyze the densification process during sintering, including the use of analytical models, scaling laws, and numerical simulations. The analytical models have been criticized from time to time because the drastic simplifications assumed in the models make them unsuitable for quantitatively predicting the sintering behavior of real powder systems. At best, they provide only a qualitative understanding of sintering. In spite of these shortcomings, the role of the analytical models in the development of our understanding of sintering should not be overlooked. Numerical simulations provide a powerful tool for gaining deeper insight into sintering, and the techniques are expected to see wider use in the future. Other approaches, such as phenomenological equations and sintering maps, attempt to represent sintering data in terms of equations or diagrams but provide very little insight into the process.

2.2 MECHANISMS OF SINTERING

Sintering of polycrystalline materials occurs by diffusional transport of matter along definite paths that define the mechanisms of sintering. Matter is transported from regions of higher chemical potential (referred to as the *source* of matter) to regions of lower chemical potential (referred to as the *sink*). There are at least six different mechanisms of sintering in polycrystalline materials, as shown schematically in Figure 2.1 for a system of three sintering particles. All six lead to bonding and growth of necks between the particles, so the strength of the powder compact increases during sintering. Only certain of the mechanisms, however, lead to shrinkage or densification, and a distinction is commonly made between *densifying* and *nondensifying* mechanisms. Surface diffusion, lattice diffusion from the particle surfaces to the neck, and vapor transport (mechanisms 1, 2, and 3) lead to neck growth without densification and are referred to as nondensifying mechanisms. Grain boundary diffusion and lattice diffusion from the grain boundary to the pore (mechanisms 4 and 5) are the most important densifying mechanisms in polycrystalline ceramics. Diffusion from the grain boundary to the pore permits neck growth as well as densification. Plastic flow by dislocation motion (mechanism 6) also leads to neck growth and densification but is more common in the sintering of metal powders. The nondensifying mechanisms cannot simply be ignored, because when they occur, they reduce the curvature of the neck surface (i.e., the driving force for sintering) and so reduce the rate of the densifying mechanisms.

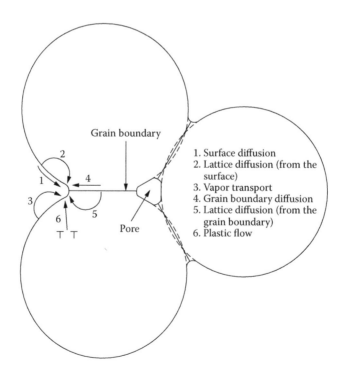

FIGURE 2.1 Six distinct mechanisms can contribute to the sintering of a consolidated mass of crystalline particles: (1) surface diffusion, (2) lattice diffusion from the surface, (3) vapor transport, (4) grain boundary diffusion, (5) lattice diffusion from the grain boundary, and (6) plastic flow. Only mechanisms 4–6 lead to densification, but all cause the necks to grow and so influence the rate of densification.

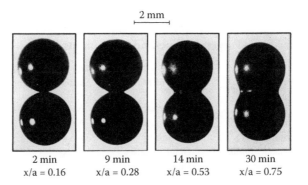

2 mm

| 2 min | 9 min | 14 min | 30 min |
| x/a = 0.16 | x/a = 0.28 | x/a = 0.53 | x/a = 0.75 |

FIGURE 2.2 Two-particle model made of glass spheres (3 mm diameter) sintered at 1000°C. (x and a represent the radius of the neck and the sphere, respectively.)

In addition to the alternative mechanisms, further complications arise from the diffusion of the different ionic species that make up the compound. As discussed in Chapter 1, the flux of the different ionic species is coupled to preserve the stoichiometry and electroneutrality of the compound. As a result, it is the slowest diffusing species along the fastest path that controls the rate of densification.

For glasses and other amorphous materials, which cannot have grain boundaries, neck growth and densification occur by viscous flow involving deformation of the particles. Figure 2.2 shows as an example the sintering of two glass spheres by viscous flow. The path by which matter flows is not clearly defined. The geometrical changes that accompany viscous flow are fairly complex, and, as we shall see later, severe simplifying assumptions are made in formulating the equations for matter transport. For the sintering of spheres, Figure 2.3 shows a schematic of two possible flow fields for viscous sintering [1]. While the form shown on the left-hand side may be expected in real systems, the results of recent numerical simulations show good agreement with the flow field on the right. Table 2.1 summarizes the sintering mechanisms in polycrystalline and amorphous solids.

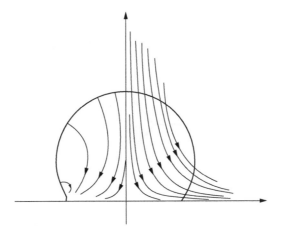

FIGURE 2.3 Flow fields for viscous sintering. *Right-hand side:* uniaxial contraction assumed in standard model; *left-hand side:* form expected in real situations. (From Uhlmann, D.R., Klein, L.C., and Hopper, R.W., Sintering, crystallization, and breccia formation, in *The Moon*, Vol. 13, D. Reidel Publishing Co., Dordrecht, Holland, 1975. With permission.)

TABLE 2.1
Mechanisms of Sintering in Polycrystalline and Amorphous Solids

Type of Solid	Mechanism	Source of Matter	Sink of Matter	Densifying	Nondensifying
Polycrystalline	Surface diffusion	Surface	Neck		×
	Lattice diffusion	Surface	Neck		×
	Vapor transport	Surface	Neck		×
	Grain boundary diffusion	Grain boundary	Neck	×	
	Lattice diffusion	Grain boundary	Neck	×	
	Plastic flow	Dislocations	Neck	×	
Amorphous	Viscous flow	Unspecified	Unspecified	×	

2.3 EFFECTS OF GRAIN BOUNDARIES

One important respect in which polycrystalline and amorphous materials differ is the presence of grain boundaries in polycrystalline material. The presence of the grain boundaries dictates the equilibrium shapes of the pores and the grains in a polycrystalline material. Consider a hypothetical pore surrounded by three grains, pictured in Figure 2.4. The forces must balance at the junction where the surfaces of the pores meet the grain boundary. They are normally represented by the tension in the interface, i.e., the tension in the solid–vapor interface and the tension in the grain boundary. By analogy with the surface tension of a liquid, a tension arises because an increase in the area of the interface leads to an increase in energy. At the junction, the tension γ_{sv} in the solid–vapor interface is tangential to that interface, while that in the grain boundary, γ_{gb}, is in the plane of the boundary. The balance of forces leads to

$$\gamma_{gb} = 2\gamma_{sv} \cos(\psi/2) \tag{2.1}$$

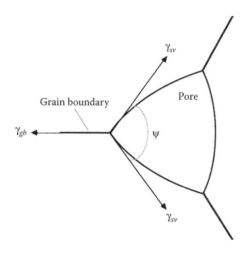

FIGURE 2.4 The equilibrium shapes of the pores in polycrystalline solids are governed by the balance between the surface and interfacial forces at the point where the grain boundary intersects the pore. γ_{sv} is the surface tension, γ_{gb} is the grain boundary tension, and ψ is the dihedral angle.

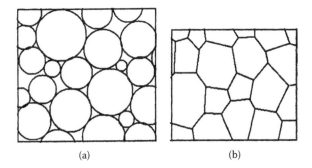

FIGURE 2.5 Schematic diagram illustrating the replacement of free surfaces (a) by grain boundaries (b) during the densification of polycrystalline systems.

where ψ is the dihedral angle. A few sintering models take the dihedral angle into account [2], but most assume that the pores have a circular cross-section, which is equivalent to assuming that $\psi = 180°$ or $\gamma_{gb} = 0$.

In the sintering of polycrystalline materials, part of the energy decreases due to the elimination of free surface area goes into creating new grain boundary area, so the driving force for sintering will be somewhat lower than that given by Equation 1.5, where the grain boundaries were neglected. Let us consider a powder compact in which the particles, assumed to be spherical, are in point contact (Figure 2.5a). The upper limit of the free-energy reduction during sintering can be expressed as

$$\Delta E_{max} = -\Delta A_s \gamma_{sv} \tag{2.2}$$

where ΔA_s is the total free surface area of the powder. When the powder compact densifies, the free surfaces are replaced by grain boundaries, as shown in Figure 2.5b. If full density is achieved, then the change in free energy associated with the densification is

$$\Delta E_d = -(\gamma_{sv} \Delta A_{sv} - \gamma_{gb} \Delta A_{gb}) \tag{2.3}$$

where ΔA_{gb} is the total grain boundary area of the dense solid. During densification, approximately two free surfaces coalesce into one grain boundary, as shown schematically in Figure 2.5, so $\Delta A_{gb} \approx \Delta A_s/2$. The total change of free energy resulting from densification is, therefore,

$$\Delta E_d \approx -A_s(\gamma_{sv} - \gamma_{gb}/2) \tag{2.4}$$

The free energy reduction given by Equation 2.4 drives the densification process. The remaining part of the free energy reduction, given by

$$\Delta E_g \approx -A_s \gamma_{gb}/2 \tag{2.5}$$

can be considered to drive the grain growth process. For most materials, γ_{sv} is larger than γ_{gb}, so ΔE_d is normally negative, and there is always a thermodynamic driving force for elimination of the pores.

One of the most important consequences of the presence of the grain boundaries is that grain growth provides an alternative process by which the powder system can decrease its free energy.

In sintering, grain growth is normally accompanied by pore growth, a process described as *coarsening*. Coarsening, therefore, occurs concurrently with sintering. A theory of sintering that rigorously analyzes the three-dimensional arrangement of particles, interaction between the particles, concurrent densification, and coarsening by the active transport mechanisms does not exist yet. The common approach, therefore, is to analyze sintering and grain growth separately. The understanding gained from the separate analyses is then used to explore the consequences of their interaction.

2.4 THEORETICAL ANALYSIS OF SINTERING

The main approaches that have been used in the theoretical analysis of sintering are summarized in Table 2.2. The development of the *analytical models,* which began about 1945–1950 [3–5], represents the first real attempt at a quantitative modeling of the sintering process. Whereas the analytical models assume an unrealistically simple geometry and, often, the occurrence of only a single mechanism, they have received the most attention and provide the basis for the present understanding of sintering. The *scaling law,* formulated by Herring [6], provides a reliable guide for understanding the particle size (i.e., length scale) dependence of the sintering mechanisms and how the relative rates of the different mechanisms are influenced by the particle size. *Numerical simulations* offer a powerful method for elucidating many of the complexities of sintering, and the approach is expected to see increasing use in the future. These simulations, which normally require advanced numerical methods, provide good insight into how matter is transported, as well as the ability to analyze more realistic particle geometries and the occurrence of simultaneous mechanisms. The *topological models,* outlined in Chapter 3, make limited predictions of the sintering kinetics and are more appropriate to the understanding of the microstructure evolution [7]. The *statistical models* [8] have received little attention since they were originally put forward and will not be considered further in this book. *Phenomenological equations* are used to fit sintering data but add almost no insight into the process. *Sintering maps* show changes in the sintering behavior and mechanisms under different conditions of temperature and particle size. Because the maps are based on the predictions of the analytical models, they suffer from the same limitations.

TABLE 2.2
Main Approaches Used for the Theoretical Analysis of Sintering

Approach	Comments	Ref.
Scaling law	Not dependent on specific geometry. Effects of change of scale on the rate of single mechanism derived.	6
Analytical models	Oversimplified geometry. Analytical equations for dependence of sintering rate on primary variables derived for single mechanism.	3–5, 11–14, 21–23, 29
Numerical simulations	Equations for matter transport solved numerically. Complex geometry and concurrent mechanisms analyzed.	31–45
Topological models	Analysis of morphological changes. Predictions of kinetics limited. More appropriate to microstructural evolution.	7
Statistical models	Statistical methods applied to the analysis of sintering. Simplified geometry. Semi-empirical analysis.	8
Phenomenological equations	Empirical or phenomenological derivation of equations to describe sintering data. No reasonable physical basis.	46, 47

2.5 HERRING'S SCALING LAW

The scaling law, formulated by Herring [6], considers the effect of change of length scale on microstructural phenomena during sintering. For a powder compact undergoing sintering, perhaps the most fundamental scaling parameter is the particle size. The scaling law attempts to answer the following important question: How does the change in scale (i.e., the particle size) influence the rate of sintering?

The scaling law does not assume a specific geometrical model. Instead, the main assumptions in the model are that during sintering: (1) the particle size of a given powder system remains the same, (2) the geometrical changes in different powder systems remain similar, and (3) the composition of the powder systems are the same. Two systems are defined as being *geometrically similar* if the linear dimensions of all of the features (grains, pores, etc.) of one system (system 1) are equal to a numerical factor times the linear dimensions of the corresponding features in the other system (system 2):

$$(Linear\ dimension)_1 = \lambda\,(Linear\ dimension)_2 \qquad (2.6)$$

where λ is a numerical factor. Geometrically similar systems, therefore, involve simply a magnification of one system relative to the other (Figure 2.6).

2.5.1 DERIVATION OF THE SCALING LAW

To illustrate the derivation of the scaling law, let us now consider a simple system consisting of two spheres in contact (Figure 2.7). We are not restricted to this geometry; it is chosen to simplify the illustration. Suppose it takes a time Δt_1 to produce a certain microstructural change in system 1 (e.g., the growth of a neck between the particles to a certain size X_1). The question we must attempt to answer is: How long (t_2) does it take to produce a geometrically similar change in system 2? For geometrically similar changes, the initial radius of the particle and the neck radius of the two systems are related by

$$a_2 = \lambda a_1; \quad X_2 = \lambda X_1 \qquad (2.7)$$

where a_1 and a_2 are the radii of the spheres in system 1 and system 2, respectively. The time required to produce a certain change by diffusional flow of matter can be expressed as

$$\Delta t = \frac{V}{JA} \qquad (2.8)$$

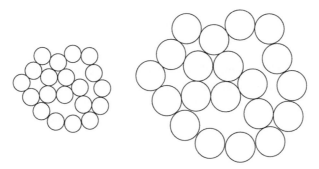

FIGURE 2.6 An example of two geometrically similar systems consisting of a random arrangement of circles. The systems differ only in scale and involve a simple magnification of one relative to the other.

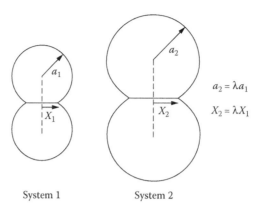

FIGURE 2.7 Geometrically similar models consisting of two spheres in contact. The linear dimensions of system 2 are a numerical factor times those of system 1.

where V is the volume of matter transported, J is the flux, and A is the cross-sectional area over which the matter is transported. We can, therefore, write

$$\frac{\Delta t_2}{\Delta t_1} = \frac{V_2 J_1 A_1}{V_1 J_2 A_2} \tag{2.9}$$

As an example of the application of Equation 2.9, consider matter transport by volume diffusion.

2.5.1.1 Scaling Law for Lattice Diffusion

For a sphere of radius a, the volume of matter transported is proportional to a^3. Therefore, V_2 is proportional to $(\lambda a_1)^3$, or $V_2 = \lambda^3 V_1$. For lattice diffusion, the area over which matter diffuses is proportional to a^2. Therefore, A_2 is proportional to $(\lambda a_1)^2$, or $A_2 = \lambda^2 A_1$. The flux J is proportional to $\nabla \mu$, the gradient in the chemical potential (Chapter 1). For a curved surface with a radius of curvature r, μ varies as $1/r$. Therefore, J varies as $\nabla(1/r)$ or as $1/r^2$. Since J_2 is proportional to $1/(\lambda r)^2$, we have $J_2 = J_1/\lambda^2$. To summarize, the parameters for lattice diffusion are as follows:

$$V_2 = \lambda^3 V_1; \quad A_2 = \lambda^2 A_1; \quad J_2 = J_1/\lambda^2 \tag{2.10}$$

Substituting into Equation 2.9 gives

$$\frac{\Delta t_2}{\Delta t_1} = \lambda^3 = \left(\frac{a_2}{a_1}\right)^3 \tag{2.11}$$

According to Equation 2.11, the time taken to produce geometrically similar changes by a lattice diffusion mechanism increases as the cube of the particle (or grain) size.

We can derive the scaling law for the other mass transport mechanisms through a procedure similar to that described above for lattice diffusion, as described in detail by Herring [6]. The law can be written in the general form

$$\frac{\Delta t_2}{\Delta t_1} = \lambda^m = \left(\frac{a_2}{a_1}\right)^m \tag{2.12}$$

TABLE 2.3
Exponents for Herring's Scaling Law
Described by Equation 2.12

Sintering Mechanism	Exponent m
Surface diffusion	4
Lattice diffusion	3
Vapor transport	2
Grain boundary diffusion	4
Plastic flow	1
Viscous flow	1

---

where m is an exponent that depends on the mechanism of sintering. Table 2.3 gives the values for m for the different sintering mechanisms.

2.5.2 APPLICATION AND LIMITATION OF THE SCALING LAW

An important application of the scaling law is the determination of how the relative rates of sintering by the different mechanisms depend on the particle size of the powder system. This type of information is useful, for example, in the fabrication of ceramics with controlled microstructure. As we observed earlier, some mechanisms lead to densification while others do not, so the achievement of high density requires that the rates of the densifying mechanisms be enhanced over those for the nondensifying mechanisms.

2.5.2.1 Relative Rates of Sintering Mechanisms

To determine the relative rates of the different mechanisms, it is useful to write Equation 2.12 in terms of a rate. For a given change, the rate is inversely proportional to the time, so Equation 2.12 can be written

$$\frac{(Rate)_2}{(Rate)_1} = \lambda^{-m} \tag{2.13}$$

In a given powder system, let us suppose that grain boundary diffusion and vapor transport (evaporation/condensation) are the dominant mass transport mechanisms. Then the rates of sintering by these two mechanisms vary with the scale of the system according to

$$(Rate)_{gb} \sim \lambda^{-4} \tag{2.14}$$

and

$$(Rate)_{ec} \sim \lambda^{-2} \tag{2.15}$$

The variation of the rates of sintering with λ for the two mechanisms is illustrated in Figure 2.8. The crossover point for the two lines is arbitrary, but this does not affect the validity of the results. We see that for small λ, i.e., as the particle size becomes smaller, the rate of sintering by grain

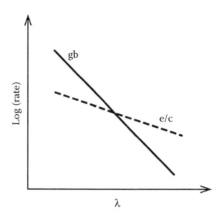

FIGURE 2.8 Schematic diagram of the relative rates of sintering by grain boundary diffusion and by evaporation/condensation as a function of the length scale (i.e., particle size) of the system.

boundary diffusion is enhanced compared to that for vapor transport. Conversely, the rate of sintering by vapor transport dominates for larger λ, i.e., for larger particle sizes. According to the scaling laws, smaller particle size is beneficial for densification when grain boundary diffusion and vapor transport are the dominant mechanisms. For the case where surface diffusion and lattice diffusion are the dominant mechanisms, we can use a similar procedure to show that surface diffusion is enhanced as the particle size decreases. We leave as an exercise for the reader to consider other combinations, for example, (1) lattice diffusion versus grain boundary diffusion and (2) surface diffusion versus grain boundary diffusion.

2.5.2.2   Limitation of the Scaling Law

Because of its general approach and the simple physical principles employed in its derivation, the scaling law approach overcomes some limitations of the analytical models. Because the geometric details of the powder system do not enter into the derivation, the law can be applied to particles of any shape and to all stages of the sintering process. On the other hand, we must remember the conditions that govern the validity of the law. The derivation assumes that the particle size of each powder system does not change during sintering and that the microstructural changes remain geometrically similar in the two systems. This requirement of geometrically similar microstructural changes is a key limitation of the scaling law, because it is difficult to maintain in real powder systems. The scaling law also requires that the two systems be identical in chemical composition so that the mass transport coefficients (e.g., the diffusion coefficients) are the same.

Since the exponent m in Equation 2.12 and Equation 2.13 depends on the mechanism of sintering, it may seem that the measurement of m would provide information on the mechanism. In practice, several factors can complicate the task of determining the mechanism. One problem is the simultaneous occurrence of more than one mechanism, in which case the measured exponent may correspond to a mechanism that is entirely different from those operating. In addition to the difficulty of maintaining a geometrically similar microstructure, assumed in the derivation of the scaling law, the sintering mechanism may change with the particle size. While the scaling law appears to be obeyed reasonably well in simple metallic systems, e.g., nickel wires, copper spheres, and silver spheres, the application of the law to the sintering of Al_2O_3 produced unexpected results [9]. For powders with a narrow size distribution, high nonintegral values of m were found. The value of m also varied with the extent of densification, especially in the early stages of sintering.

2.6 ANALYTICAL MODELS

The analytical models assume a relatively simple, idealized geometry for the powder system, and for each mechanism the mass transport equations are solved analytically to provide equations for the sintering kinetics. A problem is that the microstructure of a real powder compact changes continuously as well as drastically during sintering. It is, therefore, difficult to find a single geometrical model that can adequately represent the entire process, yet provide the degree of simplicity for the mass transport equations to be solved analytically. To overcome this problem, the sintering process is conceptually divided into separate *stages,* and for each stage an idealized geometry that has a rough similarity with the microstructure of the real powder system is assumed.

2.6.1 STAGES OF SINTERING

Sintering is normally thought to occur in three sequential stages, referred to as the *initial* stage, the *intermediate* stage, and the *final* stage. In some analyses of sintering, an extra stage, stage 0, is considered, which describes the instantaneous contact between the particles, when they are first brought together, due to elastic deformation in response to surface energy reduction at the interface [10]. However, we shall not consider this refinement. A stage represents an interval of time or density over which the microstructure is considered to be reasonably well defined. For polycrystalline materials, Figure 2.9 shows the idealized geometrical structures that were suggested by Coble [11]

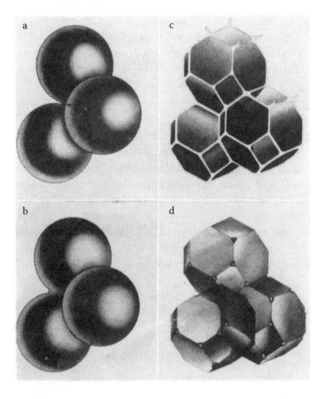

FIGURE 2.9 Idealized models for the three stages of sintering. (a) Initial stage: model structure represented by spheres in tangential contact. (b) Near the end of the initial stage: spheres have begun to coalesce. The neck growth illustrated is for center-to-center shrinkage of 4%. (c) Intermediate stage: dark grains have adopted the shape of a tetrakaidecahedron, enclosing white pore channels at the grain edges. (d) Final stage: pores are tetrahedral inclusions at the corners where four tetrakaidecahedra meet. (From Coble, R.L., Sintering crystalline solids. I. Intermediate and final state diffusion models. *J. Appl. Phys.*, 32, 787, 1961. With permission.)

as representative of the three stages. For amorphous materials, the geometrical model assumed for the initial stage is similar to that for the polycrystalline system. However, the models for the intermediate and final stages are very different from those for the polycrystalline case. We discuss this topic further below.

2.6.1.1 Initial Stage

The initial stage consists of fairly rapid interparticle neck growth by diffusion, vapor transport, plastic flow, or viscous flow. The large initial differences in surface curvature are removed in this stage, and shrinkage (or densification) accompanies neck growth for the densifying mechanisms. For a powder system consisting of spherical particles, the initial stage is represented as the transition between Figure 2.9a and Figure 2.9b. It is assumed to last until the radius of the neck between the particles has reached a value of 0.40–0.50 of the particle radius. For a powder system with an initial relative density of 0.50–0.60, this corresponds to a linear shrinkage of 3%–5%, or an increase in relative density to ~0.65 when the densifying mechanisms dominate.

2.6.1.2 Intermediate Stage

The intermediate stage begins when the pores have reached their equilibrium shapes as dictated by the surface and interfacial tensions (see Section 2.3). The pore phase is still continuous. In the sintering models, the structure is usually idealized in terms of a spaghetti-like array of porosity sitting along the grain edges, as illustrated in Figure 2.9c. Densification is assumed to occur by the pores simply shrinking to reduce their cross-section. Eventually, the pores become unstable and pinch off, leaving isolated pores; this constitutes the beginning of the final stage. The intermediate stage normally covers the major part of the sintering process and is taken to end when the density is ~0.90 of the theoretical.

2.6.1.3 Final Stage

The microstructure in the final stage can develop in a variety of ways, and we shall consider this in detail in Chapter 3. In one of the simplest descriptions, the final stage begins when the pores pinch off and become isolated at the grain corners, as shown by the idealized structure pictured in Figure 2.9d. In this simple description, the pores are assumed to shrink continuously, and they may disappear altogether. The removal of almost all of the porosity has been achieved in the sintering of several real powder systems.

Some of the main parameters associated with the three idealized stages of sintering are summarized in Table 2.4, and examples of the microstructures (planar section) of real powder compacts in the initial, intermediate, and final stages are shown in Figure 2.10.

TABLE 2.4
Parameters Associated with the Stages of Sintering for Polycrystalline Solids

Stage	Typical Microstructural Feature	Relative Density Range	Idealized Model
Initial	Rapid interparticle neck growth	Up to ~0.65	Two monosize spheres in contact
Intermediate	Equilibrium pore shape with continuous porosity	~0.65–0.90	Tetrakaidecahedron with cylindrical pores of the same radius along the edges
Final	Equilibrium pore shape with isolated porosity	>~0.90	Tetrakaidecahedron with spherical monosize pores at the corners

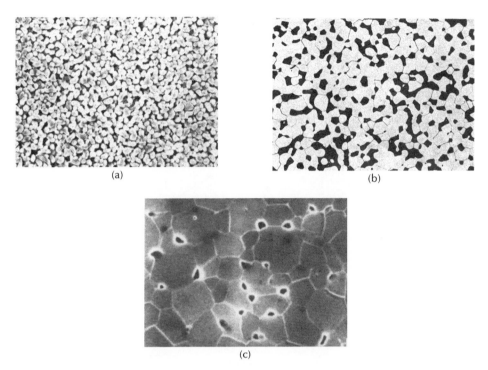

(a)

(b)

(c)

FIGURE 2.10 Examples of real microstructures (planar sections) for (a) initial stage of sintering, (b) intermediate stage, and (c) final stage.

2.6.2 MODELING THE SINTERING PROCESS

The analytical models commonly assume that the particles in the initial powder compact are spherical and of the same size, and that they are uniformly packed. With these assumptions, a unit of the powder system, called the *geometrical model*, can be isolated and analyzed. By imposing the appropriate boundary conditions, we can consider the remainder of the powder system as a continuum having the same macroscopic properties (e.g., shrinkage and densification rate) as the isolated unit. The derivation of the equations for the sintering kinetics follows a simple procedure: for the assumed geometrical model, the mass transport equations are formulated and solved under the appropriate boundary conditions.

2.6.3 INITIAL STAGE MODELS

2.6.3.1 Geometrical Model

The model for the initial stage consists of two equal-sized spheres in contact; hence, it is referred to as the *two-sphere model*. Two slightly different geometries can be considered, depending on whether the mechanisms are nondensifying (Figure 2.11a) or densifying (Figure 2.11b). The two-sphere model for the densifying mechanisms accounts for interpenetration of the spheres (i.e., shrinkage) as well as neck growth. The neck formed between the particles is assumed to be circular with a radius X and with a surface having a circular cross-section with a radius r. Assuming a circular cross-section for the neck surface is equivalent to assuming that the grain boundary energy is zero. The main geometrical parameters of the model are the principal radii of curvature of the neck surface, r and X, the area of the neck surface, A, and the volume of material transported into the neck, V. These parameters are summarized in Figure 2.11, and we leave as an exercise for the reader to derive them. It will be noticed that the parameters for the densifying model differ by only a small numerical factor from those for the nondensifying model.

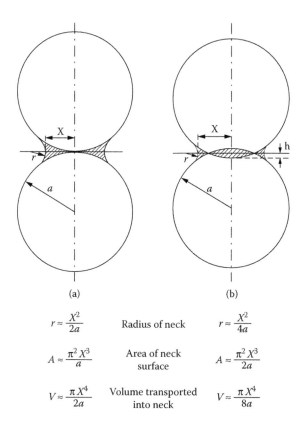

$$r \approx \frac{X^2}{2a} \qquad \text{Radius of neck} \qquad r \approx \frac{X^2}{4a}$$

$$A \approx \frac{\pi^2 X^3}{a} \qquad \begin{array}{c}\text{Area of neck}\\\text{surface}\end{array} \qquad A \approx \frac{\pi^2 X^3}{2a}$$

$$V \approx \frac{\pi X^4}{2a} \qquad \begin{array}{c}\text{Volume transported}\\\text{into neck}\end{array} \qquad V \approx \frac{\pi X^4}{8a}$$

FIGURE 2.11 Geometrical parameters for the two-sphere model used in the derivation of the initial-stage sintering equations for crystalline particles. The geometries shown correspond to those for (a) the nondensifying mechanisms and (b) the densifying mechanisms.

2.6.3.2 Kinetic Equations

Diffusional transport of matter, as described in Chapter 1, can be analyzed in terms of the flux of atoms or, equivalently, in terms of the counterflow of vacancies. In the early development of the sintering theories, the approach based on the counterflow of vacancies driven by a vacancy concentration gradient was used predominantly. We adopt this approach in the present section, but later in the chapter we will outline a more general approach based on the flux of atoms driven by a chemical potential gradient. As an illustration of the derivation of the initial-stage sintering equations, let us start with the mechanism of grain boundary diffusion.

According to Equation 1.127, the flux of atoms into the neck is

$$J_a = \frac{D_v}{\Omega} \frac{dC_v}{dx} \tag{2.16}$$

where D_v is the vacancy diffusion coefficient, Ω is the volume of an atom or vacancy, dC_v/dx is the vacancy concentration gradient (in one dimension), and C_v is the *fraction* of sites occupied by the vacancies. The volume of matter transported into the neck per unit time is

$$\frac{dV}{dt} = J_a A_{gb} \Omega \tag{2.17}$$

where A_{gb} is the cross-sectional area over which diffusion occurs. Grain boundary diffusion is assumed to occur over a constant thickness δ_{gb}, so that $A_{gb} = 2\pi X \delta_{gb}$, where X is the radius of the neck. Combining Equation 2.16 and Equation 2.17 and substituting for A_{gb} gives

$$\frac{dV}{dt} = D_v 2\pi X \delta_{gb} \frac{dC_v}{dx} \tag{2.18}$$

Since the neck radius increases radially in a direction orthogonal to a line joining the centers of the spheres, a one-dimensional solution is adequate. Assuming that the vacancy concentration between the neck surface and the center of the neck is constant, then $dC_v/dx = \Delta C_v/X$, where ΔC_v is the difference in vacancy concentration between the neck surface and the center of the neck. The vacancy concentration C_{vo} at the center of the neck is assumed to be equal to that under a flat, stress-free surface, so, according to Equation 1.113, we can write

$$\Delta C_v = C_v - C_{vo} = \frac{C_{vo} \gamma_{sv} \Omega}{kT} \left(\frac{1}{r_1} + \frac{1}{r_2} \right) \tag{2.19}$$

where r_1 and r_2 are the two principal radii of curvature of the neck surface. From Figure 2.11, $r_1 = r$ and $r_2 = -X$, and it is assumed that $X \gg r$. Substituting into Equation 2.18 gives

$$\frac{dV}{dt} = \frac{2\pi D_v C_{vo} \delta_{gb} \gamma_{sv} \Omega}{kTr} \tag{2.20}$$

Using the relations given in Figure 2.11b for V and r, and putting the grain boundary diffusion coefficient D_{gb} equal to $D_v C_{vo}$, we obtain

$$\frac{\pi X^3}{2a} \frac{dX}{dt} = \frac{2\pi D_{gb} \delta_{gb} \gamma_{sv} \Omega}{kT} \left(\frac{4a}{X^2} \right) \tag{2.21}$$

Rearranging Equation 2.21 gives

$$X^5 dX = \frac{16 D_{gb} \delta_{gb} \gamma_{sv} \Omega a^2}{kT} dt \tag{2.22}$$

After integration and application of the boundary conditions $X = 0$ at $t = 0$, Equation 2.22 becomes

$$X^6 = \frac{96 D_{gb} \delta_{gb} \gamma_{sv} \Omega a^2}{kT} t \tag{2.23}$$

We may also write Equation 2.23 in the form

$$\frac{X}{a} = \left(\frac{96 D_{gb} \delta_{gb} \gamma_{sv} \Omega}{kT a^4} \right)^{1/6} t^{1/6} \tag{2.24}$$

Equation 2.24 predicts that the ratio of the neck radius to the sphere radius increases as $t^{1/6}$.

For this densifying mechanism, the linear shrinkage, defined as the change in length ΔL divided by the original length L_0 of the geometrical model, can also be found. As a good approximation, we can write

$$\frac{\Delta L}{L_0} = -\frac{h}{a} = -\frac{r}{a} = -\frac{X^2}{4a^2} \tag{2.25}$$

where h is half the interpenetration distance between the spheres. Using Equation 2.24, we obtain

$$\frac{\Delta L}{L_0} = -\left(\frac{3D_{gb}\delta_{gb}\gamma_{sv}\Omega}{2kTa^4}\right)^{1/3} t^{1/3} \tag{2.26}$$

so the shrinkage is predicted to increase as $t^{1/3}$.

As another example, let us consider the mechanism of viscous flow. For this mechanism, matter transport is assumed to be governed by an energy balance concept, first put forward by Frenkel [3], which can be stated as follows:

Rate of energy dissipation by viscous flow = rate of energy gained by reduction in surface area

$$\tag{2.27}$$

The equation for neck growth between two spheres by viscous flow was derived by Frenkel. The original derivation contained an extra factor of π that is omitted in the version given here. For the parameters shown in Figure 2.12 and assuming that the radius of the sphere remains roughly constant during the viscous flow, the decrease in the surface area of the two spheres is

$$S_o - S = 8\pi a^2 - 4\pi a^2(1 + \cos\theta) \tag{2.28}$$

Note that this means that the material removed from the plane of contact is uniformly distributed over the surface of the sphere, rather than accumulating at the neck. For small values of θ, i.e., for small neck radius, $\cos\theta \approx 1 - \theta^2/2$, so Equation 2.28 becomes

$$S_o - S = 2\pi a^2\theta^2 \tag{2.29}$$

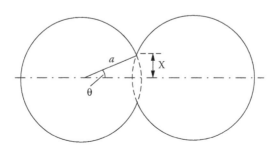

FIGURE 2.12 Geometrical parameters of the two-sphere model used by Frenkel in the derivation of the initial stage equation for viscous sintering.

The rate of change of energy due to the reduction in surface area can be written

$$\dot{E}_s = -\gamma_{sv}\frac{dS}{dt} = 4\pi a^2\gamma_{sv}\frac{d}{dt}\left(\frac{\theta^2}{2}\right) \tag{2.30}$$

where γ_{sv} is the specific surface energy of the solid–vapor interface. According to Frenkel, the rate of energy dissipation by viscous flow between the two spheres is

$$\dot{E}_v = \frac{16}{3}\pi a^3\eta\dot{u}^2 \tag{2.31}$$

where η is the viscosity of the glass and $\dot{u}$ is the velocity of motion for the viscous flow given by

$$\dot{u} = \frac{1}{a}\frac{d}{dt}\left(\frac{a\theta^2}{2}\right) = \frac{d}{dt}\left(\frac{\theta^2}{2}\right) \tag{2.32}$$

This equation is based on the assumption that flow occurs uniformly along the axis joining the centers of the spheres, rather than being concentrated near the neck. Substituting for $\dot{u}$ in Equation 2.31 and putting $\dot{E}_s = \dot{E}_v$ gives

$$\frac{16}{3}\pi a^3\eta\dot{u}\frac{d}{dt}\left(\frac{\theta^2}{2}\right) = 4\pi a^2\gamma_{sv}\frac{d}{dt}\left(\frac{\theta^2}{2}\right) \tag{2.33}$$

Rearranging Equation 2.33 gives

$$\dot{u} = \frac{3}{4}\left(\frac{\gamma_{sv}}{\eta a}\right) \tag{2.34}$$

Substituting for $\dot{u}$ in Equation 2.32 and integrating, subject to the boundary conditions of $\theta = 0$ at $t = 0$, we obtain

$$\theta^2 = \frac{3}{2}\left(\frac{\gamma_{sv}}{\eta a}\right)t \tag{2.35}$$

Since $\theta = X/a$, where X is the neck radius, Equation 2.35 gives

$$\frac{X}{a} = \left(\frac{3\gamma_{sv}}{2\eta a}\right)^{1/2}t^{1/2} \tag{2.36}$$

It is left as an exercise for the reader to determine the equation for the shrinkage by this densifying mechanism.

2.6.3.3 Summary of the Initial Stage Sintering Equations

The original derivations of the initial stage sintering equations can be found in publications by Kuczynski [5], Kingery and Berg [12], Coble [13], and Johnson and Cutler [14]. The equations

TABLE 2.5
Plausible Values for the Constants Appearing in Equation 2.37
and Equation 2.38 for the Initial Stage of Sintering

Mechanism	m	n	H^{b}
Surface diffusion[a]	7	4	$56D_{s}\delta_{s}\gamma_{sv}\Omega/kT$
Lattice diffusion from the surface[a]	4	3	$20D_{l}\gamma_{sv}\Omega/kT$
Vapor transport[a]	3	2	$3p_{o}\gamma_{sv}\Omega/(2\pi mkT)^{1/2}kT$
Grain boundary diffusion	6	4	$96D_{gb}\delta_{gb}\gamma_{sv}\Omega/kt$
Lattice diffusion from the grain boundary	5	3	$80\pi D_{l}\gamma_{sv}\Omega/kT$
Viscous flow	2	1	$3\gamma_{sv}/2\eta$

[a] Denotes nondensifying mechanism, i.e., shrinkage $L/L_{0} = 0$.
[b] D_{s}, D_{l}, D_{gb}, diffusion coefficients for surface, lattice, and grain boundary diffusion. δ_{s}, δ_{gb}, thickness for surface and grain boundary diffusion; γ_{sv}, specific surface energy; p_{o}, vapor pressure over a flat surface; m, mass of atom; k, Boltzmann constant; T, absolute temperature; η, viscosity.

for neck growth and, for the densifying mechanisms, shrinkage, can be expressed in the general form

$$\left(\frac{X}{a}\right)^{m} = \frac{H}{a^{n}}t \tag{2.37}$$

$$\left(\frac{\Delta L}{L_{0}}\right)^{m/2} = -\frac{H}{2^{m}a^{n}}t \tag{2.38}$$

where m and n are numerical exponents that depend on the mechanism of sintering, and H is a function that contains the geometrical and material parameters of the powder system. Depending on the assumptions made in the models, a range of values for m, n, and the numerical constant in H have been obtained. The values given in Table 2.5 represent the most plausible values for each mechanism [15,16].

2.6.3.4 Limitations of the Initial Stage Sintering Equations

The form of the neck growth equations indicates that a plot of log (X/a) versus log t yields a straight line with a slope equal to $1/m$, so by fitting the theoretical predictions to experimental data, the value of m can be found. A similar procedure can be applied to the analysis of shrinkage, if it occurs during sintering. Data for validating the models are commonly obtained by measuring the neck growth in simple systems (e.g., two spheres, a sphere on a plate, or two wires) or the shrinkage in a compacted mass of spherical particles. Since m is dependent on the mechanism of sintering, at first sight it may seem that the measurement of m would provide information on the mechanism of sintering. A problem is that the basic assumption in the model of a single dominant mass transport mechanism is not valid in most powder systems. As we have mentioned previously, when more than one mechanism operates simultaneously, the measured exponent may correspond to an entirely different mechanism. As an example, consider the initial sintering of copper. Kingery and Berg [12] showed that the neck growth and shrinkage data of copper spheres gave exponents characteristic of lattice diffusion as the dominant mechanism (Figure 2.13). Later analysis showed that surface diffusion was the dominant mechanism, with a significant contribution from lattice diffusion (which gave shrinkage). In the case of glass, sintering is dominated by viscous flow, and Frenkel's neck growth equation is well verified by the data of Kuczynski [17].

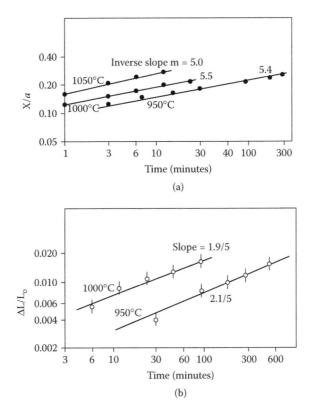

FIGURE 2.13 Data for (a) neck growth and (b) shrinkage of copper spheres. The exponents are characteristic of those for lattice diffusion. (From Kingery, W.D., and Berg, M., Study of the initial stages of sintering solids by viscous flow, evaporation–condensation, and self-diffusion, *J. Appl. Phys.*, 26, 1205, 1955. With permission.)

The other simplifying assumptions of the models must also be remembered. The extension of the two-sphere geometry to real powder compacts is valid only if the particles are spheres of the same size arranged in a uniform pattern. In practice, this system is, at best, approached only by the uniform consolidation, by colloidal methods, of monodisperse powders. Coble [18] considered the effect of a particle size distribution on the initial stage of sintering by considering a linear array of spheres.

Johnson [2] included the dihedral angle in his sintering model, but most analytical models assume that the cross-section of the neck surface is circular, which is equivalent to assuming that the dihedral angle $\psi = 180°$ or that the grain boundary energy $\gamma_{gb} = 0$. Numerical simulations, described later, indicate that this circular neck cross-section is grossly simplified. Johnson's results indicate that the neglect of the grain boundary energy is insignificant if ψ is greater than 150°, a value that is much higher than the average values of ψ (110°–120°) measured for Al_2O_3 and MgO.

The models also make simplifying assumptions about the way in which matter transported into the neck is redistributed over the surface. Matter transported into the neck by grain boundary diffusion must be redistributed over the neck surface to prevent buildup on the grain boundary groove [19]. The models assume that surface diffusion is fast enough to cause redistribution, but this assumption has been questioned from time to time in the sintering literature.

2.6.4 Intermediate Stage Models

The geometrical model for the intermediate stage sintering of polycrystalline systems differs from that proposed for amorphous systems, so we will consider the two systems separately.

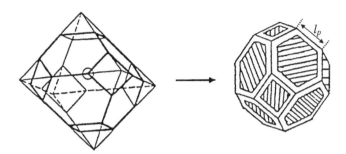

FIGURE 2.14 Sketch illustrating the formation of a tetrakaidecahedron from a truncated octahedron.

2.6.4.1 Geometrical Model for Solid-State Sintering

The geometrical model commonly used for the intermediate stage was proposed by Coble [11]. The powder system is idealized by considering it to consist of a space-filling array of equal-sized tetrakaidecahedra, each of which represents one particle. The porosity is cylindrical, with the axis of the cylinder coinciding with the edge of the tetrakaidecahedra (Figure 2.9c). The unit cell of the structure is taken as a tetrakaidecahedron with cylindrical pores along its edges.

A tetrakaidecahedron is constructed from an octahedron by trisecting each edge and joining the points to remove the six edges (Figure 2.14). The resulting structure has 36 edges, 24 corners, and 14 faces (eight hexagonal and six square). The volume of the tetrakaidecahedron is

$$V_t = 8\sqrt{2}\,l_p^3 \tag{2.39}$$

where l_p is the edge length of the tetrakaidecahedron. If r is the radius of the pore, then the total volume of the porosity per unit cell is

$$V_p = \frac{1}{3}(36\pi r^2 l_p) \tag{2.40}$$

The porosity of the unit cell, V_p/V_t, is, therefore,

$$P_c = \frac{3\pi}{2\sqrt{2}}\left(\frac{r^2}{l_p^2}\right) \tag{2.41}$$

2.6.4.2 Mechanisms

Since the model assumes that the pore geometry is uniform, the nondensifying mechanisms cannot operate. This is because the chemical potential is the same everywhere on the pore surface. We are left with the densifying mechanisms: lattice diffusion and grain boundary diffusion. Plastic flow, another densifying mechanism in metals, is not expected to operate to any significant extent for ceramic systems.

2.6.4.3 Sintering Equations

Following the analysis by Coble [11], we outline the derivation of the kinetic equations for sintering by lattice diffusion and grain boundary diffusion.

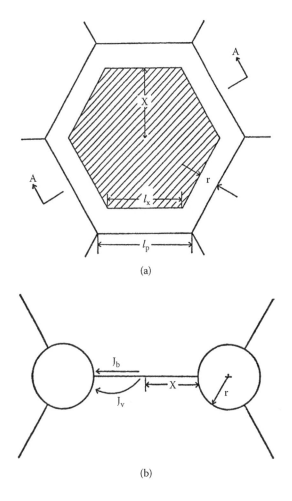

(a)

(b)

FIGURE 2.15 (a) The intermediate stage sintering equations for polycrystalline solids are based on a hexagonal neck. (b) The section (on A–A) shows a cut through the neck with the atomic flux paths for grain boundary diffusion and lattice diffusion.

2.6.4.3.1 Lattice Diffusion

Figure 2.15a shows that the cylindrical pores along the edges enclose each face of the tetrakaidecahedron. Because the vacancy flux from the pores terminates on the boundary faces (Figure 2.15b), Coble assumed radial diffusion from a circular vacancy source, and the shape effects on the corner of the tetrakaidecahedron were neglected. For the boundary to remain flat, the vacancy flux per unit area of the boundary must be the same over the whole boundary. The diffusion flux field can be approximated to that of the temperature distribution in a surface-cooled, electrically heated cylindrical conductor. The flux per unit length of the cylinder is given by

$$J/l = 4\pi D_v \Delta C \tag{2.42}$$

where D_v is the vacancy diffusion coefficient and C is the difference in vacancy concentration between the pore (source) and the boundary (sink). Coble also made several other assumptions, including the following:

1. The convergence of the flux to the boundary does not qualitatively change the flux equation with regard to its dependence on the pore radius.
2. The width of the flux field, i.e., the equivalence of l in Equation 2.42, is equal to the pore diameter.
3. The flux is increased by a factor of 2 due to the freedom of the vacancy diffusion flux to diverge initially, thereby providing additional available area.

With these assumptions, Equation 2.42 becomes

$$J = 2(4\pi D_v \Delta C)2r \tag{2.43}$$

Since there are fourteen faces in a tetrakaidecahedron and each face is shared by two grains, the volume flux per unit cell is

$$\frac{dV}{dt} = \frac{14}{2}J = 112\pi D_v \Delta C \tag{2.44}$$

For the cylindrical pore in the intermediate stage, the two principal radii of curvature are r and ∞, so from Equation 2.19, ΔC is given by

$$\Delta C = \frac{C_{vo}\gamma_{sv}\Omega}{kTr} \tag{2.45}$$

Substituting into Equation 2.44 and putting the lattice diffusion coefficient $D_l = D_v C_{vo}$, we obtain

$$dV = \frac{112\pi D_l \gamma_{sv}\Omega}{kT}dt \tag{2.46}$$

The integral of dV is equal to the porosity given by Equation 2.40, that is,

$$\int dV = 12\pi r^2 l_p \Big]_{r_o}^{r} \tag{2.47}$$

Combining with Equation 2.46 gives

$$r^2 \Big]_r^0 \approx -10\frac{D_l \gamma_{sv}\Omega}{l_p kT}t \Big]_t^{t_f} \tag{2.48}$$

where t_f is the time when the pore vanishes. Dividing both sides of this equation by l_p^2 and evaluating the integrand yields

$$P_c \approx \frac{r^2}{l_p^2} \approx \frac{10D_l \gamma_{sv}\Omega}{l_p^3 kT}(t_f - t) \tag{2.49}$$

In view of the many approximations used by Coble, this equation can only be considered as an order-of-magnitude calculation. The model applies until the pores pinch off and become isolated.

The sintering equations are commonly expressed in terms of the densification rate. Using the relation between porosity P and relative density ρ, that is, $P = 1 - \rho$, and differentiating Equation 2.49 with respect to time, we obtain

$$\frac{d}{dt}(P_c) = -\frac{d\rho}{dt} \approx -\frac{10D_l\gamma_{sv}\Omega}{l_p^3 kT} \tag{2.50}$$

If we take l_p as approximately equal to the grain size G and write the densification rate in the form of a volumetric strain rate, Equation 2.50 becomes

$$\frac{1}{\rho}\frac{d\rho}{dt} \approx \frac{10D_l\gamma_{sv}\Omega}{\rho G^3 kT} \tag{2.51}$$

According to this equation, the densification rate at a fixed density is predicted to depend inversely as the cube of the grain size, in agreement with the prediction of Herring's scaling law.

2.6.4.3.2 Grain Boundary Diffusion

Using the same geometrical model described above for lattice diffusion and modifying the flux equations to account for grain boundary diffusion, Coble derived the following equation:

$$P_c \approx \frac{r^2}{l_p^2} \approx \left(\frac{2D_{gb}\delta_{gb}\gamma_{sv}\Omega}{l_p^4 kT}\right)^{2/3} t^{2/3} \tag{2.52}$$

Following the procedure outlined above, we can express Equation 2.52 as

$$\frac{1}{\rho}\frac{d\rho}{dt} \approx \frac{4}{3}\left[\frac{D_{gb}\delta_{gb}\gamma_{sv}\Omega}{\rho(1-\rho)^{1/2}G^4 kT}\right] \tag{2.53}$$

At a fixed density, the densification rate for grain boundary diffusion is predicted to vary inversely as the fourth power of the grain size, as predicted by Herring's scaling law. An experimental test of Equation 2.51 and Equation 2.53 indicated that when the sintering kinetics are corrected for grain growth, the data for A_2O_3 (particle size of 0.3 μm) are well explained by the equation for the lattice diffusion mechanism [20].

Models for the intermediate stage of sintering have also been developed by Johnson [21] and Beeré [22]. These models can be considered as refinements of Coble's model. Johnson derived equations for the shrinkage in terms of the average values of the neck radius and the pore radius. His model cannot be used to predict the rate of sintering. Instead, the equations are meant to help in the analysis of sintering data. From the measured values of the average neck radius and pore radius, we can infer the boundary and lattice diffusion coefficients and the relative flux of matter due to the two mechanisms. Beeré extended Coble's model by allowing the pores to relax to a minimum free energy configuration. The pores have a complex curvature and meet the grain boundary at a constant dihedral angle satisfying the balance of interfacial tensions (see Section 2.3). Beeré's model shows the same dependence on grain size and temperature as Coble's model, but it contains additional terms that involve the dihedral angle and the grain boundary area.

2.6.4.4 Geometrical Model for Viscous Sintering

The structure shown in Figure 2.16 has been proposed by Scherer [23] as a model for the intermediate stage sintering of amorphous materials. It consists of a cubic array formed by intersecting cylinders, and densification is assumed to be brought about by the cylinders getting shorter and thicker. The model can be viewed as an idealized structure in which the cylinders represent strings of spherical particles joined together by necks.

The structure is represented by a unit cell consisting of twelve quarter cylinders (Figure 2.16a). The volume of the solid phase in the unit cell is

$$V_s = 3\pi a^2 l - 8\sqrt{2}a^3 \tag{2.54}$$

where l is the length of the side of the unit cell and a is the radius of the cylinder. Since the total volume of the cell is l^3, the density of the cell, d, is equal to $d_s V_s / l^3$, where d_s is the theoretical density of the solid phase. The relative density ρ, defined as d/d_s, is given by

$$\rho = 3\pi x^2 - 8\sqrt{2}x^3 \tag{2.55}$$

where $x = a/l$. According to this equation, ρ is a function of a/l only. The inverse of Equation 2.55 is

$$x = \frac{\pi\sqrt{2}}{8}\left[\frac{1}{2} + \cos\left(\Theta + \frac{4\pi}{3}\right)\right] \tag{2.56}$$

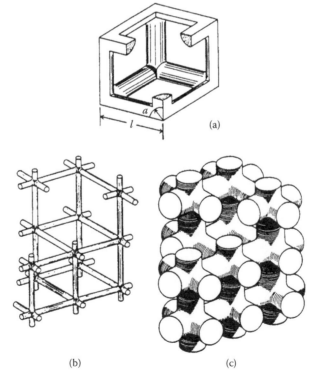

(a)

(b)   (c)

FIGURE 2.16 Scherer's model for viscous sintering. (a) Unit cell of the structure with edge length l and cylinder radius a. (b) Microstructure with a relative density of ~0.1. (c) Microstructure with a relative density of ~0.5.

where

$$\Theta = \frac{1}{3} \cos^{-1} \left[1 - \left(\frac{4}{\pi} \right)^3 \rho \right] \qquad (2.57)$$

The volume of solid phase in the unit cell is ρl^3. This volume does not change and can also be put equal to $\rho_o l_o^3$, where ρ_o and l_o are the initial values of the relative density and length, respectively, of the unit cell. Each cell contains one pore, so the number of pores per unit volume of the solid phase is

$$N = 1/\rho_o l_o^3 \qquad (2.58)$$

The model should be valid until the adjacent cylinders touch, thereby isolating the pores. This occurs when $a/l = 0.5$, i.e., when $\rho = 0.94$.

The derivation of the sintering equations for Scherer's model is quite similar to that outlined earlier for Frenkel's initial-stage model. The result is [23]:

$$\int_{t_o}^{t} \frac{\gamma_{sv} N^{1/3}}{\eta} dt = \int_{x_o}^{x} \frac{2dx}{(3\pi - 8\sqrt{2}x)^{1/3} x^{2/3}} \qquad (2.59)$$

where γ_{sv} is the specific surface energy of the solid–vapor interface and η is the viscosity of the solid phase. By making the substitution

$$y = \left(\frac{3\pi}{x} - 8\sqrt{2} \right)^{1/3} \qquad (2.60)$$

we can evaluate the integral on the right-hand side of Equation 2.59. The result can be written as

$$\frac{\gamma_{sv} N^{1/3}}{\eta} (t - t_o) = F_S(y) - F_S(y_o) \qquad (2.61)$$

where

$$F_S(y) = -\frac{2}{\alpha} \left[\frac{1}{2} \ln \left(\frac{\alpha^2 - \alpha y + y^2}{(\alpha + y)^2} \right) + \sqrt{3} \arctan \left(\frac{2y - \alpha}{\alpha\sqrt{3}} \right) \right] \qquad (2.62)$$

and $\alpha = (8\sqrt{2})^{1/3}$.

The predictions of the model are normally expressed in terms of a curve of ρ versus $(\gamma_{sv} N^{1/3}/\eta)(t - t_o)$, referred to as the *reduced time*. The curve is obtained as follows: For a chosen value of ρ the parameter y is found from Equation 2.56, Equation 2.57, and Equation 2.60. The function $F_S(y)$ is then found from Equation 2.62, and the reduced time is obtained from Equation 2.61. The procedure is repeated for other values of ρ. The predictions of the model are shown by the full curve in Figure 2.17. The curve has the characteristic sigmoidal shape observed for the

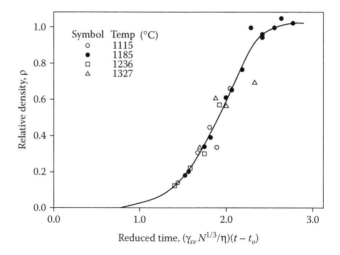

FIGURE 2.17 Comparison of the sintering data for silica soot preform sintered in air with the predictions of Scherer's model. (Courtesy of G.W. Scherer.)

density versus time data in many sintering experiments. The model has been extended by considering a Gaussian or a bimodal distribution of pore sizes [24,25].

The predictions of Scherer's model have been well validated by the data for many amorphous materials, such as glass powder compacts, colloidal gels, and polymeric gels [26,27]. In the comparison with experimental data, the predictions of the model are plotted as a function of reduced time while the data are obtained as a function of measured time (e.g., seconds). To construct such a plot, one must first find the reduced time corresponding to the measured density as described above. A plot of the reduced time versus the measured time would be a straight line with a slope of $\gamma_{sv} N^{1/3}/\eta$. Multiplying that slope by the measured time gives the average reduced time interval for each density. The data points in Figure 2.17 show a comparison of the sintering results for low-density bodies formed from SiO_2 soot with the predictions of Scherer's model [26].

2.6.5 FINAL STAGE MODELS

2.6.5.1 Geometrical Model for Solid-State Sintering

For the final-stage sintering of polycrystalline materials, the powder system is idealized in terms of an array of equal-sized tetrakaidecahedra with spherical pores of the same size at the corners (Figure 2.9d). The tetrakaidecahedron has 24 pores (one at each corner), and each pore is shared by 4 tetrakaidecahedra, so the pore volume associated with a single tetrakaidecahedron is $V_p = (24/4)(4/3)\pi r^3$, where r is the radius of a pore. Making use of Equation 2.39, the porosity in a single tetrakaidecahedron is

$$P_s = \frac{8\pi r^3}{8\sqrt{2}l_p^3} = \frac{\pi}{\sqrt{2}}\left(\frac{r^3}{l_p^3}\right) \tag{2.63}$$

A more convenient unit cell of the idealized structure can be chosen as a thick-walled spherical shell of solid material centered on a single pore of radius r (Figure 2.18). The outer radius of the spherical shell, b, is defined such that the average density of the unit cell is equal to the density of the powder system, that is,

$$\rho = 1 - (r/b)^3 \tag{2.64}$$

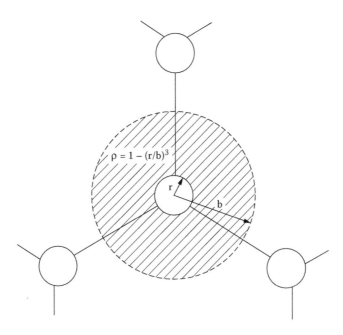

FIGURE 2.18 A porous solid during the final stage of sintering can be modeled by constructing a spherical shell centered on a single pore. The outer radius b is chosen such that the density of the shell matches that of the porous solid.

The volume of the solid phase in the unit cell is $(4/3)\pi(b^3 - r^3)$. Since a unit cell contains a single pore, the number of pores per unit volume of the solid phase is

$$N = \frac{3}{4\pi}\left(\frac{1-\rho}{\rho r^3}\right) \tag{2.65}$$

2.6.5.2 Sintering Equations

As in the intermediate stage, the uniform pore geometry assumed in the models precludes the occurrence of nondensifying mechanisms. Final-stage sintering models have been developed by Coble [11] and by Coleman and Beeré [28].

2.6.5.2.1 Lattice Diffusion

Coble [11] used a procedure similar to that outlined earlier for his intermediate-stage equations, but the atomic flux equation was approximated to that corresponding to diffusion between concentric spherical shells. The final result is

$$P_s = \frac{6\pi}{\sqrt{2}}\left(\frac{D_l \gamma_{sv} \Omega}{l_p^3 kT}\right)(t_f - t) \tag{2.66}$$

where P_s is the porosity at a time t, D_l is the lattice diffusion coefficient, γ_{sv} is the specific energy of the solid-vapor interface, Ω is the atomic volume, l_p is the edge length of the tetrakaidecahedron (taken as approximately equal to the grain size), k is the Boltzmann constant, T is the absolute temperature, and t_f is the time when the pore vanishes.

Because of the approximations made in the flux field, Equation 2.66 is believed to be valid for porosities less than about 2%. For higher porosities (2%–5%), a more complex equation was derived by Coble, but in view of the drastic approximations made, little benefit may be gained from the more complex expression. Except for a small difference in the value of the numerical constant, Equation 2.66 is identical to the corresponding equation for the intermediate stage (Equation 2.49).

2.6.5.2.2 Grain Boundary Diffusion

A final-stage sintering equation for grain boundary diffusion was not derived by Coble, but, apart from a difference in the numerical constant, the equation may be expected to be identical to Equation 2.52. Coble later developed models for diffusional sintering with an applied pressure from which sintering equations for both the intermediate and final stages can be extracted (see Section 2.10).

2.6.5.3 Geometrical Model for Viscous Sintering

A model consisting of the spherical shell shown in Figure 2.18 but without the grain boundaries was used by Mackenzie and Shuttleworth [29] to derive final-stage sintering equations for an amorphous solid by viscous flow. The use of the model means that the real system is idealized in terms of a structure consisting of spherical pores of the same size in a solid matrix. The concentric shell maintains its spherical geometry during sintering. Unlike for the initial stage, exact equations can be derived in Frenkel's energy balance concept for the rate of reduction of surface area and the rate of dissipation of energy by viscous flow. The result is

$$\int_{t_0}^{t} \frac{\gamma_{sv} N^{1/3}}{\eta} dt = \frac{2}{3} \left(\frac{3}{4\pi} \right)^{1/3} \int_{\rho_o}^{\rho} \frac{d\rho}{(1-\rho)^{2/3} \rho^{1/3}} \tag{2.67}$$

where ρ is the relative density at time t, ρ_o is the initial relative density at time t_o, γ_{sv} is the specific energy of the solid–vapor interface, N is the number of pores per unit volume (given by Equation 2.65), and η is the viscosity of the solid–phase. Equation 2.67 can be written

$$\frac{\gamma_{sv} N^{1/3}}{\eta} (t - t_o) = F_{MS}(\rho) - F_{MS}(\rho_o) \tag{2.68}$$

where

$$F_{MS}(\rho) = \frac{2}{3} \left(\frac{3}{4\pi} \right)^{1/3} \left[\frac{1}{2} \ln \left(\frac{1+\rho^3}{(1+\rho)^3} \right) - \sqrt{3} \arctan \left(\frac{2\rho - 1}{\sqrt{3}} \right) \right] \tag{2.69}$$

The form of Equation 2.68 is similar to that for Scherer's intermediate stage model (Equation 2.61), and the predictions for the sintering kinetics can be analyzed in a way similar to that outlined for Scherer's model.

Figure 2.19 gives a comparison of the predictions of Mackenzie and Shuttleworth's model with those of Scherer's model [23]. Despite the large difference in the geometry of the models, the agreement is excellent over a wide density range. Significant deviations begin to occur only when ρ falls below ~0.2. Although the Mackenzie and Shuttleworth model is strictly valid for > ~0.9, the predictions are applicable over a much wider range. A similar situation exists for Scherer's model. The predictions show excellent agreement not only with Mackenzie and Shuttleworth's model but also with Frenkel's initial stage model. The conclusion that we can draw is that, unlike the case of polycrystalline materials, the sintering behavior of amorphous materials is not sensitive to structural details.

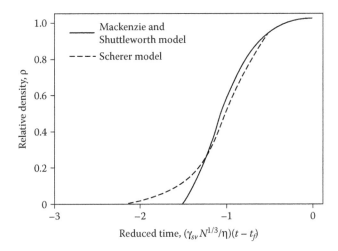

FIGURE 2.19 Relative density versus reduced time for the Mackenzie and Shuttleworth model and the Scherer model. For relative densities >0.942, the Mackenzie and Shuttleworth model applies. The curves have been shifted to coincide at the time t_f when sintering is complete. (Courtesy of G.W. Scherer.)

2.6.6 LIMITATIONS OF THE ANALYTICAL MODELS

A feature that arises from our consideration of the analytical models is the difference in complexity between the sintering phenomena in polycrystalline materials and in amorphous materials. The analysis of viscous sintering on the basis of Frenkel's energy balance concept appears relatively simple in principle. The idealization of the structure of amorphous materials leads to analytical solutions that describe the sintering behavior in a very satisfactory manner.

For polycrystalline materials, the sintering phenomena are considerably more dependent on the structural details of the powder system. Because of the dramatic simplifications made in the models, they do not provide an adequate *quantitative* representation of the sintering behavior of real powder systems. The models do, however, provide a good *qualitative* understanding of the different sintering mechanisms and the dependence of the sintering kinetics on key processing parameters such as particle size, temperature, and, as we shall see later, applied pressure.

The assumptions made in the models must be remembered. The models assume a geometry that is a dramatic simplification of a real powder system. They also assume that each mechanism operates separately. Although attempts have been made to develop analytical models with more realistic neck geometries, e.g., a catenary (a hyperbolic cosine function), and with simultaneous mechanisms (e.g., surface diffusion and grain boundary diffusion), these more complex analyses do not provide any significant advances in the understanding of sintering when compared to the simple models. The analytical models assume that the powder particles are spherical and of the same size, that there is a regular packing arrangement in the green body, and that there is no grain growth. These assumptions are almost never reproduced in real powder systems.

2.7 NUMERICAL SIMULATION OF SINTERING

In view of the dramatic simplifications made in the analytical models, numerical simulations provide a powerful tool for investigating some of the complexities of sintering, such as more realistic geometrical models and the occurrence of simultaneous mechanisms. Generally, the models developed over the last several decades can be divided into three categories, depending on the length scale of the simulations: atomic, particle, and continuum [30]. At the atomic level, each particle is modeled as an assembly of a number of atoms (or molecules), and the computer simulation follows the motion of each individual atom as the cluster of particles sinters. Simulations of this type, referred to as *molecular dynamics (MD) simulations*, are in their early stages of development for modeling the sintering process. Because of the considerable computing time required, the simulations are limited to particles with sizes smaller than a few hundred nanometers and sintering times shorter than a few nanoseconds. The MD simulations are, therefore, not relevant to practical sintering.

At the particle level, considerable progress has been made over the last ten to twenty years in the numerical simulation of sintering. In particular, *finite element (FE) analysis* has been used very effectively to analyze many of the complexities of sintering. The representative unit in the simulations is an assembly of particles with the required shape and arrangement, and the analysis can consist of up to thousands of particles, depending on the computer power. The dramatic oversimplifications assumed in the analytical models can be dropped. However, most FE simulations have been performed in two dimensions because of the enormous effort required for three-dimensional simulations.

At the continuum level, the powder compact is treated as a continuum solid, and the FE method is used to predict the macroscopic behavior of the porous body during sintering. The approach requires constitutive laws for the porous sintering body, and it gives predictions for the density, grain size, stress field, and strain field of the compact during the entire sintering process, as well as the shape and dimensions of the component. Many modern FE software packages allow the input of constitutive relationships as well as their parameters. Continuum-level modeling is particularly useful for analyzing the sintering of complex systems, such as composites and systems subjected to nonuniform stresses, and for providing guidance in industrial product development.

2.7.1 NUMERICAL SIMULATION OF SOLID-STATE SINTERING

At the particle scale, numerical simulations have been carried out by Nichols and Mullins [31], who used a finite difference method to analyze the sintering of two cylinders or a row of cylinders by the surface diffusion mechanism. Later, Bross and Exner [32] used a similar approach to analyze the sintering of a row of cylinders by surface diffusion, and by the simultaneous occurrence of surface diffusion and grain boundary diffusion. Figure 2.20 shows the results of Bross and Exner for the neck contours. The results for the surface diffusion mechanism (Figure 2.20a) are in agreement with those of Nichols and Mullins. The circle approximation used in the analytical models for the neck geometry differs from the contours found by the numerical simulation approach, which predicts undercutting and a continuous change in the curvature of the neck surface. The region of the neck surface influenced by matter transport also extends far beyond that given by the circle approximation, but the extension is less pronounced when surface diffusion and grain boundary diffusion occur simultaneously (Figure 2.20b).

More recently, various finite difference schemes have been used to study solid-state sintering [33–36]. These simulations reexamined the details of neck growth and shrinkage between two particles for a distribution of particle sizes and under complex loading conditions. As an example, Svoboda and Riedel [36] analyzed the formation of a neck between spherical particles by the simultaneous occurrence of surface diffusion and grain boundary diffusion, using an accurate numerical solution valid for short times in the limiting cases of small and large values of D_s/D_{gb}, and an approximate analytical solution based on variational calculus. Figure 2.21 shows the predictions of the model for the normalized neck radius (X/a) versus normalized time when the dihedral angle ψ is equal to 120°. Although the analysis confirms the occurrence of a power law dependence of the neck radius on time, the exponent m in Equation 2.37 increases from 11/2 for high values of D_s/D_{gb} to 7 for low D_s/D_{gb} values. This trend in the value of m is opposite to that predicted by the analytical models (Table 2.5). As outlined in Chapter 3, the variational

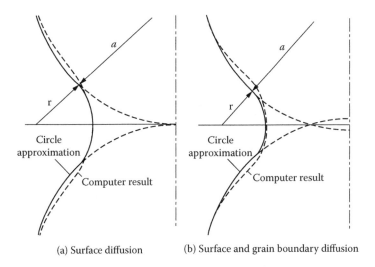

(a) Surface diffusion   (b) Surface and grain boundary diffusion

FIGURE 2.20 Contours of necks between cylinders for sintering by (a) surface diffusion and (b) simultaneous surface diffusion and grain boundary diffusion. (From Bross, P., and Exner, H.E., Computer simulation of sintering processes, *Acta Metall.*, 27, 1013, 1979. With permission.)

principle can also be used to construct FE schemes for computer simulation of microstructural evolution.

In the continuum FE modeling of sintering, constitutive equations are used to relate the strain rates, $\dot{\varepsilon}_{ij}$, of the porous sintering body with the stresses σ_{ij} imposed on the body. There are several constitutive equations in the literature, and it is difficult to determine which model works best for

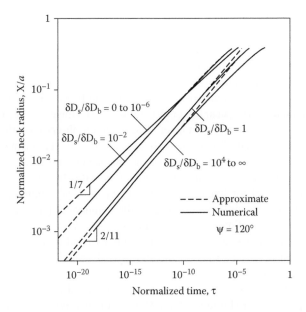

FIGURE 2.21 Results of numerical simulations for the normalized neck radius versus reduced time for sintering of two spherical particles by the simultaneous occurrence of surface diffusion and grain boundary diffusion. (From Svoboda, J., and Riedel, H., New solutions describing the formation of interparticle necks in solid-state sintering, *Acta Metall. Mater.*, 43, 1, 1995. With permission.)

a given material. Recent reviews of the constitutive equations are given by Cocks [37] and Olevsky [38]. As an example, Du and Cocks [39] developed a constitutive equation that can be expressed in the form

$$\dot{\varepsilon}_{ij} = \frac{1}{\eta_o}\left(\frac{G_o}{G}\right)^m\left[\frac{3}{2}f_1(\rho)s_{ij} + 3f_2(\rho)(\sigma_{kk}/3 - \Sigma)\delta_{ij}\right] \tag{2.70}$$

where η_o is the shear viscosity of the fully dense material with a grain size G_o, G is the grain size of the porous sintering body, m is an exponent that reflects the dependence of the deformation rate on the grain size, s_{ij} are the deviatoric stresses, σ_{kk} is the stress tensor, Σ is the sintering stress, δ_{ij} is the Kroneker delta, and $f_1(\rho)$ and $f_2(\rho)$ are functions of the relative density ρ. As defined in more detail later, Σ is the equivalent externally applied stress that has the same effect on sintering as the curved surfaces of the pores and grain boundaries. The functions $f_1(\rho)$ and $f_2(\rho)$, as well as Σ, can be determined using idealized models of sintering [37], the so-called mechanistic constitutive equations, or by fitting experimental data, the so-called empirical constitutive equations [40]. The use of Equation 2.70 also requires a grain growth relationship, which can also be obtained from mechanism-based grain growth equations (see Chapter 3) or empirically by curve fitting.

Equation 2.70 gives a general constitutive equation for a porous sintering solid that contains a shear term and a densification term, so it is applicable to sintering as well as to pressure sintering. In general, the densification and grain growth equations can be integrated numerically to produce curves of density and grain size as functions of time. These curves can be compared with experimental data to verify the constitutive models or to obtain fitting parameters for the models.

2.7.2 NUMERICAL SIMULATION OF VISCOUS SINTERING

Ross, Miller, and Weatherly [41] used finite element modeling to simulate numerically the viscous sintering of an infinite line of cylinders, but the results are not of direct interest to real powder compacts because of the two-dimensional geometry and the constraint of a fixed center-to-center distance used in the model. The FE method has been used more effectively by Jagota and Dawson [42,43] to simulate the viscous sintering of two spherical particles that are constrained to be axisymmetric. The simulations show that the viscous flow field assumed in the Frenkel model is qualitatively correct. The flow is axially downward and radially outward near the neck, with most of the energy dissipation occurring near the neck (Figure 2.22). However, the simulations also show that the Frenkel model is quantitatively in error. The common observation that the Frenkel model does not match experimental data for linear shrinkage above ~5% is attributed more to the inaccuracy of the model than to the inapplicability of the two-sphere model as such. According to the simulations, the two-sphere model is in good agreement with the Mackenzie and Shuttleworth model and the Scherer model for linear shrinkages of up to 15%. Finite element simulations in three dimensions (3D) for the viscous sintering of three spherical particles revealed effects that cannot be predicted by the two-dimensional (2D) models [44]. Viscous flow in the system of three particles produced both center-to-center approach (shrinkage) and asymmetric neck growth, which resulted in rearrangement. The rearrangement also caused nonuniform shrinkage of the system along different axes, i.e., anisotropic shrinkage.

Jagota [45] has used finite element modeling to simulate the sintering of a pair of rigid particles coated with an amorphous layer (Figure 2.23a). The use of such *coated particles* can alleviate the difficulties encountered in the sintering of mixtures consisting of rigid inclusions and sinterable particles (see Chapter 5). Figure 2.23b shows the predicted change in the relative density of a packing of coated particles (initial relative density of 0.60) as a function of reduced time for different coating thicknesses, and for the Mackenzie and Shuttleworth model. With sufficiently thick coatings

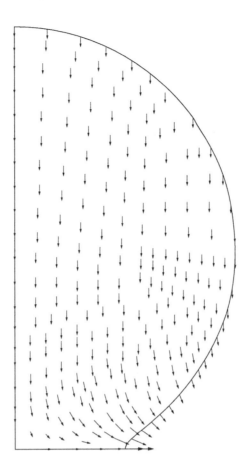

FIGURE 2.22 Viscous flow velocity field at neck radius to particle radius $X/a = 0.5$. (Courtesy of A. Jagota.)

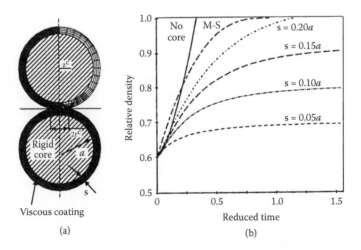

FIGURE 2.23 Finite element simulation of viscous sintering of two rigid particles coated with an amorphous layer. (a) Rigid particles of radius a coated with a viscous layer of thickness s and showing the finite element mesh. (b) Predictions for the relative density versus reduced time for different coating thicknesses, and for the Mackenzie and Shuttleworth (M–S) model. (From Jagota, A., simulation of the viscous sintering of coated particles, *J. Am. ceram. soc.*, 77, 2237, 1994. With permission.)

(in this case $s > \sim 0.2$ of the particle radius), full density is predicted at a rate comparable to a packing that consists of particles without a rigid core.

2.8 PHENOMENOLOGICAL SINTERING EQUATIONS

In the phenomenological approach, empirical equations are developed to fit sintering data, usually in the form of density (or shrinkage) versus time. Although the equations provide little or no help in understanding the process of sintering, they may serve a useful function in some numerical models that require the incorporation of empirical equations for the densification of a powder system. A simple expression that is found to be very successful in fitting sintering and hot pressing data is

$$\rho = \rho_o + K \ln(t/t_o) \tag{2.71}$$

where ρ_o is the density at an initial time t_o, ρ is the density at time t, and K is a temperature-dependent parameter. An attempt was made by Coble [11] to provide some theoretical justification for this semi-logarithmic expression. Using the rate equation for Coble's intermediate- or final-stage model for sintering by lattice diffusion (e.g., Equation 2.51), we can write

$$\frac{d\rho}{dt} = \frac{AD_l \gamma_{sv} \Omega}{G^3 kT} \tag{2.72}$$

where A is a constant that depends on the stage of sintering. Assuming that the grains grow according to a cubic law of the form

$$G^3 = G_o^3 + \alpha t \tag{2.73}$$

where G_o is the initial grain size and α is a constant, and that $G^3 \gg G_o^3$, then Equation 2.72 becomes

$$\frac{d\rho}{dt} = \frac{K}{t} \tag{2.74}$$

where $K = AD_l \gamma_{sv} \Omega /(\alpha kT)$. On integrating Equation 2.74, we obtain Equation 2.71, which is expected to be valid in both the intermediate and final stages because Equation 2.72 has the same form in both stages.

When grain growth is limited, shrinkage data over a large part of the sintering process can usually be fitted by the equation

$$\frac{\Delta L}{L_o} = K t^{1/\beta} \tag{2.75}$$

where K is a temperature-dependent parameter and β is an integer. This equation has the same form as the initial stage shrinkage equations for the analytical models.

Other empirical equations in the sintering literature include one due to Tikkanen and Makipirtti [46]:

$$\frac{V_o - V_t}{V_o - V_f} = K t^n \tag{2.76}$$

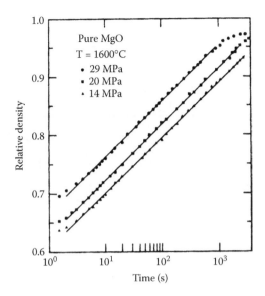

FIGURE 2.24 Semi-logarithmic dependence of the density on time during the hot pressing of high purity MgO.

where V_o is the initial volume of the powder compact, V_t is the volume after sintering for time t, V_f is the volume of the fully dense solid, and K is a temperature-dependent parameter. Depending on the material, n has values between 0.5 and 1.0. Another equation, due to Ivensen [47], is

$$\frac{V_t^P}{V_o^P} = (1 + C_1 mt)^{-1/m} \qquad (2.77)$$

where V_o^P is the initial pore volume of the body, V_t^P is the pore volume after sintering for time t, and C_1 and m are constants.

Attempts have sometimes been made to attach physical significance to parameters in the empirical equations, but the difficulties of doing so are well recognized. For example, it has been shown that the sintering data for UO_2 could be well fitted by any of the following four equations [48]: Equation 2.71, Equation 2.76, Equation 2.77, and a simple hyperbolic equation of the form $V_t^P/V_o^P = K_1/(K_2 + t) + K_3$. The conclusion is that more than one empirical equation can provide a good fit to any given set of sintering data. In practice, the choice of any one of the equations appears to be somewhat arbitrary. Coble's semi-logarithmic relationship has the advantage of simplicity and is very successful in fitting many sintering and hot-pressing data, an example of which is shown in Figure 2.24.

2.9 SINTERING DIAGRAMS

As discussed earlier, more than one mechanism commonly operates simultaneously during the sintering of polycrystalline systems. Numerical simulations provide a theoretical framework for the analysis of sintering with simultaneous mechanisms, but a more practical approach, involving the construction of *sintering diagrams*, has been developed by Ashby [49] and Swinkels and Ashby [50]. The earlier diagrams show, for a given temperature and neck size, the dominant mechanism of sintering and the net rate of neck growth or densification. Later, a second type of diagram was put forward in which the density rather than the neck size was evaluated, but the principles used for the construction of the density diagrams are similar to those for the neck size diagrams.

The form that a sintering diagram can take is shown in Figure 2.25 for the sintering of copper spheres with a radius of 57 μm. The axes are the neck radius X, normalized to the radius of the

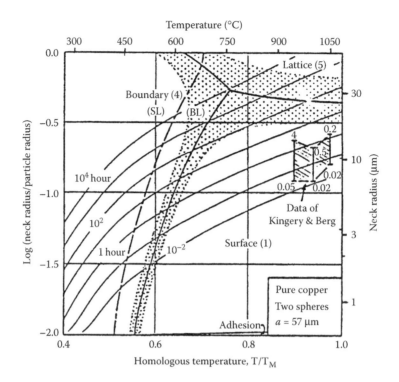

FIGURE 2.25 Neck size sintering diagram for two copper spheres with a radius of 57 mm. The data of Kingery and Berg [12] for the sintering of copper spheres with the same radius are also shown. (From Ashby, M.F., A first report on sintering diagrams, *Acta Metall.*, 22, 275, 1974. With permission.)

sphere a, and the homologous temperature T/T_M, where T_M is the melting temperature of the solid. The diagram is divided into various fields, and within each field a single mechanism is dominant, i.e., it contributes most to neck growth. Figure 2.25, for example, is divided into three fields corresponding to surface diffusion, grain boundary diffusion, and lattice diffusion (from the grain boundary). At the boundary between two fields (shown as solid lines), two mechanisms contribute equally to the sintering rate. Superimposed on the fields are contours of constant sintering time.

Some diagrams, especially those developed more recently, may contain additional information. For example, broken lines roughly parallel to the temperature axis (not shown in Figure 2.25) represent the transition between the stages of sintering. On either side of the field boundaries is a shaded band, and outside this band a single mechanism contributes more than 55% of the total neck growth. Within the shaded band, two or more mechanisms contribute significantly, but none contributes more than 55%. The shaded band provides an idea of the width of the field boundaries. A dashed line running roughly parallel to the field boundary between surface diffusion and grain boundary diffusion shows where redistribution of the matter transported by grain boundary diffusion controls the rate of this mechanism. As discussed earlier, matter transported into the neck by grain boundary diffusion requires another mechanism (e.g., surface diffusion) for redistribution over the neck surface. On the side labeled SL, redistribution by surface diffusion controls the rate; on the other side, labeled BL, boundary diffusion controls the rate.

2.9.1 CONSTRUCTION OF THE DIAGRAMS

The construction of the diagrams is described in detail by Ashby [49]. It requires neck growth equations for specified geometrical models and data for the material constants (e.g., diffusion coefficients, surface energy, and atomic volume) that appear in the equations. The geometrical models are commonly the analytical models that were discussed earlier. The diagrams are constructed by numerical methods, and it is assumed that the total neck growth rate is the sum of all the neck growth rates for the individual mechanisms. The field boundary at which one mechanism contributes 50% of the neck growth rate is then calculated. The procedure is carried out incrementally for small increases in X/a and T/T_M. The time interval between the steps is calculated and added to give the total time, which is used to plot the time contours. Refinements, such as the width of the boundaries between the fields, are added by modification of the same procedure.

2.9.2 LIMITATIONS OF THE DIAGRAMS

Because the diagrams are based on approximate geometrical models and on data for the material parameters that appear in them, they are no better than the analytical models discussed earlier. Furthermore, the material parameters (particularly for ceramics) are often not known with sufficient accuracy, so small changes in the values of the parameters can produce significant changes in the nature of the diagrams. Sintering diagrams have so far been constructed for several pure metals and a few simple ceramics (e.g., NaCl and Al_2O_3). For practical ceramic systems, small changes in the characteristics of the powders (e.g., purity) can produce significant changes in the material parameters (e.g., diffusion coefficients), so diagrams would have to be constructed for each system, requiring an enormous amount of work. In spite of these limitations, the diagrams have proved to be useful in visualizing conceptual relationships between the various mechanisms and changes in sintering behavior under different temperature and particle size regimes.

2.10 SINTERING WITH AN EXTERNALLY APPLIED PRESSURE

A difficulty that often arises in solid-state sintering of ceramics is that of inadequate densification. One solution to this difficulty is the application of an external stress or pressure to the powder system during sintering, giving rise to the techniques of hot pressing, hot isostatic pressing, and hot forging. The theory and principles of such pressure-assisted sintering are considered here; the practical applications of the techniques are described in Chapter 6.

2.10.1 HOT PRESSING

2.10.1.1 Hot Pressing Models

Models for hot pressing by diffusional mass transport under the driving forces of surface curvature and applied stress were formulated by Coble [51,52]. In one approach, the analytical models for sintering were modified to include the effects of an applied stress, while in a second approach, hot pressing is viewed in a manner analogous to that of creep in dense solids, and the creep equations were modified to account for the porosity and the surface curvature appropriate to porous powder systems.

2.10.1.1.1 Modification of the Analytical Sintering Models
Consider the idealized models for the three stages of sintering (Figure 2.9). The vacancy concentration under the neck surface ΔC_{vn} is not affected by the applied stress, so it is still given by Equation 2.19; that is,

$$\Delta C_{vn} = \frac{C_{vo}\gamma_{sv}\Omega}{kT}\left(\frac{1}{r_1}+\frac{1}{r_2}\right)=\frac{C_{vo}\gamma_{sv}\Omega K}{kT} \tag{2.78}$$

where K is the curvature of the pore surface. For the initial stage of sintering, $K = 1/r = 4a/X^2$, whereas for the intermediate and final stages $K = 1/r$ and $2/r$, respectively, where r is the pore radius, a is the particle radius, and X is the neck radius. The stress p_a applied to the powder system leads to a stress p_e on the grain boundary, and, because of the porosity, p_e is $> p_a$. Let us assume that

$$p_e = \phi p_a \tag{2.79}$$

where ϕ is a factor, referred to as the *stress intensification factor*, that we shall define in more detail later. The compressive stress on the grain boundary means that the vacancy concentration is less than that of a flat, stress-free boundary. The vacancy concentration can be found from Equation 1.109, or simply from Equation 2.78 by replacing the stress $\gamma_{sv} K$ due to the curvature by the effective stress on the boundary. It is given by

$$\Delta C_{vb} = -\frac{C_{vo} p_e \Omega}{kT} = -\frac{C_{vo} \phi p_a \Omega}{kT} \tag{2.80}$$

For the two-sphere model appropriate to the initial stage of sintering (Figure 2.11), Coble assumed that ϕ is equal to the area of the sphere projected onto the punch of the hot pressing die divided by the cross-sectional area of the neck, i.e., $\phi = 4a^2/(\pi X^2)$, whereas for both the intermediate and final stages Coble argued that $\phi = 1/\rho$, where ρ is the relative density of the body.

Using the parameters for K and ϕ, the variation of ΔC_{vn} and ΔC_{vb} is shown schematically in Figure 2.26. For hot pressing, the difference in the vacancy concentration between the neck surface and the grain boundary is given by $\Delta C = \Delta C_{vn} - \Delta C_{vb}$, so for the initial stage

$$\Delta C = \frac{C_o \Omega 4a}{kTX^2}\left(\gamma_{sv} + \frac{p_a a}{\pi}\right) \tag{2.81}$$

Equation 2.81 shows that for the initial stage, ΔC for hot pressing is identical to that for sintering except that γ_{sv} is replaced by $(\gamma_{sv} + p_a a/\pi)$. Since p_a and a are constant, it follows that the hot pressing equations can be obtained from the sintering equations by simply replacing γ_{sv} by $(\gamma_{sv} + p_a a/\pi)$.

Coble did not specifically adapt the intermediate- and final-stage sintering models to account for hot pressing. However, as we will show later, the hot pressing equations for any stage can be obtained simply by replacing the $\gamma_{sv} K$ in the sintering equations by $(\gamma_{sv} K + p_e)$.

2.10.1.1.2 Modification of the Creep Equations

Let us consider matter transport during creep of a dense solid and the form that the creep equations can take. We start with a single crystal of a pure solid with the cubic structure that has the shape of a rod with a cross-section of length L. Let normal stresses p_a act on the two sides of the rod as shown in Figure 2.27a. Nabarro [53] and Herring [54] argued that self-diffusion within the crystal will cause the solid to creep (i.e., deform slowly) in an attempt to relieve the stresses. The creep is caused by atoms diffusing from interfaces subjected to a compressive stress (where they have a higher chemical potential) toward those subjected to a tensile stress (lower chemical potential). Extending this idea of creep to a polycrystalline solid, self-diffusion within the individual grains will cause atoms to diffuse from grain boundaries under compression toward those boundaries under tension (Figure 2.27b). Creep by lattice diffusion is commonly referred to as *Nabarro–Herring creep*. An analysis of the atomic flux by Herring gave the following equation for the creep rate:

$$\dot{\varepsilon}_c = \frac{1}{L}\frac{dL}{dt} = \frac{40}{3}\frac{D_l \Omega p_a}{G^2 kT} \tag{2.82}$$

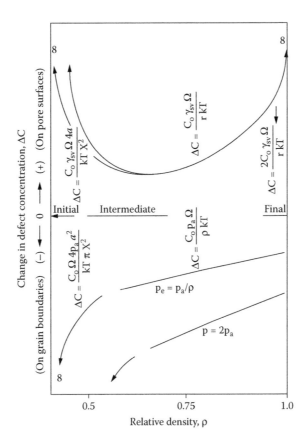

FIGURE 2.26 The driving force for hot pressing as a function of relative density.

where D_l is the lattice diffusion coefficient, Ω is the atomic volume, p_a is the applied stress, G is the grain size, k is the Boltzmann constant, and T is the absolute temperature. The creep rate is, by definition, a linear strain rate, equal to $(1/L)dL/dt$, where L is the length of the solid and t is the time.

Creep in a polycrystalline solid can also occur by diffusion along the grain boundaries (Figure 2.27c), commonly referred to as *Coble creep*, and for this mechanism the creep equation is [55]:

$$\dot{\varepsilon}_c = \frac{95}{2} \frac{D_{gb}\delta_{gb}\Omega p_a}{G^3 kT} \tag{2.83}$$

where D_{gb} is the grain boundary diffusion coefficient and δ_{gb} is the grain boundary width. Equation 2.82 and Equation 2.83 have the same linear dependence on p_a but differ in terms of the grain size dependence and the numerical constants.

The application of a high enough stress may, in some ceramics, activate matter transport by dislocation motion. For this mechanism, the creep rate is given by

$$\dot{\varepsilon}_c = \frac{AD\mu b}{kT}\left(\frac{p_a}{\mu}\right)^n \tag{2.84}$$

where A is a numerical constant, D is a diffusion coefficient, μ is the shear modulus, and b is the Burgers vector. The exponent n depends on the mechanism of the dislocation motion and has values in the range of 3–10.

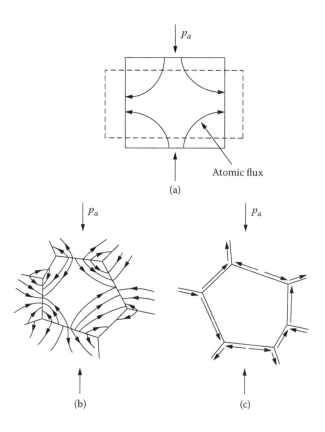

(a)

(b)   (c)

FIGURE 2.27 (a) A single crystal subjected to a uniaxial stress, showing the direction of the atomic flux, (b) a typical grain in a polycrystalline solid, showing the expected atomic flux by lattice diffusion, and (c) by grain boundary diffusion.

2.10.1.2 Densification Rate in Hot Pressing

In the hot pressing of powders, the data are normally acquired in the form of density versus time, from which the densification rate can be determined. Considerable porosity is present over a large part of the hot pressing process. The modification of the creep equations to account for hot pressing of porous powder systems, therefore, attempts to incorporate these two factors: (1) relating the creep rate (a linear strain rate) to the densification rate (a volumetric strain rate) and (2) compensating for the presence of porosity.

In hot pressing, the mass of the powder, m, and the cross-sectional area of the die, A, are approximately constant. As the density of the sample, d, increases, the sample thickness, L, decreases. The variables d and L are related by

$$\frac{m}{A} = Ld = L_o d_o = L_f d_f \tag{2.85}$$

where the subscripts o and f refer to the initial and final values. Differentiating this equation with respect to time gives

$$L\frac{d(d)}{dt} + d\frac{dL}{dt} = 0 \tag{2.86}$$

Rearranging this equation, we obtain

$$-\frac{1}{L}\frac{dL}{dt} = \frac{1}{d}\frac{d(d)}{dt} = \frac{1}{\rho}\frac{d\rho}{dt} \tag{2.87}$$

where ρ is the relative density. Equation 2.87 relates the linear strain rate of the body during hot pressing to its densification rate. As discussed in Chapter 6, the linear strain rate is commonly obtained from measurements of the distance traveled by the punch of the hot pressing die as a function of time.

In compensating for the presence of porosity in the powder system, we recall from our discussion of Coble's initial stage hot pressing model that the effective stress on the grain boundaries p_e is related to the externally applied stress p_a by Equation 2.79. To account for the effects of the applied stress and the surface curvature on the densification rate, Coble argued that the total driving force, DF, is a linear combination of the two effects; that is,

$$DF = p_e + \gamma_{sv}K = p_a\phi + \gamma_{sv}K \tag{2.88}$$

where K is the curvature of the pore, equal to $1/r$ for the intermediate-stage sintering model and $2/r$ for the final-stage model. To get the equation appropriate to hot pressing, DF as given by Equation 2.88 is substituted for the applied stress p_a in the creep equations for dense solids. The hot pressing equations obtained by the modification of the creep equations are given in Table 2.6.

Coble's modification of the creep equations can provide only an approximation to the densification rate during hot pressing. A more rigorous analysis would require additional modifications to the creep models to better represent the situation present in powder systems, such as differences in the atomic flux field and in the path length for diffusion. In the creep models, the atomic flux terminates at the boundaries under tension, whereas in hot pressing, the flux terminates at the pore surfaces. The grain boundary area, which is related to the grain size, remains constant during creep, but both the grain boundary area and the path length for diffusion increase during hot pressing.

2.10.1.3 Hot Pressing Mechanisms

The mechanisms discussed earlier for sintering also operate during hot pressing, but the nondensifying mechanisms can be neglected because they are not enhanced by the applied stress, whereas,

TABLE 2.6
Hot Pressing Equations Obtained by Modification of the Creep Equations

Mechanism	Intermediate Stage	Final Stage
Lattice diffusion	$\frac{1}{\rho}\frac{d\rho}{dt} = \frac{40}{3}\left(\frac{D_l\Omega}{G^2kT}\right)\left(p_a\phi + \frac{\gamma_{sv}}{r}\right)$	$\frac{1}{\rho}\frac{d\rho}{dt} = \frac{40}{3}\left(\frac{D_l\Omega}{G^2kT}\right)\left(p_a\phi + \frac{2\gamma_{sv}}{r}\right)$
Grain boundary diffusion	$\frac{1}{\rho}\frac{d\rho}{dt} = \frac{95}{2}\left(\frac{D_{gb}\delta_{gb}\Omega}{G^3kT}\right)\left(p_a\phi + \frac{\gamma_{sv}}{r}\right)$	$\frac{1}{\rho}\frac{d\rho}{dt} = \frac{15}{2}\left(\frac{D_{gb}\delta_{gb}\Omega}{G^3kT}\right)\left(p_a\phi + \frac{2\gamma_{sv}}{r}\right)$
Dislocation motion[a]	$\frac{1}{\rho}\frac{d\rho}{dt} = A\left(\frac{D\mu b}{kT}\right)\left(\frac{p_a\phi}{\mu}\right)^n$	$\frac{1}{\rho}\frac{d\rho}{dt} = B\left(\frac{D\mu b}{kT}\right)\left(\frac{p_a\phi}{\mu}\right)^n$

[a] A and B are numerical constants; n is an exponent that depends on the mechanism of dislocation motion.

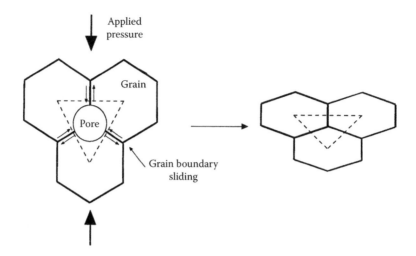

FIGURE 2.28 Sketch illustrating the change in grain shape that occurs during hot pressing. The grains are flattened in the direction of the applied pressure. When matter transport occurs by diffusion, grain boundary sliding is necessary to accommodate the change in grain shape.

for most hot pressing experiments, the densifying mechanisms are significantly enhanced. Because of this enhancement of the densifying mechanisms relative to the nondensifying mechanisms, hot pressing provides an effective technique for determining the mechanism of densification. However, new mechanisms can be activated by the applied stress and may provide complications, so their occurrence needs to be recognized.

Particle rearrangement contributes to the densification during the initial stage of hot pressing, but it is difficult to analyze. Grain boundary sliding is necessary to accommodate the diffusion-controlled shape changes that occur during the intermediate and final stages. As shown in Figure 2.28 for a cross-section, a representative element of the system (e.g., three idealized grains with a hexagonal shape) must mimic the overall shape change of the powder compact. Since the diameter of the hot pressing die is fixed, powder compaction occurs predominantly in the direction of the applied pressure, so the grains will be flattened. This shape change must be accommodated through sliding of the grains over one another. Grain boundary sliding and diffusional mass transport are not independent mechanisms. They occur sequentially, so that the slower mechanism controls the rate of densification. The major mechanisms that commonly operate during hot pressing are lattice diffusion, grain boundary diffusion, plastic deformation by dislocation motion, viscous flow (for amorphous systems), rearrangement, and grain boundary sliding.

According to the equations given in Table 2.6, when the applied stress is much greater than the driving force due to surface curvature (which corresponds to the common practical situation), the densification rate during hot pressing can be written

$$\frac{1}{\rho}\frac{d\rho}{dt} = \frac{HD\phi^n}{G^m kT} p_a^n \tag{2.89}$$

where H is a numerical constant, D is the diffusion coefficient of the rate-controlling species, ϕ is the stress intensification factor, G is the grain size, k is the Boltzmann constant, T is the absolute temperature, and the exponents m and n depend on the mechanism of densification. Table 2.7 gives the values of m and n, as well as the appropriate diffusion coefficient, for the various mechanisms.

TABLE 2.7
Hot Pressing Mechanisms and the Associated Exponents and Diffusion Coefficients

Mechanism	Grain Size Exponent, m	Stress Exponent, n	Diffusion Coefficient[a]
Lattice diffusion	2	1	D_l
Grain boundary diffusion	3	1	D_{gb}
Plastic deformation	0	≥3	D_l
Viscous flow	0	1	—
Grain boundary sliding	1	1 or 2	D_l or D_{gb}
Particle rearrangement			

[a] D_l = lattice diffusion coefficient; D_{gb} = grain boundary diffusion coefficient.

According to Equation 2.89, a plot of the densification rate (at a fixed density) versus p_a allows the exponent n and thus the mechanism of densification to be determined. For the commonly used hot pressing pressures (typically 10–50 MPa), data for many ceramics give $n \approx 1$, suggesting that a diffusion mechanism is responsible for densification [56,57]. This finding is not surprising in view of the strong bonding in ceramics (which limits dislocation motion) and the tendency for the use of fine powders (which favor diffusion mechanisms). Higher values of n, characteristic of dislocation mechanisms, have been reported for a few ceramics [58].

The dominant densification mechanism may change depending on the applied pressure, temperature, grain size, and composition. For example, because of the larger grain size exponent ($m = 3$) and, in general, the lower activation energy, fine powders and lower temperatures favor grain boundary diffusion over lattice diffusion ($m = 2$). With careful selection of the process variables, the dominant mechanism under a given set of conditions can be identified, and the results are commonly displayed in the form of a map, referred to as a *hot pressing map* or a *deformation mechanism map*, analogous to a sintering map, discussed earlier. Figure 2.29 shows a hot pressing

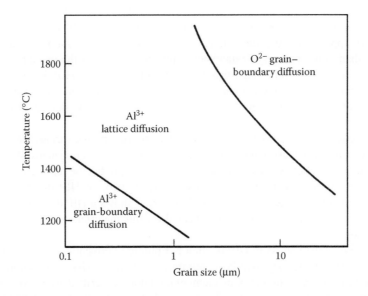

FIGURE 2.29 Hot pressing (deformation) map for pure α-alumina.

map for α-Al_2O_3, plotted as temperature versus grain size, showing the relationships between the mechanisms and hot pressing parameters [56].

2.10.2 SINTER FORGING

Sinter forging, also referred to as hot forging, is similar to hot pressing, but the sample is not confined in a die. Typically, a uniaxial stress is applied to a powder compact or partially densified compact at the sintering temperature. If a uniaxial stress equal to p_z is applied to the compact, then the hydrostatic component of the stress $p_z/3$ provides an additional driving force for densification. The hot pressing equations (Table 2.6) can, therefore, be used to predict the densification rate if p_a is replaced by $p_z/3$. There is also a shear component of the applied stress, which causes creep of the compact. As for the case of hot isostatic pressing, continuum finite element modeling provides a useful method for predicting the stresses, densification, and deformation of the compact. Because the uniaxial strains in the direction of the applied stress are significantly larger than those in hot pressing, sinter forging provides an effective method for producing some ceramics with aligned grain microstructure (see Chapter 6). As will be described later, researchers have used sintering under a uniaxial load to investigate the simultaneous creep and deformation of powder compacts and to determine the constitutive properties of porous sintering compacts.

2.10.3 HOT ISOSTATIC PRESSING

In hot isostatic pressing, sometimes abbreviated HIP, a hydrostatic pressure p_a is applied by means of a gas to the powder compact, which is often tightly enclosed in a glass or metal container (see Chapter 6). The applied pressures used in HIP (150–200 MPa) are typically much higher than those used in hot pressing (20–50 MPa). Because the nondensifying mechanisms are not enhanced by the applied pressure, they can be neglected. The densification of powder systems can be treated by equations similar to those for hot pressing (Table 2.6), but kinetic data are not easily obtained in HIP. The data for most ceramics are limited to values for the final density after a given time at an isothermal temperature and under a given applied pressure.

Particle rearrangement, which contributes to densification during the initial stage, is usually ignored because of the transient nature of the contribution and the difficulty of analyzing the process. The much higher applied pressure means that, in metals, plastic deformation plays a more important role in HIPing than in hot pressing. Instantaneous plastic yielding of the metal powder particles at their contact points will be more significant in the early stages of HIP. Plastic yielding is still expected to be unlikely for many ceramic powders, so the possible mechanisms that need to be considered are lattice diffusion, grain boundary diffusion, and plastic deformation by dislocation motion (power-law creep).

Constitutive equations have been developed for densification by the various mechanisms [59,60]. Using a procedure similar to that outlined for the construction of the sintering maps, along with available data for the material parameters, these equations were used to predict the relative contributions of the different mechanisms to densification. The results are plotted on HIP diagrams that show the conditions of dominance for each mechanism. Figure 2.30 shows a HIP diagram for Al_2O_3 powder with a particle size of 2.5 µm. As might be expected for this material, densification is predicted to occur by diffusion.

In practice, the applied stress in HIP is often not purely hydrostatic, and a shear component is also present. The shrinkage may also be nonuniform, leading to shape distortion of the powder compact (see Chapter 6). Continuum finite element modeling, described earlier, provides a valuable approach for predicting the stresses, densification, and shape change during HIP [40].

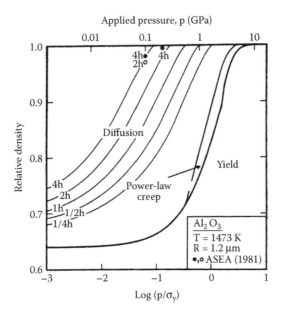

FIGURE 2.30 Hot isostatic pressing (HIP) map for α-Al_2O_3 powder with a particle size of 2.5 μm at 1473 K.

2.11 STRESS INTENSIFICATION FACTOR AND SINTERING STRESS

2.11.1 STRESS INTENSIFICATION FACTOR

In the hot pressing models, we found that the factor ϕ, defined by Equation 2.79, arose every time we needed to relate the mean stress on the grain boundary, p_e, to the externally applied stress, p_a. The factor ϕ is referred to as the *stress intensification factor* (or the stress multiplication factor). It should not be confused with a similarly named factor, the stress intensity factor, which has a different meaning in fracture mechanics. The significance of ϕ is such that while p_a is the stress that is measured, p_e is its counterpart that influences the rate of matter transport. The factor ϕ is geometrical in origin and would be expected to depend on the porosity and the shape of the pores.

Consider a hydrostatic pressure p_a applied to the external surface of a powder system, a model of which is shown schematically in Figure 2.31a. The applied pressure exerts on the surface of the solid an applied load $F_a = A_T p_a$, where A_T is the total external cross-sectional area of the solid, including such areas as may be occupied by pores. The presence of porosity located at the grain boundaries makes the actual grain boundary area, A_e, lower than the total external area. Assuming a force balance across any plane of the solid, we obtain

$$p_a A_T = p_e A_e \tag{2.90}$$

Therefore,

$$\phi = \frac{p_e}{p_a} = \frac{A_T}{A_e} \tag{2.91}$$

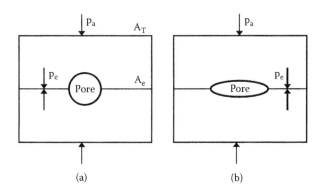

FIGURE 2.31 (a) The effective stress on the grain boundary, p_e, is greater than the externally applied stress p_a due to the presence of pores on the grain boundary. The stress intensification factor ϕ is defined as the ratio of the total external area to the actual grain boundary area. (b) ϕ depends on the porosity as well as the pore shape, having a stronger dependence on porosity for pores with smaller dihedral angles.

For spherical pores randomly distributed in a porous solid, ϕ can be easily found. Taking a random plane through the solid, the area fraction of porosity in the plane is equal to the volume fraction of porosity. If A_T is assumed to be unity, then A_e is equal to $(1 - P)$, where P is the porosity of the solid. Since $(1 - P)$ is also equal to the relative density, we can write

$$\phi = 1/\rho \tag{2.92}$$

This expression would be applicable to a glass containing isolated, nearly spherical pores or to a polycrystalline solid in which the equilibrium shapes of the isolated pores are roughly spherical (i.e., for dihedral angles > ~150°). When the pores deviate significantly from a spherical shape, this simple expression will no longer hold, and, in fact, ϕ may be expected to be quite complex. As illustrated in Figure 2.31 for pores with the same volume but different shapes, we may expect ϕ to depend not only on the porosity but also on the shape of the pores. As the pore shape deviates from a spherical geometry (i.e., as the dihedral angle decreases), the actual area of the grain boundary decreases, so ϕ would be expected to increase.

Computer simulation of the equilibrium shapes of a continuous network of pores has been performed by Beeré [22]. An analysis of Beeré's results by Vieira and Brook [58] showed that ϕ can be approximated by the expression

$$\phi = \exp(\alpha P) \tag{2.93}$$

where α is a factor that depends on the dihedral angle and P is the porosity. A semi-logarithmic plot of Beeré's results is shown in Figure 2.32 for several dihedral angles from which the factor α can be determined. As will be outlined later, Equation 2.93 has been verified by experimental data for several systems. However, this does not imply that it is the most accurate or appropriate model. As summarized in Table 2.8, other expressions have been proposed for ϕ, and, depending on the powder characteristics and packing, one of these expressions may be appropriate to fit the data [61–63]. A plot of ϕ versus ρ for several expressions (Figure 2.33) shows large differences between the values [64]. Except for the model based on monosize spheres, all of the results can be represented by an exponential function of the form given by Equation 2.93.

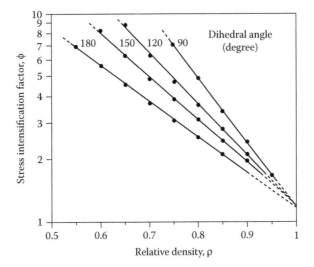

FIGURE 2.32 Stress intensification factor represented by an exponential function of the porosity.

2.11.2 SINTERING STRESS

According to Table 2.6, the densification rate for the diffusion-controlled mechanisms may be written in the general form

$$\frac{1}{\rho}\frac{d\rho}{dt} = -\frac{3}{L}\frac{dL}{dt} = \frac{3}{\eta_\rho}(p_a\phi + \sigma) \tag{2.94}$$

where $(1/L)dL/dt$ is the linear strain rate of the sintering solid, equal to 1/3 the volumetric strain rate; η_ρ has the dimensions of a viscosity and can be called the densification viscosity; and σ is the effective

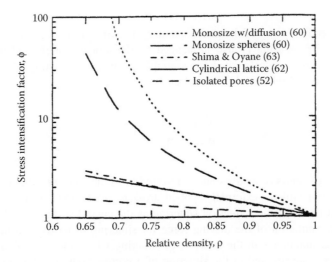

FIGURE 2.33 Stress intensification factor versus relative density for various idealized powder geometries and for the data of Shima and Oyane [63].

TABLE 2.8
Stress Intensification Factors for Various Powder Geometries

Stress Intensification Factor γ	Density Range	Model	Ref.
$(1 - \rho_0)/\rho^2(\rho - \rho_0)$	$\rho < 0.90$	Monosize spherical powder	60, 61
$(1 - \rho)^2/\rho(\rho - \rho_0)^2$	$\rho < 0.90$	Monosize spherical powder, including neck growth due to diffusion	60, 61
$1/\rho$	$\rho > 0.90$	Random distribution of isolated pores	51, 52
$\exp[\alpha(1 - \rho)]$	$\rho < 0.90$	Free energy minimization for equilibrium shapes of continuous pores	22, 58
$(3 - 2\rho)/\rho$	$\rho < 0.95$	Cubic array of intersecting cylindrical particles	62
$1/\rho^{5/2}$	$\rho < 1.0$	Empirical equation developed to fit sintering data of spherical Cu particles	63

stress on the atoms beneath the pore surface, given by the equation of Young and Laplace:

$$\sigma = \gamma_{sv}\left(\frac{1}{r_1} + \frac{1}{r_2}\right) \tag{2.95}$$

where r_1 and r_2 are the two principal radii of curvature of the pore surface. The quantity σ is the thermodynamic driving force for sintering. It has the units of a pressure (or stress) and is sometimes referred to as the *sintering pressure* or the *sintering potential*. The equation for σ is actually more complex for a polycrystalline ceramic where the pores are in contact with the grain boundaries, and it consists of two contributions, one attributed to the pores and the other attributed to the grain boundaries [65]. For example, for an idealized final stage microstructure where the pores and boundaries are assumed to have a spherical shape, σ is given by [66]

$$\sigma = \frac{2\gamma_{gb}}{G} + \frac{2\gamma_{sv}}{r} \tag{2.96}$$

where γ_{gb} is the specific energy of the grain boundary, G is the grain size, and r is the pore radius. The driving force for sintering is also influenced by the dihedral angle and the mass transport mechanism, and it has been calculated for a simple geometry consisting of a line of spherical particles [67]. Equation 2.94 can also be written

$$\frac{1}{\rho}\frac{d\rho}{dt} = \frac{3\phi}{\eta_\rho}(p_a + \Sigma) \tag{2.97}$$

where $\Sigma = \sigma/\phi$ has the units of a stress and is referred to as the *sintering stress*. Because Σ occurs in a linear combination with the externally applied stress, p_a, it is defined as the *equivalent externally applied stress* that has the same effect on sintering as the curved surfaces of the pores and grain boundaries. The formulation of the driving force for sintering in terms of a fictitious externally applied stress is advantageous in the analysis of sintering where mechanical stress effects arise, such as in pressure sintering and in the sintering of a porous body adherent on a rigid substrate (see Chapter 5). It also provides a conceptual basis for designing experiments to measure the driving force for sintering.

2.11.3 MEASUREMENT OF THE SINTERING STRESS AND STRESS INTENSIFICATION FACTOR

Although the sintering stress Σ and the stress intensification factor ϕ are key parameters in sintering, their measurement has been attempted in only a few studies.

2.11.3.1 Zero-Creep Technique

A relatively straightforward technique for measuring Σ is similar to the zero-creep technique suggested by Gibbs [68] for the determination of surface energies. The powder formed into the shape of a wire or a tape is clamped at one end, and a load is suspended from the other end. The creep of the sample is monitored, and the load required to produce zero creep is determined. The sintering stress is taken as the applied stress at the zero-creep condition, because the creep may be assumed to stop when the applied stress just balances the sintering stress. Gregg and Rhines [69] used this technique for copper powder in the form of a wire and found that the sintering stress increased with relative density ρ up to a value of ~0.95, after which it decreased (Figure 2.34). The increase in Σ with density can be attributed to a decrease in the pore size, and it is likely that significant coarsening of the microstructure (grain growth and pore growth) caused the observed decrease at values of the relative density above ~0.95 [70]. At a given density of the powder system, Gregg and Rhines also found that Σ varied inversely as the initial particle size of the powder, which is consistent with Equation 2.96 if the average pore size is assumed to be proportional to the average particle size.

2.11.3.2 Loading Dilatometry

The technique of loading dilatometry [71], in which a small uniaxial stress, p_z, is applied to a powder compact during sintering, has been used to investigate the simultaneous occurrence of densification and creep, as well as their interaction. Parameters such as Σ, ϕ, and the creep (or shear) viscosity of the porous powder compact can be determined from data obtained in the same experiment [72,73]. In the experiments, a constant uniaxial load W is applied to a cylindrical powder compact of initial length L_o and radius R_o at an isothermal sintering temperature, and the length L and the radius R of the compact are measured continuously as a function of time t in a dilatometer (Figure 2.35).

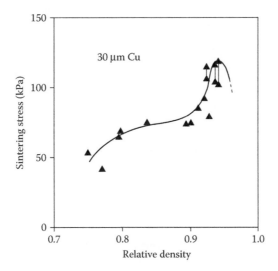

FIGURE 2.34 Data for the sintering stress versus relative density for copper particles, determined by the zero-creep technique.

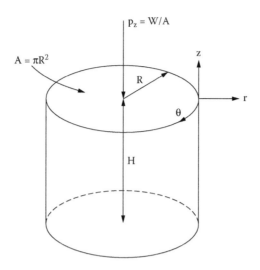

FIGURE 2.35 Geometrical parameters for the sintering of a cylindrical powder compact under a uniaxial stress.

From these data, the true strain rates in the axial direction $\dot{\varepsilon}_z$ and in the radial direction $\dot{\varepsilon}_r$ are determined using the following equations:

$$\dot{\varepsilon}_z = \frac{1}{L}\frac{dL}{dt} \tag{2.98}$$

$$\dot{\varepsilon}_r = \frac{1}{R}\frac{dR}{dt} \tag{2.99}$$

It is essential to use the true strains, rather than the engineering strains, because of the large deformations involved in sintering. The cross-sectional area A of the compact, which changes with time, is given by $A = \pi R^2$. Using Equation 2.99, then

$$A = A_o \exp(2\varepsilon_r) \tag{2.100}$$

where $A_o = \pi R_o^2$. The axial stress p_z is given by

$$p_z = \frac{W}{A} = \frac{W}{A_o}\exp(-2\varepsilon_r) \tag{2.101}$$

The densification strain rate $\dot{\varepsilon}_p$ and the creep strain rate $\dot{\varepsilon}_c$ of the sintering compact subjected to the uniaxial stress p_z are determined from the data using the following equations [74]:

$$\dot{\varepsilon}_c = (2/3)(\dot{\varepsilon}_z - \dot{\varepsilon}_r) \tag{2.102}$$

$$\dot{\varepsilon}_p = \frac{1}{3\rho}\frac{d\rho}{dt} = -(1/3)(\dot{\varepsilon}_z + 2\dot{\varepsilon}_r) \tag{2.103}$$

In the loading dilatometer experiments, the hydrostatic component of the applied uniaxial stress (equal to $p_z/3$) is commonly small compared to Σ, so the sintering mechanism of the loaded powder compact is expected to be identical to that for a compact sintered without the load. The hydrostatic component of the applied uniaxial stress can be neglected, and under these conditions Equation 2.97 can be written

$$\dot{\varepsilon}_\rho = \frac{1}{\eta_\rho} \Sigma \phi \qquad (2.104)$$

Similarly, the creep rate can be expressed as

$$\dot{\varepsilon}_c = \frac{1}{\eta_c} p_z \phi \qquad (2.105)$$

where η_c can be taken as the viscosity of the porous compact for the creep process. The ratio of the densification rate to the creep rate can be found from Equation 2.104 and Equation 2.105 as

$$\frac{\dot{\varepsilon}_\rho}{\dot{\varepsilon}_c} = \frac{F\Sigma}{p_z} \qquad (2.106)$$

where F is a parameter equal to η_c/η_ρ. As shown in Figure 2.36, measurements of the simultaneous densification and creep rates at a fixed value of p_z indicate that the ratio $\dot{\varepsilon}_\rho/\dot{\varepsilon}_c$ is approximately constant over a wide density range for several materials [75]. The ratio is also found to be approximately constant over a wide temperature range, but for a given system it decreases with the green density [76].

If it is assumed that the atomic flux fields in sintering and creep of porous ceramics at low applied stress are not significantly different, then $F \approx 1$, and the value of Σ can be determined from

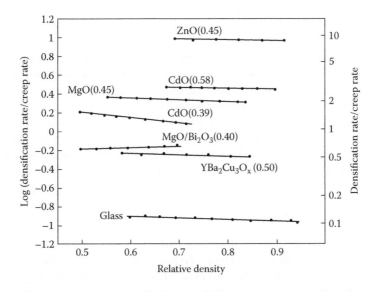

FIGURE 2.36 Ratio of the densification rate to the creep rate versus relative density at a constant uniaxial applied stress of 200 kPa. The numbers in parentheses represent the relative green density.

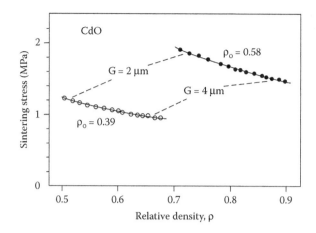

FIGURE 2.37 The sintering stress for CdO powder measured using the technique of loading dilatometry.

the measured values of $\dot{\varepsilon}_\rho/\dot{\varepsilon}_c$ and p_z. The sintering stress determined in this way is shown in Figure 2.37 for CdO powder compacts with two different initial densities ($\rho_o = 0.39$ and $\rho_o = 0.58$). It is of the order of 1–2 MPa, and, for a given initial density, it decreases slightly with increasing density. Based on the sintering model of Coble [52], it might be expected that Σ would increase with density after the initial stage of sintering. However, Coble's model assumes no grain growth or coarsening of the microstructure, so the pore size decreases with increasing density. The decrease of Σ with increasing ρ observed in Figure 2.37 is attributed largely to coarsening of the microstructure, leading to an increase in the average grain and pore sizes. As given in Figure 2.37, the average grain size increases from a value of ~2 μm to ~4 μm during the experiments. If Σ is normalized with respect to the initial grain size, then an increase in Σ with ρ would be obtained. For glass particles in which grain boundaries are absent, Σ is observed to increase with density, in agreement with theoretical predictions [77]. The measured value of Σ for CdO is of the same order of magnitude as the value estimated from the surface energy γ_{sv} and the pore radius r. Although the value of γ_{sv} for CdO is not available, assuming $\gamma_{sv} \approx 1$ J/m², a value comparable to those measured for a few oxides, and assuming that the average pore size is approximately one third the average grain size (i.e., $r \approx 0.5$–1 μm), then $\Sigma \approx \gamma_{sv}/r \approx 1$–2 MPa.

According to Equation 2.105, a plot of the creep rate determined at a constant value of p_z and η_c (i.e., at constant grain size) versus the relative density ρ gives the functional dependence of ϕ on ρ. Data for the creep rates of some porous ceramic compacts measured in loading dilatometry experiments [72,78] give values for ϕ that can be well fitted by Equation 2.93. An example of such data is shown in Figure 2.38 [72]. The lower sets of data show the measured creep rates for CdO powder compacts with two different initial densities, $\rho_o = 0.39$ and $\rho_o = 0.58$, whereas the upper set of data give the creep rates corrected for grain growth. For CdO, the stress intensification factor ϕ decreases exponentially with ρ, in accordance with Equation 2.93. Evaluation of the parameter α in Equation 2.93 from the slopes of the grain size-compensated creep rates gives $\alpha = 2.0$.

2.11.3.3 Sinter Forging Technique

The sinter forging method has also been used to study the kinetics of simultaneous deformation and densification during sintering [79]. The procedure is quite similar to the loading dilatometry technique outlined above, but the applied uniaxial stress p_z is considerably larger. By measuring the axial and radial strains as functions of time t, for given values of p_z, the creep and densification strain rates can be determined from Equation 2.98 and Equation 2.99. The hydrostatic component of the applied uniaxial stress is $p_z/3$. According to Equation 2.97, a plot of the grain size-corrected

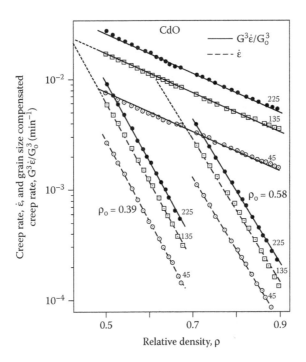

FIGURE 2.38 Dependence of the creep rate and the grain size–corrected creep rate on relative density for CdO powder compacts with initial densities of 0.39 and 0.58, subjected to the uniaxial stresses shown in kPa.

densification rate at a given compact density versus the hydrostatic component of the applied stress yields Σ for that density, by extrapolation to zero densification rate. The sintering stress determined for Al_2O_3 powder compacts (initial particle size of 0.3 μm) was found to decrease from 0.8 MPa at $\rho = 0.70 - 0.4$ MPa at $\rho = 0.95$.

2.12 ALTERNATIVE DERIVATION OF THE SINTERING EQUATIONS

As an alternative to the analytical models described in Section 2.6, one may derive the sintering equations by solving the differential equations for the atomic flux, subject to the appropriate boundary conditions. The method is outlined for grain boundary diffusion, and the concept of the sintering stress is incorporated into the derivation [80].

To simplify the derivation, we consider a geometrical model consisting of spherical particles with a diameter a arranged in a simple cubic pattern (Figure 2.39a). It is assumed that the grain boundary remains flat, with a constant width δ_{gb} (Figure 2.39b). Following Equation 1.124, the flux of atoms as a function of distance along the neck is

$$j(x) = -\frac{D_{gb}}{\Omega kT} \nabla \mu \tag{2.107}$$

where D_{gb} is the grain boundary diffusion coefficient, Ω is the atomic volume, k is the Boltzmann constant, T is the absolute temperature, and μ is the chemical potential of the atoms. The total number of atoms crossing the neck at a radius x per unit time is

$$J(x) = 2\pi x \delta_{gb} j(x) = -\frac{2\pi x D_{gb} \delta_{gb}}{\Omega kT} \nabla \mu \tag{2.108}$$

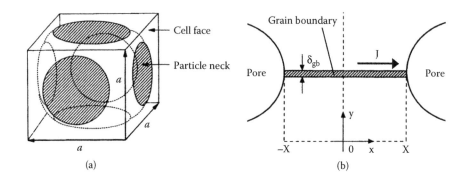

FIGURE 2.39 (a) Schematic of cell surrounding a spherical particle, showing the neck (grain boundary) section with an adjoining particle. (b) Geometrical parameters for matter transport by grain boundary diffusion.

Since the displacement of the boundary must be independent of x, the rate of approach of the particle centers, dy/dt, is related to $J(x)$ by

$$J(x) = \frac{\pi x^2}{\Omega} \frac{dy}{dt} \tag{2.109}$$

From Equation 2.108 and Equation 2.109, we obtain

$$\frac{d\mu}{dx} = -2Ax \tag{2.110}$$

where

$$A = \frac{kT}{4D_{gb}\delta_{gb}} \frac{dy}{dt} \tag{2.111}$$

Integrating Equation 2.110 gives

$$\mu(x) = -Ax^2 + B \tag{2.112}$$

where B is a constant. The chemical potential μ is related to the normal stress on the boundary σ by $\mu = \sigma\Omega$, so Equation 2.112 can be written

$$\sigma(x) = \frac{-Ax^2 + B}{\Omega} \tag{2.113}$$

The constants A and B in Equation 2.113 can be found from the boundary conditions. The first boundary condition is that the stresses must balance at $x = X$ and at $x = -X$. We will recall that according to the definition of the sintering stress, the effects of the pores and grain boundaries were artificially set at zero and replaced by an equivalent external stress equal to Σ. In this representation, the stress acting at the surface of the pore must be equal to zero. Therefore, our first boundary condition is $\sigma = 0$ at $x = \pm X$, and substituting either of these into Equation 2.113 gives

$$B = AX^2 \tag{2.114}$$

The second boundary condition is that the average stress on the grain boundary is equal to $\phi\Sigma$, where ϕ is the stress intensification factor. This condition can be expressed as

$$\frac{1}{\pi X^2} \int_0^{2\pi} \int_0^X \sigma(x) x\, dx\, d\theta = \phi\Sigma \tag{2.115}$$

Substituting for $\sigma(x)$ and integrating gives

$$-AX^2/2 + B = \Omega\phi\Sigma \tag{2.116}$$

Using Equation 2.114, we obtain

$$A = 2\Omega\phi\Sigma/X^2; \quad B = 2\Omega\phi\Sigma \tag{2.117}$$

From Equation 2.108, the total flux at the surface of the neck between two particles is given by

$$J(X) = \frac{8\pi D_{gb}\delta_{gb}\phi\Sigma}{kT} \tag{2.118}$$

To relate $J(X)$ to the shrinkage of the system, the total volume transported out of one neck in a time Δt is given by

$$\Delta V = -J(X)\Omega\Delta t = \pi X^2 \Delta a \tag{2.119}$$

where Δa is the corresponding change in the center-to-center distance between the particles. The total volumetric shrinkage in all three orthogonal directions is

$$\frac{\Delta V}{V} = \frac{3\Delta a}{a} \tag{2.120}$$

and since $V = a^3$, the rate of change of the cell volume is

$$\frac{dV}{dt} = \frac{\Delta V}{\Delta t} = 3a^2 \left(\frac{\Delta a}{\Delta t} \right) \tag{2.121}$$

Substituting for $\Delta a/\Delta t$ from Equation 2.119 gives

$$\frac{dV}{dt} = -3a^2 \frac{J(X)\Omega}{\pi X^2} \tag{2.122}$$

By definition, $\phi = a^2/(\pi X^2)$, so the instantaneous volumetric strain rate is given by

$$\frac{1}{V}\frac{dV}{dt} = -\frac{3J(X)\Omega\phi}{a^3} \tag{2.123}$$

The linear densification strain rate $\dot{\varepsilon}_p$ is equal to $(-1/3V)(dV/dt)$, and substituting for $J(X)$ from Equation 2.118, we obtain

$$\dot{\varepsilon}_p = \frac{8\pi D_{gb}\delta_{gb}\Omega}{a^3 kT}\phi^2\Sigma \tag{2.124}$$

Except for the geometrical constant (8π) and the incorporation of the density-dependent term (ϕ), Equation 2.124 has the same form as the analytical sintering equations discussed earlier.

The derivation of the sintering equation for lattice diffusion follows a procedure similar to that described above for grain boundary diffusion. It can also be obtained directly from Equation 2.124 by replacing $\pi D_{gb}\delta_{gb}$ with $2XD_l$, where D_l is the lattice diffusion coefficient [81]. Remembering that $\phi = a^2/(\pi X^2)$, these substitutions give

$$\dot{\varepsilon}_p = \frac{16D_l\Omega}{\pi^{1/2}a^2 kT}\phi^{3/2}\Sigma \tag{2.125}$$

2.12.1 GENERAL ISOTHERMAL SINTERING EQUATION

For matter transport by diffusion, a general equation for the linear densification strain rate may be written as

$$\dot{\varepsilon}_p = \frac{H_1 D\phi^{(m+1)/2}}{G^m kT}(\Sigma + p_h) \tag{2.126}$$

where H_1 is a numerical constant that depends on the geometry of the model, p_h is the hydrostatic component of an externally applied stress, G is the grain (or particle) size, k is the Boltzmann constant, T is the absolute temperature, and

$$D = D_{gb}; \quad m = 3 \quad \text{for grain boundary diffusion} \tag{2.127}$$

$$D = D_l; \quad m = 2 \quad \text{for lattice diffusion} \tag{2.128}$$

Compared with Coble's hot pressing equations (Table 2.6), Equation 2.126 has the same dependence on the material and physical parameters, except for the value of the exponent for the density-dependent parameter ϕ. The difference in the exponents arises because Coble assumed that $X = G$, whereas the present derivation relates X to G through ϕ.

2.12.2 GENERAL ISOTHERMAL CREEP EQUATION FOR POROUS SOLIDS

The procedure used in the derivation of Equation 2.124 can also be used to analyze diffusional creep of porous solids under an applied uniaxial stress p_z. In this case, the mean stress on the grain boundary is ϕp_z, and the general creep equation has the form

$$\dot{\varepsilon}_c = \frac{H_2 D\phi^{(m+1)/2}}{G^m kT}p_z \tag{2.129}$$

where H_2 is a numerical constant and the other parameters are the same as those defined by Equation 2.126.

2.13 CONCLUDING REMARKS

In this chapter, we considered the main theoretical approaches for describing solid-state sintering and viscous sintering. The models for the densification of glasses by viscous flow assume Frenkel's energy balance concept in the derivation of the sintering rate equations. The predictions of the models give good agreement with the data for real powder compacts of glass particles and for other porous amorphous systems. The sintering phenomena in polycrystalline ceramics are considerably more complex. Matter transport can occur by at least six different paths that define the mechanisms of sintering, and a distinction is often made between the densifying mechanisms and the nondensifying mechanisms. Analytical models yield explicit expressions for the dependence of the sintering rate on the primary variables, such as particle size, temperature, and applied pressure, but in view of the dramatic simplifications used in their development, the models provide only a qualitative description of sintering. Numerical simulations provide a powerful tool for analyzing many of the complexities of sintering, and the approach is expected to see wider use in the future. The stress intensification factor and the sintering stress are fundamental parameters in the understanding of sintering. The physical meaning of these parameters and methods for their measurement were discussed. The analysis of the sintering rate is, by itself, not very useful. We must also understand how the microstructure of the powder system evolves during sintering. To develop a clearer understanding of solid-state sintering, the reader must combine the results of the present chapter with those of the next chapter.

PROBLEMS

2.1   The dihedral angles for two oxides are $150°$ and $90°$. If the oxides have the same surface energy, which would you expect to densify more readily? Explain why.

2.2   Show that in Equation 2.12, relating the scaling laws, $m = 4$ for grain boundary diffusion and $m = 2$ for vapor transport.

2.3   Sintering is a continuous process, but in the models the process is divided into stages. Explain why. For the idealized solid-state sintering theory, make a table of the three stages of sintering, and for each stage give the approximate density range of applicability and the mechanisms of sintering, indicating which mechanisms lead to densification and which do not.

2.4   For the two-sphere sintering model, derive the equations given in Figure 2.11 for the neck radius X, the surface area of the neck A, and the volume of material transported into the neck V.

2.5   For the two-sphere model, derive the equations given in Table 2.5 for sintering by

(a) lattice diffusion from the grain boundary
(b) vapor transport
(c) surface diffusion

2.6   The lattice diffusion coefficient for Al^{3+} ions in Al_2O_3 is 4.0×10^{-14} cm^2/sec at $1400°C$, and the activation energy is 580 kJ/mol. Assuming that sintering is controlled by lattice diffusion of Al^{3+} ions, estimate the initial rate of sintering for an Al_2O_3 powder compact of 1 μm particles at $1300°C$.

2.7   Consider a pore with constant volume. Assuming that the pore takes on the shape of an ellipsoid of revolution, calculate the surface area of the pore as a function of the ratio of the short axis to the long axis.

2.8   Derive Equation 2.54 for the volume of the solid phase in a unit cell of Scherer's model for viscous sintering.

2.9   Consider the sintering of a spherical pore in an infinite viscous matrix. Derive an expression for the radius of the pore as a function of the sintering time. For a borosilicate glass, the viscosity at 800°C is measured to be 10^9 Pa s. Assuming that the surface energy of the glass is ~0.35 J/m^2, estimate the time it will take for an isolated pore of diameter 10 μm to be removed from a large block of the glass by sintering at 800°C.

2.10 A ZnO powder compact is formed from particles with an average size of 3 μm. Assuming that densification occurs by a lattice diffusion mechanism with an activation energy of 250 kJ/mol, estimate the factor by which the densification rate will change if

(a) the particle size is reduced to 0.3 μm.
(b) the compact is hot pressed under an applied pressure of 40 MPa.
(c) the sintering temperature is raised from 1000°C to 1200°C.
    The specific surface energy of ZnO can be assumed to be 1 J/m^2.

2.11 Zinc oxide has a fairly high vapor pressure for commonly used sintering temperatures above half the melting point, so that coarsening due to vapor transport can reduce the densification rate. Discuss how the changes in the particle size, applied pressure, and temperature described in Problem 10 will influence the rate of vapor transport.

2.12 For MgO-doped Al$_2$O$_3$, the average dihedral angle is found to be approximately 120°. Assuming that the stress intensification factor ϕ is given by Equation 2.93, determine the value of the parameter α in Equation 2.93 by using the values for ϕ given in Figure 2.32.

2.13 Explain the meaning of the term *sintering stress*. Briefly describe the methods that can be used to measure the sintering stress, indicating the principles on which each method is based, and list the advantages and disadvantages of each method.

REFERENCES

1. Uhlmann, D.R., Klein, L.C., and Hopper, R.W., Sintering, crystallization, and breccia formation, in *The Moon*, Vol. 13, D. Reidel Publishing Co., Dordrecht-Holland, 1975, p. 277.
2. Johnson, D.L., New method of obtaining volume, grain boundary, and surface diffusion coefficients from sintering data, *J. Appl. Phys.*, 40, 192, 1969.
3. Frenkel, J., Viscous flow of crystalline bodies under the action of surface tension, *J. Phys. (Moscow)*, 5, 385, 1945.
4. Shaler, A.J., Seminar on kinetics of sintering, *Metals Trans.*, 185, 796, 1949.
5. Kuczynski, G.C., Self-diffusion in sintering of metal particles, *Trans. AIME*, 185, 169, 1949.
6. Herring, C., Effect of change of scale on sintering phenomena, *J. Appl. Phys.*, 21, 301, 1950.
7. Rhines, F.N., and DeHoff, R.T., Channel network decay in sintering, in *Sintering and Heterogeneous Catalysis*, Mater. Sci. Res. Vol. 16, Kuczynski, G.C., Miller, A.E., and Sargent, G.A., Eds., Plenum Press, New York, 1984, p. 49.
8. Kuczynski, G.C., Statistical theory of sintering, *Z. Metallkunde*, 67, 606, 1976.
9. Song, H., Coble, R.L., and Brook, R.J., The applicability of Herring's scaling law to the sintering of powders, in *Sintering and Heterogeneous Catalysis*, Mater. Sci. Res. Vol. 16, Kuczynski, G.C., Miller, A.E., and Sargent, G.A., Eds., Plenum Press, New York, 1984, p. 63.
10. Johnson, K.L., Kendall, K., and Roberts, A.D., Surface energy and the contact of elastic solids, *Proc. Roy. Soc. Lond. A*, 324, 301, 1971.
11. Coble, R.L., Sintering crystalline solids. I. Intermediate and final state diffusion models. *J. Appl. Phys.*, 32, 787, 1961.
12. Kingery, W.D., and Berg, M., Study of the initial stages of sintering solids by viscous flow, evaporation–condensation, and self-diffusion, *J. Appl. Phys.*, 26, 1205, 1955.
13. Coble, R.L., Initial sintering of alumina and hematite, *J. Am. Ceram. Soc.*, 41, 55, 1958.
14. Johnson, D.L., and Cutler, I.B., Diffusion sintering. I. Initial stage sintering models and their application to shrinkage of powder compacts, *J. Am. Ceram. Soc.*, 46, 541, 1963.
15. Coblenz, W.S., Dynys, J.M., Cannon, R.M., and Coble, R.L., Initial stage solid state sintering models. A critical analysis and assessment, in Sintering Processes, *Mater. Sci. Res.*, Vol. 13, Kuczynski, G.C., Ed., Plenum Press, New York, 1980, p. 141.

16. Exner, H.E., Principles of single phase sintering, *Rev. Powder Metall. Phys. Ceram.*, 1, 7, 1979.

17. Kuczynski, G.C., Study of the sintering of glass, *J. Appl. Phys.*, 20, 1160, 1949.

18. Coble, R.L., Effect of particle size distribution in initial stage sintering, *J. Am. Ceram. Soc.*, 56, 461, 1973.

19. Swinkels, F.B., and Ashby, M.F., Role of surface redistribution in sintering by grain boundary transport, *Powder Metall.*, 23, 1, 1980.

20. Coble, R.L., Sintering crystalline solids. II. Experimental test of diffusion models in powder compacts, *J. Appl. Phys.*, 32, 793, 1961.

21. Johnson, D.L., A general model for the intermediate stage of sintering, *J. Am. Ceram. Soc.*, 53, 574, 1970.

22. Beeré, W., The second stage sintering kinetics of powder compacts, *Acta Metall.*, 23, 139, 1975.

23. Scherer, G.W., Sintering of low-density glasses: I, theory, *J. Am. Ceram. Soc.*, 60, 236, 1977.

24. Scherer, G.W., Sintering of low-density glasses: III, effect of a distribution of pore sizes, *J. Am. Ceram. Soc.*, 60, 243, 1977.

25. Scherer, G.W., Viscous sintering of a bimodal pore size distribution, *J. Am. Ceram. Soc.,* 67, 709, 1984.

26. Scherer, G.W., and Bachman, D.L., Sintering of low-density glasses: II, experimental study, *J. Am. Ceram. Soc.*, 60, 239, 1977.

27. Brinker, C.J., and Scherer, G.W., *Sol–Gel Science*, Academic, New York, 1990, Chap. 11.

28. Coleman, S.C., and Beeré, W., The sintering of open and closed porosity in UO_2, *Phil. Mag.*, 31 1403, 1975.

29. Mackenzie, J.K., and Shuttleworth, R., A phenomenological theory of sintering, *Proc. Phys. Soc. (London)*, 62, 833, 1949.

30. Pan, J., Modeling sintering at different length scales, *Intl. Mater. Rev.*, 48, 69, 2003.

31. Nichols, F.A., and Mullins, W.W., Morphological changes of a surface of revolution due to capillary-induced surface diffusion, *J. Appl. Phys.*, 36, 1826, 1965.

32. Bross, P., and Exner, H.E., Computer simulation of sintering processes, *Acta Metall.*, 27, 1013, 1979.

33. Bouvard, D., and McMeeking, R.M., Deformation of interparticle necks by diffusion-controlled creep, *J. Am. Ceram. Soc.*, 79, 666, 1996.

34. Zhang, W., and Schneibel, J.H., The sintering of two particles by surface and grain boundary diffusion – a two-dimensional numerical study, *Acta Metall. Mater.*, 43, 4377, 1995.

35. Pan, J., Le, H., Kucherenko, S., and Yeomans, J.A., A model for the sintering of particles of different sizes, *Acta Mater.*, 46, 4671, 1998.

36. Svoboda, J., and Riedel, H., New solutions describing the formation of interparticle necks in solid-state sintering, *Acta Metall. Mater.*, 43, 1, 1995.

37. Cocks, A.C.F., The structure of constitutive laws for the sintering of fine-grained materials, *Acta Metall. Mater.*, 42, 2191, 1994.

38. Olevsky, E.A., Theory of sintering: from discrete to continuum, *Mater. Sci. Eng.*, R23, 41, 1998.

39. Du, Z.Z., and Cocks, A.C.F., Constitutive models for the sintering of ceramic components – I. Material models, *Acta Metall. Mater.*, 40, 1969, 1992.

40. Besson, J., and Abouaf, M., Rheology of porous alumina and simulation of hot isostatic pressing, *J. Am. Ceram. Soc.,* 75, 2165, 1992.

41. Ross, J.W., Miller, W.A., and Weatherly, G.C., Dynamic computer simulation of viscous flow sintering kinetics, *J. Appl. Phys.*, 52, 3884, 1981.

42. Jagota, A., and Dawson, P.R., Micromechanical modeling of powder compacts – I. Unit problems for sintering and traction-induced deformation, *Acta Metall.*, 36, 2551, 1988.

43. Jagota, A., and Dawson, P.R., Simulation of the viscous sintering of two particles, *J. Am. Ceram. Soc.*, 73, 173, 1990.

44. Zhou, H., and Derby, J.J., Three-dimensional finite-element analysis of viscous sintering, *J. Am. Ceram. Soc.*, 81, 533, 1998.

45. Jagota, A., Simulation of the viscous sintering of coated particles, *J. Am. Ceram. Soc.*, 77, 2237, 1994.

46. Tikkanen, M.H., and Makipirtti, S.A., A new phenomenological sintering equation, *Int. J. Powder Metall.*, 1, 15, 1965.

47. Ivensen, V.A., *Densification of Metal Powders during Sintering*, Consultants Bureau, New York, 1973.

48. Pejovnik, S., Smolej, V., Susnik, D., and Kolar, D., Statistical analysis of the validity of sintering equations, *Powder Metall. Int.*, 11, 22, 1979.

49. Ashby, M.F., A first report on sintering diagrams, *Acta Metall.*, 22, 275, 1974.
50. Swinkels, F.B., and Ashby, M.F., A second report on sintering diagrams, *Acta Metall.*, 29, 259, 1981.
51. Coble, R.L., Mechanisms of densification during hot pressing, in *Sintering and Related Phenomena*, Kuczynski, G.C., Hooton, N.A., and Gibbon, C.F., Eds., Gordon and Breach, New York, 1967, p. 329.
52. Coble, R.L., Diffusion models for hot pressing with surface energy and pressure effects as driving forces, *J. Appl. Phys.*, 41, 4798, 1970.
53. Nabarro, F.R.N., Deformation of crystals by the motion of single ions, in *Report of a Conference on the Strength of Solids*, Physical Society, London, 1948, p. 75.
54. Herring, C., Diffusional viscosity of a polycrystalline solid, *J. Appl. Phys.*, 21, 437, 1950.
55. Coble, R.L., A model for boundary diffusion controlled creep in polycrystalline materials, *J. Appl. Phys.*, 34, 1679, 1963.
56. Harmer, M.P., and Brook, R.J., The effect of MgO additions on the kinetics of hot pressing in Al_2O_3, *J. Mater. Sci.*, 15, 3017, 1980.
57. Beeré, W., Diffusional flow and hot pressing: a study of MgO, *J. Mater. Sci.*, 10, 1434, 1975.
58. Vieira, J.M., and Brook, R.J., Kinetics of hot pressing: the semi-logarithmic law, *J. Am. Ceram. Soc.*, 67, 245, 1984.
59. Swinkels, F.B., Wilkinson, D.S., Arzt, E., and Ashby, M.F., Mechanisms of hot isostatic pressing, *Acta Metall.*, 31, 1829, 1983.
60. Artz, E., Ashby, M.F., and Easterling, K.E., Practical applications of hot isostatic pressing diagrams, *Metall. Trans.*, 14A, 211, 1983.
61. Helle, A.S., Easterling, K.E., and Ashby, M.F., Hot isostatic pressing diagrams: new developments, *Acta Metall.*, 33, 2163, 1985.
62. Scherer, G.W., Sintering of inhomogeneous glasses: application to optical waveguides, *J. NonCryst. Solids*, 34, 239, 1979.
63. Shima, S., and Oyane, M., Plasticity theory for porous metals, *Int. J. Mech. Sci.*, 18, 285, 1976.
64. Dutton, R.E., Shamasundar, S., and Semiatin, S.L., Modeling the hot consolidation of ceramic and metal powders, *Metall. Mater. Trans. A*, 26, 2041, 1995.
65. De Jonghe, L.C., and Rahaman, M.N., Sintering stress of homogeneous and heterogeneous powder compacts, *Acta Metall.*, 36, 223, 1988.
66. Raj, R., Analysis of the sintering pressure, *J. Am. Ceram. Soc.*, 70, C210, 1987.
67. Cannon, R.M., and Carter, W.C., Interplay of sintering microstructures, driving forces, and mass transport mechanisms, *J. Am. Ceram. Soc.*, 72, 1550, 1989.
68. Gibbs, J.W., in *The Scientific Papers of J. Willard Gibbs*, Dover Press, New York, 1961.
69. Gregg, R.A., and Rhines, F.N., Surface tension and the sintering force in copper, *Metall. Trans.*, 4, 1365, 1973.
70. Aigeltinger, E.H., Relating microstructure and sintering force, *Int. J. Powder Metall. Powder Technol.*, 11, 195, 1975.
71. De Jonghe, L.C., and Rahaman, M.N., A loading dilatometer, *Rev. Sci. Instrum.*, 55, 2007, 1984.
72. Rahaman, M.N., De Jonghe, L.C., and Brook, R.J., Effect of shear stress on sintering, *J. Am. Ceram. Soc.*, 69, 53, 1986.
73. Scherer, G.W., Viscous sintering under a uniaxial load, *J. Am. Ceram. Soc.*, 69, C206, 1986.
74. Raj, R., Separation of cavitation strain and creep strain during deformation, *J. Am. Ceram. Soc.*, 65, C46, 1982.
75. Chu, M.Y., De Jonghe, L.C., and Rahaman, M.N., Effect of temperature on the densification/creep viscosity during sintering, *Acta Metall.*, 37, 1415, 1989.
76. Rahaman, M.N., De Jonghe, L.C., and Chu, M.Y., Effect of green density on densification and creep during sintering, *J. Am. Ceram. Soc.*, 74, 514, 1991.
77. Rahaman, M.N., and De Jonghe, L.C., Sintering of spherical glass powder under a uniaxial stress, *J. Am. Ceram. Soc.*, 73, 707, 1990.
78. Rahaman, M.N., and De Jonghe, L.C., Creep sintering of zinc oxide, *J. Mater. Sci.*, 22, 4326, 1987.
79. Venkatachari, K.R., and Raj, R., Shear deformation and densification of powder compacts, *J. Am. Ceram. Soc.*, 69, 499, 1986.
80. Chu, M.Y., Rahaman, M.N., De Jonghe, L.C., and Brook, R.J., Effect of heating rate on sintering and coarsening, *J. Am. Ceram. Soc.*, 74, 1217, 1991.
81. Raj, R., and Ashby, M.F., Intergranular fracture at elevated temperature, *Acta Metall.*, 23, 653, 1975.

3 Grain Growth and Microstructure Control

3.1 INTRODUCTION

The engineering properties of ceramics are strongly dependent on the microstructure, the important features of which are the size and shape of the grains, the amount of porosity, the pore size, the distribution of the pores in the body, and the nature and distribution of any second phases. For most applications, microstructural control usually means the achievement of as high a density, as small a grain size, and as homogeneous a microstructure as possible. The microstructure of the fabricated article is influenced strongly by the microstructure of the green body, which in turn is dependent largely on the powder characteristics and the forming method [1]. However, even if proper procedures are employed in the production of the green body, which is rarely the case, significant manipulation of the microstructure is still necessary during the sintering step.

The densification of a polycrystalline powder compact is normally accompanied by coarsening of the microstructure: both the grains and the pores increase in average size while decreasing in number. In Chapter 2 we considered the densification process. We now address the coarsening process and the interplay between densification and coarsening in order to develop an understanding of microstructural control.

The treatment of microstructural control in ceramics often starts with an analysis of grain growth in fully dense, single-phase solids where the complicating effects of pores and second phases are absent. Grain growth in ceramics is divided into two main types: normal grain growth and abnormal grain growth. In *normal* grain growth, the grain sizes and shapes occur within a fairly narrow range and, except for a magnification factor, the grain size distribution at a later time is similar to that at an earlier time. *Abnormal* grain growth is characterized by the rapid growth of a few larger grains at the expense of the smaller ones. Simple models have been developed to predict the kinetics of normal grain growth, but many of them analyze an isolated grain boundary or a single grain and neglect the topological requirements of space filling. In recent years, computer simulation has played an important role in exploring many of the complexities of grain growth.

Sintering of porous powder compacts involves the interaction between pores and grain boundaries that migrate as a result of grain growth. A common approach is to analyze how the average sizes of the grains and the pores increase with time, but a broader understanding of microstructural development also requires an understanding of how the pore network evolves. The interaction between spherical, isolated pores and the grain boundaries can be analyzed in terms of simple models, and, since grain growth is more pronounced in the final stage of sintering, this interaction plays an important role in determining the limits of densification. A key result is that the separation of the boundaries from the pores, a condition that is symptomatic of abnormal grain growth in porous solids, must be prevented if high density is to be achieved during sintering. Coarsening also occurs in the earlier stages of sintering, and this process influences microstructural evolution in the later stages, so the coarsening of more porous compacts is also important.

The densification process and the coarsening process are by themselves very complex, so a detailed theoretical analysis of the interplay between the two processes is a daunting task. Computer simulation techniques are only now beginning to address some of the complexities of microstructural evolution in simple model systems. Qualitative analyses based on simple models indicate that the achievement of high density and controlled grain size during sintering is dependent on reducing the grain growth rate, increasing the densification rate, or some combination of these two. Several fabrication approaches that achieve one or both of these goals are available.

3.2 GENERAL FEATURES OF GRAIN GROWTH

3.2.1 GRAIN GROWTH AND COARSENING

Grain growth is the term used to describe the increase in the grain size of a single-phase solid, or in the matrix grain size of a solid containing a second phase. Grain growth occurs in both dense and porous polycrystalline solids at sufficiently high temperatures. For the conservation of matter, the sum of the individual grain sizes must remain constant, so an increase in the average grain size is accompanied by the disappearance of some grains, usually the smaller ones. In porous solids, both the grains and the pores commonly increase in size while decreasing in number. Because of the considerable interaction between the grains and the pores, microstructural evolution is considerably more complex than for dense solids. Frequently, the term *coarsening* is used to describe the process of grain growth coupled with pore growth. At a later time the microstructure appears to be roughly a coarser version of that at an earlier time.

3.2.2 OCCURRENCE OF GRAIN GROWTH

In the widely accepted picture, the grain boundary is considered to be a region of disorder between two crystalline regions (the grains), as sketched in Figure 3.1 for a section across two grains. High-resolution transmission electron microscopy indicates that the thickness of the grain boundary

FIGURE 3.1 Classical picture of a grain boundary. The boundary migrates downward as atoms move less than an interatomic spacing from the convex side of the boundary to the concave side.

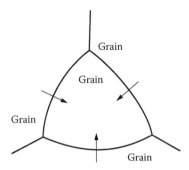

FIGURE 3.2 Movement of the grain boundary toward its center of curvature.

region is 0.5–1 nm. Grain growth occurs as atoms (or ions) diffuse less than an interatomic distance from one side of the boundary to new positions on the other side, resulting in one grain growing at the expense of another. The atoms move from the "convex" surface on one side of the grain boundary to the "concave" surface on the other side more readily than in the reverse direction, because the chemical potential of the atoms under the "convex" surface is higher than that for the atoms under the "concave" surface (see Chapter 1). The result of this net flux is that the boundary moves toward its center of curvature (Figure 3.2).

3.2.3 DRIVING FORCE FOR GRAIN GROWTH

It will be recalled that the atoms in the grain boundary have a higher energy than those in the bulk of the crystalline grain, so the grain boundary is characterized by a specific energy γ_{gb}, typically on the order of 0.2–1.0 J/m^2. The *driving force* for grain growth is the decrease in free energy that accompanies the reduction in the total grain boundary area (Equation 2.5).

3.2.4 NORMAL AND ABNORMAL GRAIN GROWTH

Grain growth in ceramics is generally divided into two types: *normal* grain growth and *abnormal* grain growth, which is sometimes referred to as exaggerated grain growth, discontinuous grain growth, or, in the case of metals, secondary crystallization. In normal grain growth, the average grain size increases, but the grain sizes and shapes remain within a fairly narrow range, so the grain size distribution at a later time is fairly similar to that at an early time (Figure 3.3a). The form of the grain size distribution is therefore said to be time invariant, and this characteristic is sometimes expressed as the grain size distribution having the property of *scaling* or *self-similarity*. In abnormal grain growth, a few large grains develop and grow relatively faster than the surrounding matrix of smaller grains. The grain size distribution may change significantly, giving rise to a bimodal distribution (Figure 3.3b), and in this case the property of time invariance of the distribution is lost. Eventually, the large grains impinge, and the grain size distribution may revert to a normal distribution.

Grain growth in *porous* ceramics is also described as being normal or abnormal, but the interaction of the pores with the grains is also taken into account. Thus, normal grain growth in porous ceramics is, in addition, characterized by the pores remaining at the grain boundaries. When the boundaries break away from the pores, leaving them inside the grains, the situation is usually indicative of abnormal grain growth. Although ceramic microstructures commonly show nearly *equiaxial* grain shapes, it is sometimes observed that the normal or abnormal grains grow in an anisotropic manner, resulting in grains that have an elongated or plate-like morphology. In this case, the growth is described as *anisotropic* grain growth.

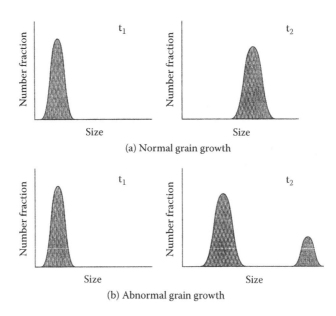

FIGURE 3.3 For an initial microstructure consisting of a unimodal distribution of grain sizes, (a) normal grain growth results in an increase in the average grain size, whereas the grain size distribution remains almost self-similar; (b) abnormal grain growth is characterized by a few large grains growing rapidly at the expense of the surrounding matrix grains and may lead to a bimodal distribution of grain sizes.

Figure 3.4 shows some examples of normal and abnormal grain growth in dense and porous ceramics. A dense Al_2O_3 structure with normal grains is shown in Figure 3.4a. If the structure is heated further, a few abnormal grains may start to develop (Figure 3.4b), subsequently growing rapidly and eventually consuming the smaller grains. Figure 3.4c shows a combination of normal and abnormal grain growth in a nickel–zinc ferrite. The pores deep inside the large, abnormal grain at the top are characterized by long diffusion distances for matter transport into them from the grain boundary. As discussed below, these pores are difficult to remove, so they limit the final density that can be achieved during sintering. In comparison, the pores in the fine-grained array at the bottom lie almost exclusively on the grain boundaries; these pores are less difficult to remove during sintering because of the short diffusion distances. Figure 3.4d shows a sample of Al_2O_3 that has undergone considerable abnormal grain growth, with the residual pores trapped within the grains.

3.2.5 IMPORTANCE OF CONTROLLING GRAIN GROWTH

The control of grain growth during sintering forms one of the most important considerations in the fabrication of ceramics for two main reasons [2]. First, many engineering properties of ceramics are dependent on the grain size, so grain growth control is directly related to the achievement of the desired properties. Second, grain growth increases the diffusion distance for matter transport, thereby reducing the rate of densification, so grain growth control forms an important approach for achieving the normally required high density.

3.2.5.1 Effect of Grain Size on Properties

Few properties are completely independent of grain size, and the subject of grain size effects on properties is well covered in several texts and review articles, including Kingery, Bowen, and Uhlmann [3]. Generally, most properties are enhanced by smaller grain size, a notable exception being the creep resistance, which increases with larger grain size. For example, the fracture strength

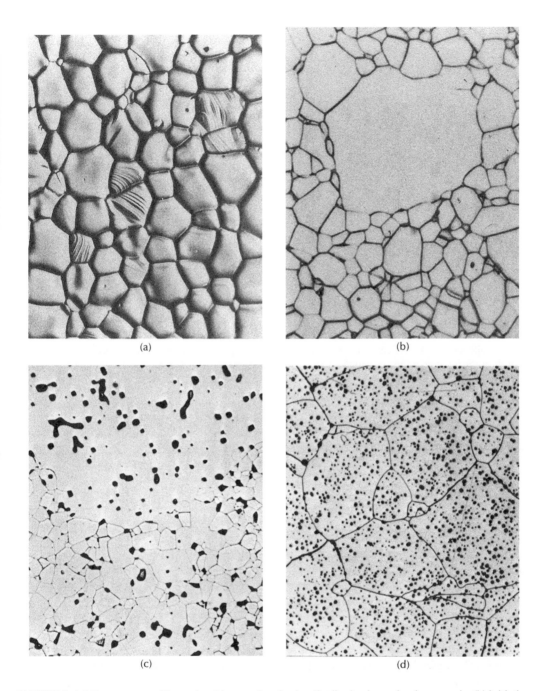

FIGURE 3.4 Microstructures illustrating (a) normal grain size distribution in an alumina ceramic; (b) initiation of abnormal grain growth in an alumina ceramic; (c) normal and abnormal grain growth in a porous nickel–zinc ferrite; (d) an alumina ceramic that has undergone considerable abnormal grain growth.

of many ceramics is often found to increase with smaller grain size G, according to $1/G^{1/2}$. A wide range of electrical and magnetic phenomena are affected by grain size, and it is in this area that grain size control has been used most effectively to produce ceramics with properties suitable for a variety of applications. For example, the electrical breakdown strength of ZnO varistors used in electrical surge suppressors increases as $1/G$, while the dielectric constant of $BaTiO_3$ capacitors is found to increase with decreasing grain size, down to ~1 μm.

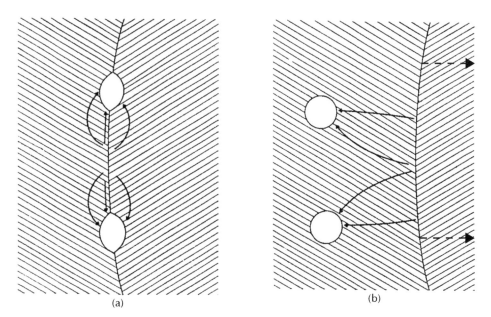

FIGURE 3.5 (a) Densification mechanisms for porosity attached to a grain boundary. The arrows indicate paths for atom diffusion. (b) Densification mechanisms for porosity separated from a grain boundary. The solid arrows indicate paths for atom diffusion, and the dashed arrows indicate the direction of boundary migration.

3.2.5.2 Attainment of High Density

As described in Chapter 2, densification occurs by the flux of matter from the grain boundaries (the source) to the pores (the sink). For sintering by diffusion mechanisms, at a given temperature and density, the dependence of the densification rate on the grain size G can be written

$$\frac{1}{\rho}\frac{d\rho}{dt} = \frac{K}{G^m} \tag{3.1}$$

where K is a temperature-dependent constant and the exponent $m = 3$ for lattice diffusion and $m = 4$ for grain boundary diffusion. Rapid densification requires that the diffusion distance between the source of matter and the sink be kept small; that is, the grain size must remain small (Figure 3.5a). According to Equation 3.1, rapid grain growth causes a drastic reduction in the densification rate, so prolonged sintering times are needed to achieve the required density, which increases the tendency for abnormal grain growth to occur. When abnormal grain growth occurs, the pores become trapped inside the grains and become difficult or almost impossible to remove because the transport paths become large, in the case of lattice diffusion, or are eliminated, in the case of grain boundary diffusion (Figure 3.5b). The attainment of high density therefore requires the control of normal grain growth as well as the avoidance of abnormal grain growth.

3.3 OSTWALD RIPENING

Ostwald ripening refers to the coarsening of precipitates (particles) in a solid or liquid medium. Its relevance to microstructural control is that many features of grain growth and pore growth during sintering are shared by the Ostwald ripening process. Let us consider a system consisting of a dispersion of spherical precipitates with different radii in a medium in which the particles have

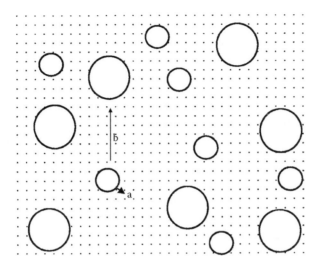

FIGURE 3.6 Coarsening of particles in a medium by matter transport from the smaller to the larger particles. Two separate mechanisms can control the rate of coarsening: (a) reaction at the interface between the particles and the medium and (b) diffusion through the medium.

some solubility (Figure 3.6). The chemical potential of the atoms under the surface of a sphere of radius a is given by (see Chapter 1)

$$\mu = \mu_o + \frac{2\gamma\Omega}{a} \tag{3.2}$$

where μ_o is the chemical potential of the atoms under a flat surface, γ is the specific energy of the interface between the sphere and the medium, and Ω is the atomic volume. Due to their higher chemical potential, the atoms under the surface of the sphere have a higher solubility in the surrounding medium than the atoms under a flat surface. Making use of the relation between chemical potential and concentration (Equation 1.94) and assuming ideal solutions, we can write

$$kT \ln\left(\frac{C}{C_o}\right) = \mu - \mu_o = \frac{2\gamma\Omega}{a} \tag{3.3}$$

where C is the concentration of solute surrounding a particle of radius a, C_o is the concentration over a flat surface, k is the Boltzmann constant, and T is the absolute temperature. If $\Delta C = C - C_o$ is small, then $\ln(C/C_o) \approx \Delta C/C_o$ and Equation 3.3 becomes

$$\frac{\Delta C}{C_o} = \frac{2\gamma\Omega}{kTa} \tag{3.4}$$

The higher solute concentration around a precipitate with smaller radius gives rise to a net flux of matter from the smaller precipitates to the larger ones, so the dispersion coarsens by a process in which the smaller precipitates dissolve and the larger ones grow. This type of coarsening process is referred to as *Ostwald ripening*, in honor of the work of Ostwald in this area around 1900. The driving force for the process is the reduction in the interfacial area between the precipitates and the medium.

3.3.1 THE LSW THEORY

The basic theory of Ostwald ripening was developed independently by Greenwood [4], Wagner [5], and Lifshitz and Slyozov [6]. It is often referred to as the LSW theory. The idealized system considered in the LSW theory is required to satisfy three conditions during the coarsening process:

1. The growth rate of a precipitate is equal to the atomic flux at the precipitate surface.
2. The distribution of precipitate sizes is continuous.
3. Mass conservation holds, in which the total volume (mass) of the precipitates is conserved.

There are also several idealized assumptions, including the following: (a) the precipitate and the medium are isotropic fluids, (b) the precipitates are spherical in shape, (c) the number of precipitates is sufficiently large for the distribution of the precipitate radii to be represented as a continuous function, (d) the solute concentration at the surface of the precipitate is determined only by the radius of the precipitate, (e) nucleation and precipitate coalescence are neglected, and (f) the total volume of the system is infinite.

As illustrated in Figure 3.6, two different mechanisms can control the rate of coarsening. In one case, the solubility of the particles into the medium or the deposition of the solute onto the particle surfaces may be the slowest step, and under these conditions Ostwald ripening is said to be controlled by the *interface reaction*. In the other case, the diffusion of atoms through the medium may be the slowest step, and the process is said to be *diffusion controlled*.

3.3.1.1 Ostwald Ripening Controlled by the Interface Reaction

In his analysis, Wagner [5] assumed that the rate of transfer of atoms is proportional to the difference between the solute concentration around a precipitate of radius a, given by Equation 3.4, and an average concentration of the solute, C^*, defined as the concentration that is in equilibrium with a precipitate of radius a^* that neither grows nor shrinks. It is also assumed that the change in C^* with increasing a^* can be neglected. The rate of change of the precipitate radius can be written

$$\frac{da}{dt} = -\alpha_T \Omega (C_a - C^*)$$

(3.5)

where α_T is a transfer constant, and the negative sign is included to indicate that for smaller precipitates $C_a - C^*$ is positive but the radius, a, decreases with time. If the total volume of the precipitates is constant, then

$$4\pi \sum_i a_i^2 \frac{da_i}{dt} = 0$$

(3.6)

where the summation is taken over all the precipitates in the system. Writing Equation 3.5 as

$$\frac{da}{dt} = \alpha_T \Omega [(C^* - C_o) - (C_a - C_o)]$$

(3.7)

and substituting into Equation 3.6 gives

$$\sum_i a_i^2 (C^* - C_o) = \sum_i a_i^2 (C_{a_i} - C_o)$$

(3.8)

Putting $(C_{a_i} - C_o) = \Delta C$ and using Equation 3.4 for ΔC, Equation 3.8 becomes

$$C^* - C_o = \frac{2\gamma\Omega C_o}{kT}\left(\frac{\sum a_i}{\sum a_i^2}\right) \tag{3.9}$$

Substituting Equation 3.9 into Equation 3.7 gives

$$\frac{da}{dt} = \frac{2\alpha_T\gamma\Omega^2 C_o}{kT}\left[\frac{\sum a_i}{\sum a_i^2} - \frac{1}{a}\right] \tag{3.10}$$

Putting $a^* = \sum a_i^2/\sum a_i$, Equation 3.10 can be written

$$\frac{da}{dt} = \frac{2\alpha_T\gamma\Omega^2 C_o}{kT}\left(\frac{1}{a^*} - \frac{1}{a}\right) \tag{3.11}$$

According to Equation 3.11, the rate of change of the precipitate radius is proportional to the difference between the critical precipitate curvature and the actual precipitate curvature.

The evolution of a system of precipitates is described by a distribution function $f(a,t)$ such that $f(a,t)da$ represents the fractional number of precipitates between a and $a + da$. The distribution function must satisfy the continuity equation

$$\frac{df}{dt} + \frac{\partial}{\partial a}\left[f\left(\frac{da}{dt}\right)\right] = 0 \tag{3.12}$$

The solution of the coupled differential equations, Equation 3.11 and Equation 3.12, shows that after a sufficient amount of time, coarsening proceeds in a steady-state manner such that the precipitate size distribution remains stationary in time. If the precipitate radius is expressed as a reduced size $s = a/a^*$, the distribution function takes the form [5]

$$f(s,t) \approx s\left(\frac{2}{2-s}\right)^5 \exp\left(\frac{-3s}{2-s}\right) \quad \text{for } 0 < s < 2$$
$$= 0 \quad\quad\quad\quad\quad\quad\quad\quad\quad \text{for } s > 2 \tag{3.13}$$

This asymptotic distribution is independent of the initial distribution at the start of coarsening. In this distribution, the average radius, $\bar{a}$, taken as the arithmetic mean radius, is given by $\bar{a} = (8/9)a^*$, and the maximum particle radius is $2a^*$. The distribution function f is shown in Figure 3.7. The critical radius in this steady-state coarsening regime increases parabolically according to

$$(a^*)^2 - (a_o^*)^2 = \left(\frac{\alpha_T C_o \gamma\Omega^2}{kT}\right)t \tag{3.14}$$

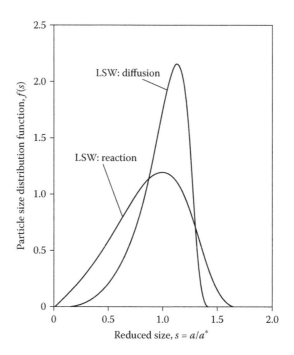

FIGURE 3.7 The particle size distribution function plotted versus the reduced size for Ostwald ripening controlled by interface reaction or diffusion.

3.3.1.2 Ostwald Ripening Controlled by Diffusion

The solution of the coarsening problem when diffusion is rate controlling follows along the same lines as that for reaction control. The rate of change of the particle radius, equated to the diffusive flux at the precipitate surface, is

$$\frac{da}{dt} = -D\Omega \frac{dC}{da} \tag{3.15}$$

where D is the diffusion coefficient for the solute atoms in the medium. For a dilute dispersion of particles in the medium, Equation 3.15 can be written

$$\frac{da}{dt} = D\Omega \frac{(C^* - C_a)}{a} = \frac{DC_o\gamma\Omega^2}{kTa}\left(\frac{1}{a^*} - \frac{1}{a}\right) \tag{3.16}$$

Solution of the coupled differential equations, Equation 3.12 and Equation 3.16, gives the distribution function for steady-state coarsening as [5]

$$f(s,t) \approx s^2 \left(\frac{3}{3+s}\right)^{7/3} \left(\frac{3/2}{3/2-s}\right)^{11/3} \exp\left(\frac{-s}{3/2-s}\right) \quad \text{for } 0 < s < 3/2 \tag{3.17}$$

$$= 0 \qquad\qquad\qquad\qquad\qquad\qquad\qquad\qquad \text{for } s > 3/2$$

The distribution function is also independent of the initial size distribution, but the average radius $\bar{a}$ is now equal to a^*, and the maximum particle radius is equal to $3a^*/2$. It is compared with the function for reaction control in Figure 3.7. In the case of diffusion control, the critical (or average) radius in the steady-state coarsening regime increases according to a cubic law:

$$(a^*)^3 - (a_o^*)^3 = \left(\frac{8DC_o\gamma\Omega^2}{9kT} \right)t \tag{3.18}$$

3.3.2 Modifications to the LSW Theory

Despite the idealized and restrictive set of assumptions, the LSW theory provides a remarkably successful description of many coarsening observations. However, for diffusion-controlled coarsening, it is often found that real systems exhibit a broader and more symmetric steady-state distribution. Furthermore, although the average precipitate size is usually observed to increase qualitatively with the predicted $t^{1/3}$ dependence, the rate constant is generally found to be different from that predicted by the LSW theory. For purely reaction-controlled coarsening where the mechanism is strictly a surface phenomenon, the process should be independent of the volume fraction of precipitates, as long as the precipitates do not touch. On the other hand, for diffusion control, the interaction of the diffusion fields of the precipitates must be taken into account, and there have been several attempts to incorporate volume fraction effects into the LSW theory [7].

A volume fraction modification to the LSW theory was investigated by Ardell [8], who assumed that a random spatial arrangement of different-sized precipitates can be related through a mean free diffusion path that is a function of the volume fraction. On repeating the LSW analysis, the results showed that as the volume fraction increases, the rate constant increases and the size distribution function broadens. However, the precipitate size increases as $t^{1/3}$ at all volume fractions. Figure 3.8 shows the

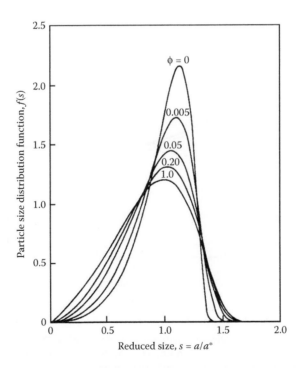

FIGURE 3.8 Dependence of the particle size distribution on the particle volume fraction for diffusion-controlled Ostwald ripening.

predicted size distribution for several values of the volume fraction ϕ. In the limit of $\phi = 0$, the results reduce to those of the LSW theory, but for $\phi = 1$, the distribution function is formally identical to the LSW function for reaction control.

Modifications to the LSW theory have also been made by Davies, Nash, and Stevens [9], who included the effect of direct contact between the precipitates, and by Brailsford and Wynblatt [10], who considered a precipitate embedded in a random array of precipitates with different sizes. Both modifications give the $t^{1/3}$ coarsening dependence and reduce to the LSW results at zero volume fraction. The analysis of Davies et al. shows that the rate constant increases as a result of the encounters between the precipitates and that the steady-state distribution function broadens and becomes more symmetric with increasing volume fraction. The results of Brailsford and Wynblatt

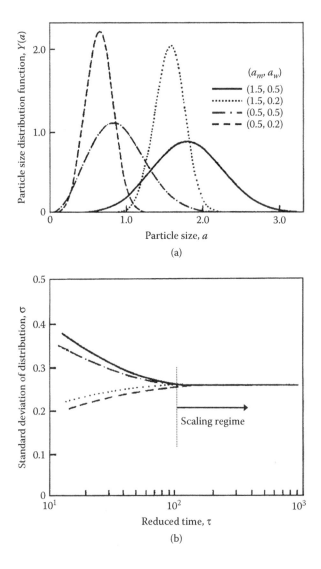

FIGURE 3.9 Effect of the initial particle size distribution shown in (a) on the time variation of the standard deviation of the distribution function (b). (From Enomoto, Y., Kawasaki, K., and Tokuyama, M., Computer modeling of Ostwald ripening, *Acta Metall.*, 35, 904, 1987. With permission.)

show that the broadening of the steady-state distribution is less than that found by Ardell and by Davies et al. The modifications by Davies et al. and by Brailsford and Wynblatt provide reasonable agreement with most experimental data.

3.3.3 TIME-DEPENDENT OSTWALD RIPENING

The LSW theory and its modifications apply to the limit of long time, sometimes referred to as the scaling regime, when the precipitate size distribution function is time invariant. Prior to reaching this scaling regime, the distribution function is time dependent, and the approach to the steady-state distribution is an interesting problem that has been addressed by Enomoto, Kawasaki, and Tokuyama [11,12]. For precipitates with a modified Gaussian distribution of sizes, given as

$$Y(a) = a^2 \exp\left[-\frac{(a - a_m)^2}{2a_w^2} \right] \tag{3.19}$$

where the parameters a_m and a_w correspond approximately to the position of the maximum and the width of the distribution, respectively. Enomoto et al. chose distributions with varying a_m and a_w (Figure 3.9a) and analyzed the changes in the properties of the distribution with time. The results show that for distributions with a broad width, the standard deviation of the distribution decreases to the characteristic steady-state value, while for systems with a narrow width, the standard deviation increases to the steady-state value (Figure 3.9b). We may interpret this behavior during the transient regime to mean that the scaling regime acts as a strong attractor for the evolution of the precipitate size distribution.

3.4 TOPOLOGICAL AND INTERFACIAL TENSION REQUIREMENTS

A dense polycrystalline solid consists of a space-filling array of grains. Whereas many theories of grain growth analyze the interfacial tension requirements of an isolated grain boundary or an isolated grain, certain topological requirements of space filling and the local requirements for equilibrium of the interfacial tensions must be satisfied, as recognized many years ago by Smith [13]. Consider a two-dimensional section through a dense polycrystalline solid (Figure 3.10). The structure consists of vertices joined by edges (also called sides) that surround faces. Provided that the face at infinity is not counted, the numbers of faces F, edges E, and vertices V obey Euler's equation

$$F - E + V = 1 \tag{3.20}$$

For stable topological structures, i.e., those for which the topological features are unchanged by small deformations, the number of edges that intersect at a vertex is equal to 3. For isotropic grain boundary energies (i.e., γ_{gb} is the same for all grain boundaries), balance of the grain boundary tensions requires that the edges meet at an angle of 120°. Let us now consider polygons in which the sides intersect at 120°. Taking N as the number of sides, under these conditions a hexagon ($N = 6$) has plane sides, whereas a polygon with $N > 6$ has "concave" sides, and one with $N < 6$ has "convex" sides (Figure 3.11). Since the grain boundary migrates toward its center of curvature, grains with $N > 6$ tend to grow, while those with $N < 6$ tend to shrink.

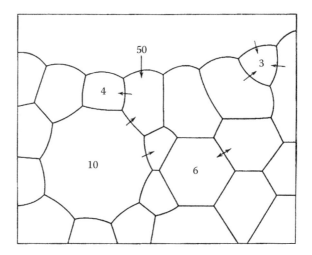

FIGURE 3.10 Sketch of a section through a dense polycrystalline solid. The sign of the curvature changes as the number of sides increases from less than six to more than six. The arrows indicate the direction in which the boundaries migrate.

In three dimensions, the faces surround cells, and, provided the cell at infinity is not counted, we have

$$F - E + V - C = 1 \tag{3.21}$$

where C is the number of cells. Equation 3.21 can be easily verified for cells with the shape of a cube or a tetrakaidecahedron. Assuming isotropic interfacial energies, balance of interfacial tension requires that the surfaces meet in groups of three at angles of 120°, along lines that themselves meet in groups of four mutually at an angle of 109.5°. This is the angle subtended by straight lines joining the corners of a regular tetrahedron. No regular polyhedron with plane sides has exactly this angle between its edges. The nearest approach to space filling by a regular plane-sided polyhedron is obtained with tetrakaidecahedra arranged on a body-centered cubic lattice (Figure 3.12a), but even with this structure, the angles are not exactly those required, and the boundaries must become curved to satisfy local equilibrium at the vertices. In general, real grains are arranged randomly and also have different sizes (Figure 3.12b).

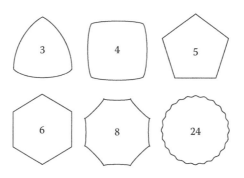

FIGURE 3.11 Polygons with curved sides meeting at 120°.

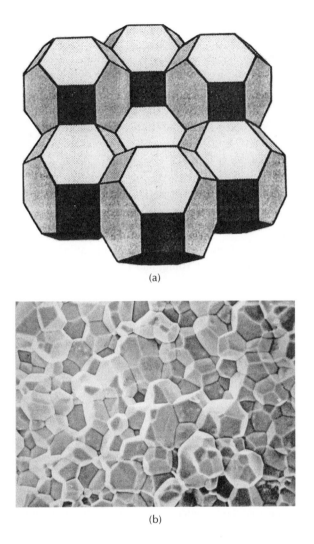

(a)

(b)

FIGURE 3.12 (a) A group of regular tetrakaidecahedra arranged in a space-filling array, and (b) grains in a real system consisting of an almost fully dense ZnO material.

3.5 NORMAL GRAIN GROWTH IN DENSE SOLIDS

Starting from the early 1950s, the theoretical analysis of normal grain growth in dense polycrystalline solids has received considerable attention. A review of the subject is given by Atkinson [14]. Here, we discuss the main approaches that have had a key impact on advancing our understanding of the grain growth process.

3.5.1 THE MODEL OF BURKE AND TURNBULL

In one of the earliest analyses, Burke and Turnbull [15] modeled the grain growth in dense solids as occurring by transport of atoms across the grain boundary under the driving force of the pressure gradient across the grain boundary. More correctly, the chemical potential gradient across the boundary should be used. Burke and Turnbull considered an isolated part of the grain boundary and assumed that the grain growth equation derived in the analysis represents the average behavior of the whole system. In addition, the grain boundary energy γ_{gb} is assumed to be isotropic and independent of the crystallographic direction, and the grain boundary width δ_{gb} is assumed to be constant.

If the average rate of grain boundary migration v_b, sometimes called the *grain boundary velocity*, is taken to be approximately equal to the instantaneous rate of grain growth, we can write

$$v_b \approx dG/dt \qquad (3.22)$$

where G is the average grain size. It is also assumed that v_b can be represented as the product of the driving force for boundary migration F_b and the boundary mobility M_b, so

$$v_b = M_b F_b \qquad (3.23)$$

where M_b includes effects arising from the mechanism of migration. The pressure difference across the boundary is given by the equation of Young and Laplace:

$$\Delta p = \gamma_{gb} \left(\frac{1}{r_1} + \frac{1}{r_2} \right) \qquad (3.24)$$

where γ_{gb} is the specific grain boundary energy (energy per unit area), and r_1 and r_2 are the principal radii of curvature of the boundary. Assuming that the radius of the boundary is proportional to G, then

$$\left(\frac{1}{r_1} + \frac{1}{r_2} \right) = \frac{\alpha}{G} \qquad (3.25)$$

where α is a geometrical constant that depends on the shape of the boundary. Taking the driving force for atomic diffusion across the boundary to be equal to the gradient in the chemical potential, we have

$$F_b = \frac{d\mu}{dx} = \frac{d}{dx}(\Omega \Delta p) = \frac{1}{\delta_{gb}} \left(\frac{\Omega \gamma_{gb} \alpha}{G} \right) \qquad (3.26)$$

where Ω is the atomic volume and $dx = \delta_{gb}$ is the grain boundary width. From Equation 1.124, the flux of atoms across the boundary is

$$J = \frac{D_a}{\Omega kT} \frac{d\mu}{dx} = \frac{D_a}{\Omega kT} \left(\frac{\Omega \gamma_{gb} \alpha}{\delta_{bg} G} \right) \qquad (3.27)$$

where D_a is the diffusion coefficient for atomic motion across the grain boundary. The boundary velocity becomes

$$v_b \approx \frac{dG}{dt} = \Omega J = \frac{D_a}{kT} \left(\frac{\Omega}{\delta_{gb}} \right) \left(\frac{\alpha \gamma_{gb}}{G} \right) \qquad (3.28)$$

In many analyses, the driving force is taken as the pressure difference across the boundary, $\alpha\gamma_{gb}/G$, and v_b is given by

$$v_b \approx \frac{dG}{dt} = M_b\left(\frac{\alpha\gamma_{gb}}{G}\right) \tag{3.29}$$

Comparing Equation 3.28 and Equation 3.29, M_b is defined as

$$M_b = \frac{D_a}{kT}\left(\frac{\Omega}{\delta_{gb}}\right) \tag{3.30}$$

Integrating Equation 3.29 gives

$$G^2 - G_o^2 = Kt \tag{3.31}$$

where G_o is the initial grain size ($G = G_o$ at $t = 0$) and K is a temperature-dependent growth factor given by

$$K = 2\alpha\gamma_{gb}M_b \tag{3.32}$$

Equation 3.31 is the parabolic grain growth law that has the same form as the LSW equation for interface reaction-controlled Ostwald ripening (Equation 3.14). The growth factor K has an Arrhenius dependence on temperature T, given by $K = K_o \exp(-Q/RT)$, where K_o is a constant, R is the gas constant, and Q is the activation energy for grain growth.

The boundary mobility M_b, which depends on the diffusion coefficient D_a for the atomic jumps across the boundary of the pure material, is termed the *intrinsic* boundary mobility. For an ionic solid, in which both cations and anions must diffuse, D_a represents the diffusion coefficient of the rate-limiting (or slowest) species. In ceramics, the experimentally determined boundary mobilities are rarely as high as M_b given by Equation 3.30. As will be described in more detail later, segregated solutes, inclusions, pores, and second-phase films can also exert a drag force on the boundary, reducing the mobility.

3.5.2 MEAN FIELD THEORIES

The next major advance in the theoretical analysis of grain growth was made by Hillert [16]. His approach was to consider the change in size of an isolated grain embedded in an environment that represented the average effect of the whole array of grains. Theories based on this type of approach are often referred to in the grain growth literature as mean field theories. Hillert's analysis formed the first significant application of the LSW theory of Ostwald ripening to grain growth. He deduced an expression for the growth rate of grains of radius a of the form

$$\frac{da}{dt} = \alpha_1\gamma_{gb}M_b\left(\frac{1}{a^*} - \frac{1}{a}\right) \tag{3.33}$$

where α_1 is a geometrical factor equal to 1/2 in two dimensions (2D) and 1 in three dimensions (3D), a is the radius of the circle (2D) or sphere (3D) having the same area or volume as the grain, and a^* is a critical grain size such that if $a > a^*$ the grain will grow and if $a < a^*$ the grain will shrink. It is expected that a^* would be related to the average grain size of the system, but the exact relation was not derived.

Equation 3.33 has the same form as the LSW equation for the rate of change of the particle radius in Ostwald ripening controlled by the interface reaction (Equation 3.11). By following the LSW analysis, Hillert derived an equation for the rate of change of the critical radius, given by

$$\frac{d(a^*)^2}{dt} = (1/2)\alpha_1\gamma_{gb}M_b \tag{3.34}$$

Assuming that the terms on the right-hand side of this equation are constant with time, then integration leads to parabolic grain growth kinetics, similar in form to the equation of Burke and Turnbull. Hillert also derived an equation for the size distribution during steady-state grain growth, given by

$$f(s,t) = (2e)^\beta \frac{\beta s}{(2-s)^{2+\beta}} \exp\left(\frac{-2\beta}{2-s}\right) \tag{3.35}$$

where s is the reduced size equal to a/a^*, e is the base of the natural logarithm (equal to 2.718), and $\beta = 2$ in two dimensions or $\beta = 3$ in three dimensions. The distribution is sharply peaked when compared, for example, to a log-normal distribution (Figure 3.13).

Other theories that employ the mean field approach include those of Feltham [17] and of Louat [18]. Feltham assumed that the grain size distribution is log-normal and time invariant if plotted as a function of the reduced size $s = a/\bar{a}$, where $\bar{a}$ is the average grain size. Using an approach similar to that of Hillert, he obtained parabolic growth kinetics. Louat assumed that the grain boundary motion can be modeled as a random walk process in which the fluctuation of the grain size is the cause of the drift to larger average sizes. He deduced that the grain growth kinetics are parabolic and derived a time-invariant grain size distribution function given by

$$f(s,t) = As \exp\left(-\frac{s^2}{2}\right) \tag{3.36}$$

where A is a constant. A plot of this distribution function is shown in Figure 3.13.

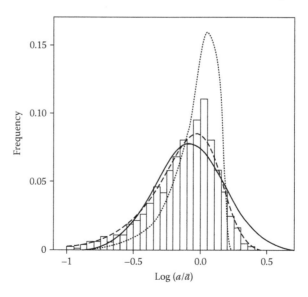

FIGURE 3.13 The grain size distribution function for three theoretical distributions and that obtained from a computer simulation employing the Monte Carlo procedure: log-normal distribution (*solid curve*); Hillert's model (*dotted curve*); Louat's model (*dashed curve*); and computer simulation (*histogram*).

To summarize at this stage, in spite of the different assumptions, the theories considered so far all predict a parabolic law for normal grain growth that is formally similar to the LSW equation for interface reaction-controlled Ostwald ripening. The grain size distribution functions that these theories predict also show the property of time invariance found in the LSW theory. Because of these similarities, normal grain growth in dense polycrystalline solids is sometimes considered as a special case of interface-controlled Ostwald ripening.

In practice, the grain growth data of dense polycrystalline solids do not follow the predicted parabolic law very well. Writing the grain growth equation as

$$G^m - G_o^m = Kt \qquad (3.37)$$

the exponent m is often found to be between 2 and 4 and may also depend on the temperature. In ceramics, the value $m = 3$ has been reported most often, but, as will be discussed later, this value can be indicative of any of five different mechanisms. Deviations from $m = 2$ are often explained away in terms of additional factors not considered in the models. One common explanation is the segregation of impurities to the grain boundaries, but the grain growth kinetics of very pure, zone-refined metals also show deviations from $m = 2$. Other explanations that have been suggested include the presence of second-phase particles at the grain boundaries and anisotropy in the grain boundary energy.

3.5.3 TOPOLOGICAL ANALYSIS OF GRAIN GROWTH

A difficulty with the grain growth theories considered so far is that they neglect the topological requirements of space filling. The most detailed analysis of the topological features of grain growth was carried out by Rhines and Craig [19], who argued that the volume of a shrinking grain must be shared with the grains throughout the whole structure. Similarly, changes in the topological parameters, e.g., the number of faces, edges, and vertices, must also be shared with the other grains. Rhines and Craig introduced two constants of grain growth, the *sweep constant* θ and the *structure gradient* σ, which have been the subject of some debate. The sweep constant is defined as the number of grains lost when grain boundaries throughout the whole grain structure sweep through the equivalent of unit volume of the solid. There is some doubt about whether the sweep constant is indeed constant [20]. The structure gradient is defined as

$$\sigma = \frac{M_V S_V}{N_V} \qquad (3.38)$$

where M_V, S_V, and N_V are the curvature, surface area, and number of grains, respectively, per unit volume of the solid. From a limited amount of experimental data, Rhines and Craig found that σ was constant.

In the analysis of the grain growth kinetics, the mean boundary velocity is taken as the product of the boundary mobility and the force on the boundary, that is,

$$v_b = M_b \gamma_{gb} M_V \qquad (3.39)$$

Considering unit volume of the solid, the volume swept out per unit time is $v_b S_V$, and, from the definition of the sweep constant, the number of grains lost per unit time is $\theta v_b S_V$. The total volume transferred from the disappearing grains to the remaining grains per unit time is $\theta v_b S_V V_G$, where V_G is the average volume of a grain. The rate of increase in the average volume per grain can be written

$$\frac{dV_G}{dt} = \frac{\theta v_b S_V V_G}{N_V} \qquad (3.40)$$

where N_V, the number of grains per unit volume, is equal to $1/V_G$. Substituting for S_V and v_b from Equation 3.38 and Equation 3.39 gives

$$\frac{dV_G}{dt} = \frac{M_b \gamma_{gb} \theta \sigma}{N_V} \tag{3.41}$$

and, assuming that the terms on the right-hand side of this equation are constant, integration leads to

$$V_G - V_{Go} = \left(\frac{M_b \gamma_{gb} \theta \sigma}{N_{Vo}} \right) t \tag{3.42}$$

where V_{Go} and N_{Vo} are the average grain volume and the average number of grains at $t = 0$. For the grain growth of aluminum at 635°C, the data in Figure 3.14 show a linear dependence of V_G (measured as $1/N_V$) on time, in agreement with the predictions of Equation 3.42. Values for N_V were measured directly by serial sectioning.

Since V_G is proportional to G^3, where G is the average grain size, and assuming self-similarity of the grain size distribution, Equation 3.42 indicates that the growth exponent m in Equation 3.37 is equal to 3, in contrast to the theories described earlier where $m = 2$. Most investigators conventionally use the mean grain intercept length from 2D sections as a measure of G. A puzzling aspect of Rhines and Craig's experiments is that when the conventional intercept lengths were determined from the samples that were previously examined by serial sectioning, the mean intercept length was found to have increased as $t^{0.43}$, giving $m = 2.3$. Rhines and Craig used this difference in their measured m values to cast doubt on the use of mean grain intercept lengths determined from 2D sections. However, a possible source for the discrepancy in the two measured m values may be the presence of anisotropy in the grain shapes.

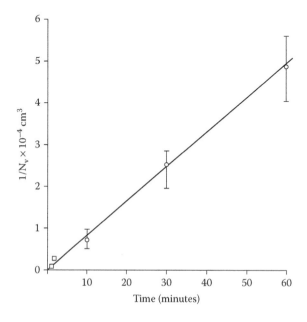

FIGURE 3.14 Plot of the grain volume ($1/N_V$) versus time of annealing for aluminum, showing a linear relationship.

3.5.4 Computer Simulation of Normal Grain Growth

Since the 1980s, computer simulations have played a key role in exploring some of the complexities of grain growth. Several models have been developed, and these are reviewed by Atkinson [14]. As summarized in Figure 3.15, the models simulate grain growth in two dimensions for a soap froth [21,22] or a polycrystal [23]. The models can employ either a *direct simulation,* where the boundary network is constructed and operated on in the computer program, or a *statistical approach,* in which ensemble probabilities for a collection of grains are determined from a series of interrelated equations so as to find the behavior of typical grain types. Whereas the statistical models do not take into account the topological constraints of space filling, the direct simulations do, and they can be divided into *deterministic* and *probabilistic* models. *Deterministic* in this context means that each subsequent configuration of the boundary network is exactly defined by a previous configuration, whereas *probabilistic,* which is used synonymously with the terms *stochastic* or *random,* relates to probabilities of events on a microscopic scale.

The probabilistic models of Anderson et al. [23]   and Srolovitz et al. [24], which employ a Monte Carlo procedure, have the advantage that both the topological requirements of space filling and the detailed local effects of grain boundary curvature can be incorporated into the simulations. To analyze the complexity of the grain boundary topology, the microstructure is mapped out onto a discrete lattice (Figure 3.16). Each lattice site is assigned a number between 1 and Q, corresponding to the orientation of the grain in which it is embedded. Commonly, a sufficiently large value of Q (>30) is chosen to limit the impingement of grains of the same orientation. The grain boundary segment is defined to lie between two sites of unlike orientation, and the grain boundary energy is specified by defining the interaction between nearest neighbor lattice sites in terms of a Hamiltonian operator. The unit of time for the simulations is defined as 1 Monte Carlo step (MCS) per site, which corresponds to N microtrials or reorientation attempts, where N is the total number of sites in the system.

In the two-dimensional analysis, the average area per grain $\bar{A}$ is monitored, and this can be related to the average grain size, G. An initial simulation of a shrinking circular grain embedded in an infinite matrix ($Q = 2$), i.e., a structure equivalent to the mean field approximation, showed that the size of the circular grain increases uniformly with time according to the equation

$$\bar{A} - \bar{A}_o = Kt \tag{3.43}$$

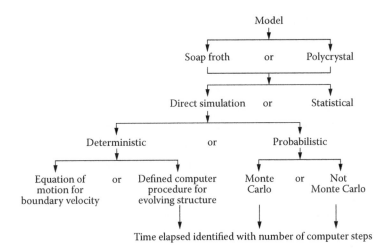

FIGURE 3.15 Classification of computer simulation models for soap froth and grain boundary evolution.

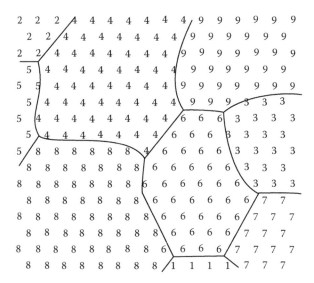

FIGURE 3.16 Sample microstructure on a triangular lattice where the integers denote orientation and the lines represent grain boundaries.

where $\bar{A}_o$ is the initial area and K is a constant. The grain growth kinetics of the isolated grain are therefore parabolic, in agreement with the mean field theories and the theory of Burke and Turnbull.

For an interconnected network of polycrystalline grains, the kinetics are not parabolic. Large Q values and long times give grain structures (Figure 3.17) that resemble those of real single-phase systems. After an initial transient, the grain growth exponent, m, obtained from the simulations is equal to 2.44. Anderson et al. [23] attempted to justify the m value found in their simulations by comparing it with measured m values determined from data for six very pure metals. Whereas the m value obtained from their simulations is in good agreement with the average experimental value for the six metals, the limited number of samples and the wide variation of measured m values (2–4) for this sampling should be recognized.

Reasons for the deviation between the grain growth exponent obtained in the simulations ($m = 2.44$) and the value predicted by the mean field theories ($m = 2$) were discussed by Anderson et al. [23]. In the mean field approach, the driving force for grain growth, as outlined earlier, is the reduction of the curvature (or area) of the boundary. In the lattice model used in the simulations, the curvature is discretely allocated as kinks on the boundary. Such kinks can be eliminated by two mechanisms: (1) the meeting and annihilation of two kinks of identical orientation (as defined

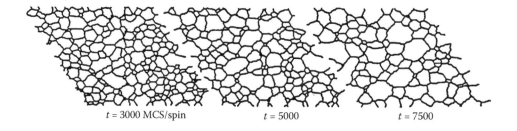

$t = 3000$ MCS/spin $t = 5000$ $t = 7500$

FIGURE 3.17 The evolution of the microstructure for a $Q = 64$ model on a triangular lattice. (From Anderson, M.P., Srolovitz, D.J., Grest, G.S., and Sahni, P.S., Computer simulation of grain growth — I. Kinetics, *Acta Metall.*, 32, 783, 1984. With permission.)

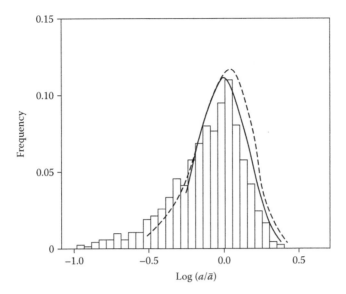

FIGURE 3.18 The grain size distribution function obtained from a computer simulation (*histogram*) compared with a log-normal fit to grain growth data for aluminum (*solid curve*) and with grain growth data for MgO (*dashed curve*). (From Srolovitz, D.J., Anderson, M.P., Sahni, P.S., and Grest, G.S., Computer simulation of grain growth — II. Grain size distribution, topology, and local dynamics, *Acta Metall.*, 32, 793, 1984. With permission.)

by the lattice) but opposite sign, and (2) adsorption of a kink at a vertex where more than two grains meet. The first mechanism corresponds to grain growth driven purely by curvature, as in the case of the simulation of a circular grain embedded in the infinite matrix. However, the second mechanism can operate only when vertices are present. By absorbing kinks, vertices are capable of decreasing the curvature without causing grain growth. Effectively, the growth is slowed relative to the case of the circular grain in the infinite matrix, resulting in a higher m value.

In a subsequent analysis, Srolovitz et al. [24] showed that when the grain size is normalized to the average grain size, then the size distribution function becomes time invariant, as also predicted by the mean field theories. Figure 3.13 shows a comparison of the distribution function obtained from the computer simulations with the log-normal distribution, along with the distribution functions derived by Hillert [16] and Louat [18]. The distribution function obtained from the simulations can also provide an excellent fit to some experimental data (Figure 3.18).

To summarize at this stage, the computer simulations employing the Monte Carlo procedure show a remarkable ability to provide realistic pictures of the normal grain growth process, and they provide a good fit to some experimental data. On the other hand, the physical significance of some of the results is not clearly understood. Srolovitz et al. [24] have suggested that the true nature of the grain growth process might lie somewhere between growth driven by the grain boundary curvature and a random walk process.

3.6 ABNORMAL GRAIN GROWTH IN DENSE SOLIDS

Microstructures of polycrystalline ceramics that have been heated for some time often show a number of very large grains in a fine-grained matrix. The large grains, referred to as abnormal grains, are said to have developed as a result of abnormal or runaway grain growth in which the large grains had a much faster growth rate relative to the surrounding fine-grained matrix. A well-known microstructure in the ceramic literature involves the growth of a relatively large single-crystal Al_2O_3 grain in a fine-grained Al_2O_3 matrix, which appears to show the single-crystal grain

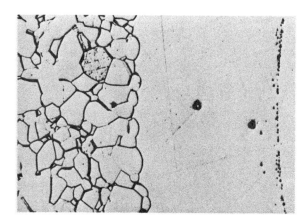

FIGURE 3.19 Example of abnormal grain growth: a large Al_2O_3 crystal growing into a matrix of uniformly sized grains. (magnification 495×).

growing much faster than the matrix grains (Figure 3.19). An understanding of abnormal grain growth and how to control it is vital for the production of ceramics with high density and controlled microstructure by sintering.

3.6.1 Causes of Abnormal Grain Growth

Abnormal grain growth has often been explained in terms of the grain size distribution in the starting material. According to the Hillert theory (Equation 3.35), a steady-state normal grain size distribution has a maximum or cutoff grain size equal to twice the critical grain size (or equal to approximately twice the average grain size). It was inferred that a grain larger than twice the average size would be outside the maximum sustainable size and would grow abnormally. Surprisingly, when the evolution of large grains in a fine-grained matrix is analyzed theoretically, it is found that large grains do not undergo abnormal or runaway grain growth.

Using the Monte Carlo procedure outlined in the previous section, Srolovitz, Grest, and Anderson [25] simulated the growth of large grains in a matrix of fine (normal) grains in two dimensions. They found that, for isotropic systems with uniform grain boundary energy and mobility, although the large grains did grow, they did not outstrip the normal grains. The normal grains grew at a faster relative rate, so the large (abnormal) grains eventually returned to the normal size distribution (Figure 3.20). The growth of large grains in a matrix of fine grains can also be treated analytically [26]. Consider a large grain of radius a in a matrix of fine, normal grains. By analogy with Equation 3.11, we can write

$$\frac{da}{dt} = 2M_b\gamma_{gb}\left(\frac{1}{a^*} - \frac{1}{a}\right)$$

(3.44)

where a^* is the critical radius at which a grain neither grows nor shrinks. The *relative* growth rate of the large grain can be defined as

$$\frac{d}{dt}\left(\frac{a}{a^*}\right) = \frac{1}{(a^*)^2}\left(a^*\frac{da}{dt} - a\frac{da^*}{dt}\right)$$

(3.45)

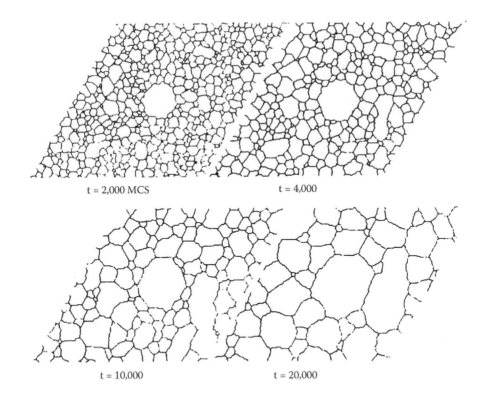

t = 2,000 MCS   t = 4,000

t = 10,000   t = 20,000

FIGURE 3.20 Computer simulation of the growth of a large grain in an otherwise normal grain growth microstructure for an isotropic system. The microstructure was produced by running the normal grain growth simulation procedure for 1000 MCS, and the large grain was then introduced as a circular grain with an initial size equal to 5 times the average grain size of the matrix grains. (From Srolovitz, D.J., Grest, G.S., and Anderson, M.P., Computer simulation of grain growth — V. Abnormal grain growth, *Acta Metall.*, 33, 2233, 1985. With permission.)

Since the number of fine, normal grains is much greater than that of the abnormal grain (one), the time dependence of a^* is still described by Equation 3.14, which, in differential form, is

$$\frac{da^*}{dt} = \frac{M_b \gamma_{gb}}{2a^*} \qquad (3.46)$$

Substituting Equation 3.44 and Equation 3.46 into Equation 3.45 gives

$$\frac{d}{dt}\left(\frac{a}{a^*}\right) = -\frac{M_b \gamma_{gb}}{2aa^*}\left(\frac{a}{a^*} - 2\right)^2 \qquad (3.47)$$

This equation shows that the relative growth rate of the large grains is always negative, except for $a = a^*$, in which case it is zero. Abnormal grains (i.e., with $a > 2a^*$), therefore, do not outstrip the normal grain population but rejoin it at the upper limit of $2a^*$. Due to irregularities in their shape and to fluctuations, they cannot remain at exactly $2a^*$ after they rejoin the population. Eventually, they continue to decrease in relative size and are incorporated into the normal distribution. Therefore, for isotropic systems, size difference alone is not a sufficient criterion for initiating abnormal grain growth.

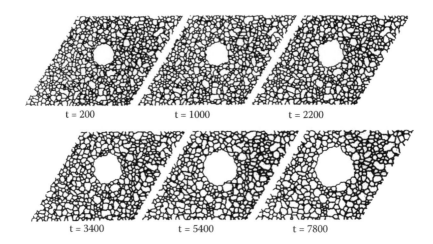

FIGURE 3.21 Computer simulation of the microstructural evolution in which a large grain embedded in a fine-grained matrix has a size that is 6 times the average size and a mobility that is 7.5 times the mobility of the fine grains. (From Rollet, A.D., Srolovitz, D.J., and Anderson, M.P., Simulation and theory of abnormal grain growth — anisotropic grain boundary energies and mobilities, *Acta Metall.*, 37, 1227, 1989. With permission.)

Computer simulations and theoretical analysis show that true abnormal grain growth can occur as a result of variable (or anisotropic) grain boundary energy or mobility [27–29]. Abnormal grain growth is favored for grains with boundaries that have a higher mobility or a lower energy than the surrounding matrix grains. Figure 3.21 shows the results of simulations in which the single abnormal grain embedded in a fine-grained matrix has a size that is 6 times the average size and a mobility that is 7.5 times the mobility of the fine grains. The abnormal grain is predicted to grow faster than the grains of average size in the surrounding matrix.

There are several ways in which variable grain boundary properties can arise in practice. One factor is the structure and misorientation of the grain boundary. *Special* or *low-angle* grain boundaries, which result when two grains have only a slight misorientation relative to one another, possess a lower energy than general boundaries that may have high misorientation angles. Transfer of matter from the surrounding grains to the low-energy boundaries results in abnormal grain growth. Alternatively, low-energy grain boundaries are generally believed to have low mobility, so growth is slow normal to and rapid parallel to the low-energy boundaries, often resulting in faceting and anisotropic abnormal growth (Figure 3.22).

Preferential segregation of dopants and impurities to different types of boundaries can also change the relative grain boundary mobility or energy. Considerable attention has been paid to the Al_2O_3 system, largely based on attempts to understand the role of MgO in suppressing abnormal grain growth in this system [30]. Although the single most important role of MgO is the reduction in the grain boundary mobility [31], it is also recognized that MgO plays an important role in reducing anisotropies in the surface and grain boundary energies and mobilities [32,33]. On the other hand, doping of Al_2O_3 with TiO_2 is observed to enhance anisotropic abnormal grain growth [34,35], presumably due to preferential segregation to some types of boundaries to change the relative boundary energies and mobilities.

The release of solutes, second-phase particles, and pores from moving grain boundaries leads to a sudden increase in the boundary mobility, and this has been suggested as a cause of abnormal grain growth. Liquid phases have long been cited as a major cause of abnormal grain growth. The presence of a liquid phase has been observed to affect the grain growth kinetics and grain morphology in Al_2O_3 [36,37]. The formation of liquid films at grain boundaries often has the effect of significantly enhancing the mobility. In the Al_2O_3 system, a small amount of SiO_2 impurity is often present in the material, and this, together with other impurities, may form a liquid phase at high temperatures. It has been observed that the liquid tends to selectively wet the low-energy boundaries,

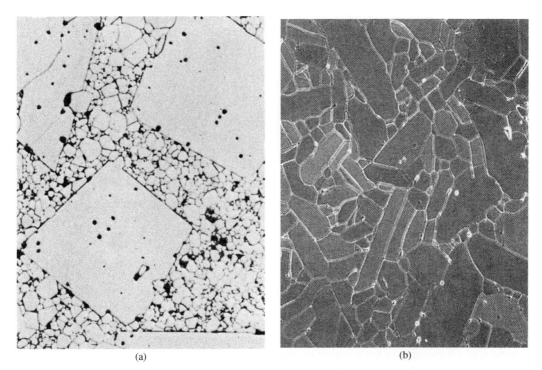

FIGURE 3.22 (a) Large faceted grains in a polycrystalline spinel. The straight boundaries of the large grains suggest a crystallographic orientation of low energy, possibly with a wetting liquid (magnification 350×). (b) Faceted grain growth in TiO_2-doped Al_2O_3 (500 ppm Ti) (magnification 650×). (Courtesy of A.M. Glaeser.)

whereas a small fraction of the other types of boundaries are not wetted [38]. The large mobility difference between the non-wetted, fast-moving boundaries and the wetted, low-energy boundaries has been claimed to predispose the system toward anisotropic abnormal grain growth.

Physical and chemical inhomogeneities, such as inhomogeneous packing and non-uniform distribution of dopants and second-phase particles, commonly lead to local variation in the microstructure [1], and this has often been considered a major cause of abnormal grain growth. The local microstructural variations produce inequalities in the boundary mobility or energy, leading to the initiation of abnormal growth.

3.6.2 APPLICATION OF CONTROLLED ABNORMAL GRAIN GROWTH

Although the suppression of abnormal grain growth is commonly a key goal in sintering, the ability to exploit abnormal grain growth in a *controlled* manner can provide significant benefits for several ceramic systems. For example, seeding a fine-grained polycrystalline ceramic with a very large grain does have the effect that this grain will become very large before the slower-growing matrix will have developed an average size even comparable to the seed grain. Some single-crystal ferrites are made this way [39]. The procedure involves consolidating and sintering nearly monodisperse ferrite particles to form a fine-grained matrix with a narrow distribution of grain sizes, bonding a large-seed crystal to the matrix, and annealing the system to make the boundary of the seed crystal migrate into the matrix, converting virtually the entire sample into a single grain. The procedure has also proved successful for $BaTiO_3$, Al_2O_3, and lead magnesium niobate–lead titanate [40–42]. An example of a $BaTiO_3$ single crystal grown by the procedure is shown in Figure 3.23.

The *in situ* growth of a controlled distribution of anisotropic abnormal grains in a fine-grained matrix (Figure 3.24a) has been used to enhance the fracture toughness of Si_3N_4, SiC, and mullite [43–45]. Preferential alignment of growing anisotropic grains leads to *texturing* of the microstructure

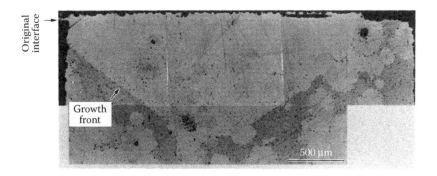

FIGURE 3.23 Optical micrograph of a barium titanate single crystal growing into a polycrystalline matrix after annealing for 20 h at 1300°C. (From Yamamoto, T., and Sakuma, T., Fabrication of barium titanate single crystals by solid-state grain growth, *J. Amer. Ceram. Soc.*, 77, 1107, 1994. With permission.)

(Figure 3.24b), which may cause the development of anisotropic properties in some ceramics. For example, the dielectric constant of bismuth titanate, a ceramic with a layered crystal structure, is significantly higher in the direction of alignment than in the perpendicular direction. Such anisotropy in dielectric properties may be beneficial for some applications in the electronic ceramic industry. Alignment can be achieved by applying a unidirectional pressure during densification of the powder compact, as for example in hot pressing or sinter forging [46].

An elegant procedure developed by Messing and coworkers [47,48], referred to as *templated grain growth*, involves the use of large, elongated seed crystals that are mixed with a fine equiaxial

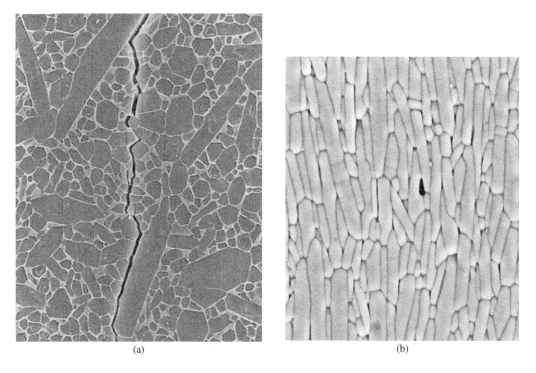

(a)   (b)

FIGURE 3.24 Exploitation of controlled anisotropic abnormal grain growth: (a) self-reinforced silicon nitride, showing the interaction of a propagating crack with the microstructure (magnification 2000×). (b) Bismuth titanate with a microstructure of aligned anisotropic grains, giving anisotropic dielectric properties (magnification 1000×).

powder and aligned during the forming of the green body by, for example, tape casting or extrusion [1]. The aligned seed crystals act as templates to pattern the growth of the matrix grains during sintering. After the production of a dense body, subsequent grain growth increases the grain orientation by preferential growth of the template particles to yield a highly aligned grain microstructure.

3.7 GRAIN GROWTH IN THIN FILMS

An understanding of grain growth in thin films is important because grain size, grain-size distribution, and grain orientation can strongly influence the properties of polycrystalline films. If the initial grain size of a dense film is smaller than the film thickness, grain growth often leads to the rapid development of a grain structure with an average grain size larger than or comparable to the film thickness. The grain boundaries fully traverse the thickness of the film, leading to the development of a columnar microstructure. The occurrence of subsequent grain growth primarily by grain boundary motion in directions lying in the plane of the film can, in principle, be modeled as a two-dimensional process. However, application to thin films of the grain growth models discussed earlier for dense solids is not straightforward. In thin films, grain growth is not truly a two-dimensional process. The film has a higher percentage of free surfaces than the bulk solid, which provide additional constraints on the shape and motion of the grain boundary.

Normal grain growth in thin films is found to stagnate when the average grain size (or grain diameter) is two to three times the film thickness [49,50]; this is often referred to as the *specimen thickness effect*. The grain-size distribution in the stagnant structure can be well fitted by a log-normal distribution. After stagnation, any further grain growth occurs by abnormal growth, whereby a small fraction of the grains grows at the expense of the majority of the grains. These abnormal grains develop a new population distinct from the original normal grain population. They can grow to impingement, resulting in a film with grain sizes that are many times larger than the film thickness.

Mullins [51] proposed that the stagnation of grain growth in films is due to grain-boundary grooving at the junctions where the grain boundaries meet the free surfaces. The action of the grain-boundary grooves to restrain grain growth may be understood by considering a flat boundary that is perpendicular to the plane of the film, running from a groove at the bottom surface to a groove on the top surface (Figure 3.25). Migration of the boundary would require an increase in its area, making the process energetically unfavorable. By incorporating the effects of grain-boundary grooving into a model for two-dimensional grain growth, Frost, Thompson, and Walton [52] accounted for the stagnation of normal grain growth and the resulting log-normal distribution of grain sizes, as well as the scaling relationship between the average grain size of the stagnant structure and the film thickness.

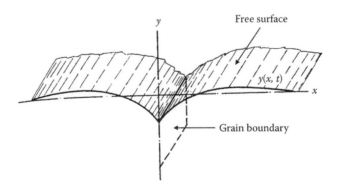

FIGURE 3.25 Illustration of the restraining effect of grain boundary groove on grain growth in thin films. A flat boundary, perpendicular to the plane of the film, is shown running from the groove on the top surface to the bottom surface. For the boundary to migrate, it must "climb" out of the groove, thereby increasing its area.

As discussed earlier, several factors can contribute to the growth of abnormal grains in dense solids, including anisotropies in the grain boundary energy and mobility. For thin films, anisotropies in surface energy (or interfacial energy) can also contribute to abnormal grain growth. When the grains become larger than the film thickness, the area of the free surface of the film becomes comparable to the area of the grain boundaries. If there is sufficient anisotropy in the surface-free energy, then the minimization of surface free energy becomes an important driving force for grain boundary migration. Grains with orientations having low surface energies are favored during growth and may escape from the pinning effects of grain-boundary grooving and become abnormal grains. The effects of variations in the driving force for grain growth due to variations in the surface-free energy of the different grains were analyzed using a two-dimensional grain-growth model [53]. The grain boundary velocity v_b is the sum of the contributions from the grain boundary curvature and the surface free energy difference, given by

$$v_b = M_b \gamma_{gb} (K + \Gamma) \qquad (3.48)$$

where M_b is the boundary mobility, γ_{gb} is the grain boundary energy (per unit area), K is the curvature of the boundary, and Γ is defined as

$$\Gamma = \frac{2 \Delta \gamma_{sv}}{h \gamma_{gb}} \qquad (3.49)$$

where $\Delta \gamma_{sv}$ is the difference in the surface free energy, and h is the film thickness. By coupling the stagnation of normal grain growth with an additional driving force due to surface free energy difference, the results for the evolution of the grain structure are found to closely resemble experimentally observed grain growth in thin films. Figure 3.26 shows the results for the evolution of the microstructure in which 5% of the grains initially present were given an additional driving

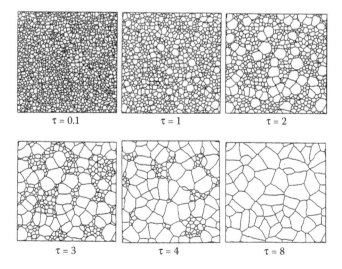

$\tau = 0.1$ $\qquad$ $\tau = 1$ $\qquad$ $\tau = 2$

$\tau = 3$ $\qquad$ $\tau = 4$ $\qquad$ $\tau = 8$

FIGURE 3.26 Evolution of the structure in a thin film in which 5% of the grains initially present were given an additional driving force of $\Gamma = 0.732$. The stagnation condition is $K_{crit} = 0.4\sqrt{A_o}$, where A_o is the initial area of the grain, and τ is the normalized time. (From Frost, H.J., Thompson, C.C., and Walton, D.T., Simulation of thin film grain structures — II. Abnormal grain growth, *Acta Metall. Mater.*, 40, 779, 1992. With permission.)

force of $\Gamma = 0.732$ (in units of the curvature $1/\sqrt{A_o}$, where A_o is initial area of the grain) at a normalized time $\tau = 0.1$, coupled with a stagnation condition in which the critical curvature for boundary motion K_{crit} is equal to $0.4\sqrt{A_o}$ (i.e., v_b is zero below K_{crit}). There is some general growth initially, but by $\tau = 2$ the normal grains are beginning to reach stagnation. Most of the abnormal grains, however, overcome the stagnation and continue to grow, giving rise to the transient development of a bimodal grain size distribution, until the entire field is taken over at $\tau = 8$. At this stage, there is no additional driving force for the boundaries between the abnormal grains, so growth of the abnormal grains stagnates.

It is found that surface free energy differences $\Delta\gamma_{sv}$ of only a few percent should be sufficient to trigger abnormal grain growth. In this case, thin film systems that develop a bimodal grain size distribution are expected to have nearly isotropic surface free energies for most grain orientations. According to Equation 3.49, the rate of abnormal grain growth should be higher for thinner films. However, the final grain size after abnormal grain growth should be greater in thicker films.

In many systems, particularly in the case of metallic films, the film is deposited at a temperature that is different from the temperature at which grain growth occurs. Because the film and the substrate will, in general, have different coefficients of thermal expansion, the film is subjected to a biaxial stress in the plane of the film during grain growth. The difference in strain energy between adjacent grains can provide an additional driving force for abnormal grain growth [54].

We see that grain boundary drag and stagnation, coupled with variations in the surface, interface, and strain energy, lead to abnormal grain growth in thin polycrystalline films. A bimodal grain size distribution results from a restraining effect on most boundaries, coupled with an energy term that favors the growth of a subpopulation of grains.

3.8 MECHANISMS CONTROLLING THE BOUNDARY MOBILITY

So far we have limited our analysis of grain growth to the case of single-phase solids in which the boundary mobility M_b, termed the *intrinsic boundary mobility*, defined by Equation 3.30, is controlled by the diffusion across the boundary of the atoms (ions) present in the pure material. While dopants, impurities, fine second-phase particles, and porosity are commonly present in ceramics, they have been assumed to produce no effect on the basic rate of atom diffusion across the boundary, so their effect on the boundary mobility has been neglected. However, boundary mobilities determined from grain growth data for dense or nearly dense ceramics are often found to be significantly lower than the calculated intrinsic mobility [55]. Figure 3.27 shows selected grain boundary mobilities for a few common ceramics, where it is assumed that $\gamma_{gb} = 0.3$ J/m² for each material and the mobilities are normalized to that for a fixed grain size of 90 μm. Although we recognize the uncertainty in the calculated intrinsic mobility due to the lack of definitive data for the diffusion coefficient of the rate-controlling ion, trends in the mobility values are clear. First, the measured mobilities of the oxides are well below the calculated intrinsic values. Second, dense samples containing dopants or second-phase precipitates have low mobilities, and the mobilities for porous samples are much lower than those for the dense samples. Third, when liquid phases are present, the mobilities are higher than those for porous samples, but they are often not much higher than for dense samples.

Because of their influence on the boundary mobility, the use of fine second-phase particles and, in particular, the use of dopants form two of the most effective approaches for controlling grain size during solid-state sintering. Prior to considering their practical use, we examine the mechanism by which they influence M_b.

3.8.1 PINNING BY FINE SECOND-PHASE PARTICLES

Let us consider a system of fine second-phase particles (also referred to as fine inclusions or precipitates) dispersed randomly in a polycrystalline solid in which they are insoluble and immobile. If a

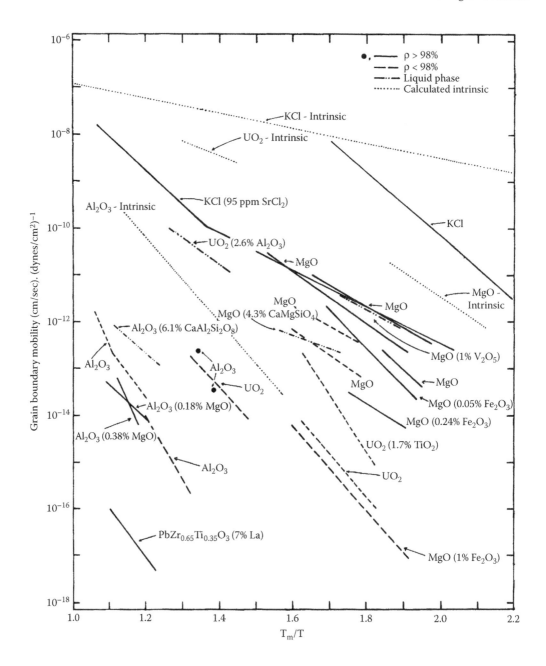

FIGURE 3.27 Selected grain boundary mobilities for some ceramics compared with the calculated intrinsic boundary mobility. T_m is the melting temperature. (From Yan, M.F., Cannon, R.M., and Bowen, H.K., Grain boundary migration in ceramics, in *Ceramic Microstructures '76*, Fulrath, R.M., and Pask, J.A., Eds., Westview Press, Boulder, CO, 1977.)

grain boundary moving under the driving force of its curvature encounters an inclusion, the boundary will be held up by the particle until the motion elsewhere has proceeded sufficiently far for it to break away. If there are a sufficient number of particles, we might expect that the boundary will be pinned when it encounters the particles, and boundary migration will, therefore, cease. The Zener model and computer simulations are the basic approaches that have been used to describe this situation.

3.8.1.1 The Zener Model

The first quantitative model to describe particle inhibited grain growth was developed by Zener, as communicated by Smith [56]. Zener assumed that the inclusion particles are monosize, spherical, insoluble, immobile, and randomly distributed in the polycrystalline solid. Taking a grain boundary with principal radii of curvature a_1 and a_2, the driving force (per unit area) for boundary motion is

$$F_b = \gamma_{gb}\left(\frac{1}{a_1} + \frac{1}{a_2}\right) \tag{3.50}$$

Assuming that a_1 and a_2 are proportional to the grain size G, then

$$F_b = \alpha\gamma_{gb}/G \tag{3.51}$$

where α is a geometrical shape factor (e.g., $\alpha = 2$ for a spherical grain). When the grain boundary intersects an inclusion, its further movement is hindered (Figure 3.28a and Figure 3.28b). A dimple is formed in the grain boundary and, compared to the inclusion-free boundary, extra work must be performed for the equivalent motion of the boundary. This extra work manifests itself as a retarding force on the boundary. If r is the radius of the inclusion (Figure 3.28c), then the retarding force exerted by the inclusion on the boundary is

$$F_r = (\gamma_{gb} \cos\theta)(2\pi r \sin\theta) \tag{3.52}$$

According to Equation 3.52, the retarding force is the grain boundary tension, resolved in the direction opposite to that of the grain boundary motion, times the perimeter of contact. The retarding force is a maximum when $\theta = 45°$, so $\sin\theta \cos\theta = 1/2$, and

$$F_r^{\max} = \pi r \gamma_{gb} \tag{3.53}$$

If there are N_A inclusions per unit area of the grain boundary, then the maximum retarding force per unit area of the boundary is

$$F_d^{\max} = N_A \pi r \gamma_{gb} \tag{3.54}$$

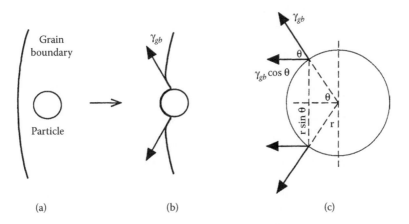

(a)   (b)   (c)

FIGURE 3.28 Interaction of a grain boundary with an immobile particle. (a) Approach of the boundary toward the particle. (b) Interaction between the grain boundary and the particle leading to a retarding force on the boundary. (c) Detailed geometry of the particle-grain boundary interaction.

N_A is difficult to determine but is related to N_V, the number of inclusions per unit volume, by the relation

$$N_A = 2rN_V \tag{3.55}$$

If the volume fraction of the inclusions in the solid is f, then

$$N_V = \frac{f}{(4/3)\pi r^3} \tag{3.56}$$

Substituting for N_A in Equation 3.54 gives

$$F_d^{max} = \frac{3f\gamma_{gb}}{2r} \tag{3.57}$$

The net driving force per unit area of the boundary is

$$F_{net} = F_b - F_d^{max} = \gamma_{gb}\left(\frac{\alpha}{G} - \frac{3f}{2r}\right) \tag{3.58}$$

When $F_{net} = 0$, boundary migration will cease, and this occurs when

$$G_L = \frac{2\alpha}{3}\frac{r}{f} \tag{3.59}$$

where G_L is the limiting grain size. Equation 3.59 is sometimes referred to as the *Zener relationship*. It indicates that a limiting grain size will be reached, the magnitude of which is proportional to the inclusion size and inversely proportional to the inclusion volume fraction. Changes in temperature would not affect this equilibrium relationship but would affect only the rate at which the system approaches the equilibrium condition. Although it may be difficult to achieve in practice, if a limiting grain size is reached, further grain growth can occur only if (1) the inclusions coarsen by Ostwald ripening, (2) the inclusions go into solid solution in the matrix, or (3) abnormal grain growth occurs.

Several refinements have been made to the Zener analysis by modifying the simple assumptions used in the model, but the approach has remained essentially the same [57–60]. The refinements lead to a different geometrical constant in Equation 3.59, but they do not affect the dependence of G_L on f.

3.8.1.2 Computer Simulations of Particle-Inhibited Grain Growth

A completely different approach was used by Srolovitz and coworkers [61–63], who modified their Monte Carlo computer simulation technique, outlined earlier for normal grain growth in single-phase solids, to include a dispersion of fine second-phase particles that are assumed to be monosize, spherical, insoluble, immobile, and randomly distributed (Figure 3.29). The temporal evolution of the microstructure for a system undergoing grain growth in the presence of second-phase particles shows that the grains grow and finally reach a pinned, limiting grain size. In the growing regime, simulated microstructures show several features that are characteristic of normal grain growth. The grain growth exponent and the grain size distribution are found to be identical with and without the presence of the inclusions.

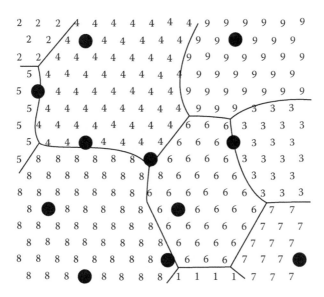

FIGURE 3.29 Sample microstructure on a triangular lattice where the integers denote crystallographic orientation, the lines represent boundaries, and the solid circles represent second-phase particles.

Figure 3.30 shows the pinned microstructures for two different volume (or areal) fractions f of the inclusions. When we compare this to the simulated microstructures for normal grain growth without the presence of inclusions (Figure 3.17), we observe a few grains with highly irregular shapes in the pinned microstructures. The simulations predict that the time taken to reach the limiting grain size, as well as the value of the limiting grain size, decreases with increasing values of f. The simulations also reveal that although the total number of inclusions remains constant, the fraction of inclusions ϕ_b that are located at the boundaries decreases with time, and the effect becomes more significant for smaller values of f. The limiting grain size G_L is predicted to follow an expression of the form

$$G_L = K_S \frac{r}{(\phi_b f)^{1/3}} \tag{3.60}$$

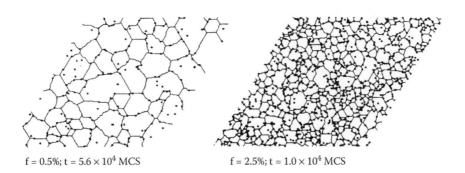

f = 0.5%; t = 5.6 × 10⁴ MCS   f = 2.5%; t = 1.0 × 10⁴ MCS

FIGURE 3.30 Pinned microstructures for two different inclusion concentrations. (From Srolovitz, D.J., Anderson, M.P., Grest, G.S., and Sahni, P.S., Computer simulation of grain growth — III. Influence of a particle dispersion, *Acta Metall.*, 32, 1429, 1984. With permission.)

where K_S is a constant and r is the radius of the inclusions. When compared to the Zener relationship, the different dependence of G_L on f given by Equation 3.60 should be noted. It is sometimes claimed that the Zener analysis overestimates the driving force for grain growth because it neglects the effect of the inclusions on the curvature of the boundary and considers only the work involved in dragging the inclusion. As a result, for a given value of f, the Zener relationship predicts more grain growth and a larger limiting grain size. On the other hand, in the computer simulations, when the boundary intersects an inclusion, the inclusion becomes a new vertex that has the effect of reducing the boundary curvature. The boundaries also deform to maximize the number of pinning inclusion intersections, so the pinning effect is strong, leading to smaller limiting grain sizes.

3.8.1.3 Comparison with Experimental Results

Alumina containing fine ZrO_2 particles is the most widely studied ceramic system for particle-inhibited grain growth [64–66], but the experimental conditions often differ from the assumptions of the models. For example, ZrO_2 particles are quite mobile at Al_2O_3 grain boundaries, and they migrate with the boundaries. The particles also tend to coalesce, so the particle size does not remain constant. Zirconia containing fine Al_2O_3 particles is another widely studied system [66,67], but the Al_2O_3 particles coarsen faster than the ZrO_2 grains grow. Thus, the model assumptions of immobile inclusions and fixed inclusion size are not met in these systems.

Grain growth in Al_2O_3 containing fine SiC particles has been investigated by Stearns and Harmer [68,69], and it is claimed that this system closely fulfills the criteria for an ideal pinning experiment. A plot of the Al_2O_3 grain size versus the SiC volume fraction after annealing dense, hot-pressed samples for 100 h at 1700°C indicates that the data deviate from the predictions of both the Zener model and the computer simulations (Figure 3.31). The fraction of SiC particles at the grain boundaries ϕ_b is found to decrease with increasing grain size G during thermal annealing in a predictable manner.

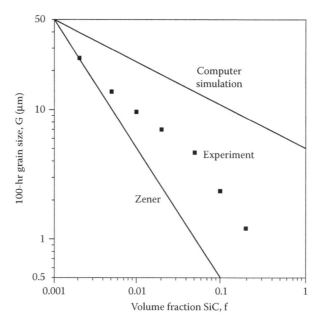

FIGURE 3.31 Data for grain size of Al_2O_3 (after annealing for 100 h at 1700°C) as a function of SiC inclusion volume fraction, compared with the predictions of the Zener model and computer simulations. (Courtesy of M.P. Harmer.)

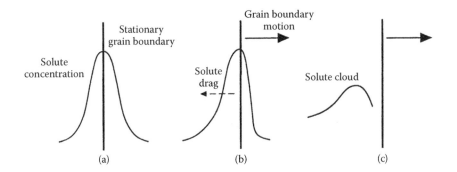

FIGURE 3.32 Sketch of the solute drag effect produced by the segregation of dopants to the grain boundaries. (a) Symmetrical distribution of the dopant in the region of a stationary grain boundary. (b) For a moving boundary, the dopant distribution becomes asymmetric if the diffusion coefficient of the dopant atoms across the boundary is different from that of the host atoms. The asymmetric distribution produces a drag on the boundary. (c) Breakaway of the boundary from the dopant leaving a solute cloud behind.

By relaxing the assumption of totally random distribution of the particles at the grain boundaries and incorporating the observed relation between ϕ_b and G into the Zener analysis, good agreement between the limiting grain size G_L and the volume fraction of particles f was achieved.

3.8.2 Solute Drag

Let us consider a system in which a small amount of a dopant (referred to as the *solute*) is dissolved in solid solution in a polycrystalline solid (referred to as the *host*). If the solute is attracted to or repelled from the grain boundary due to an interaction potential, the solute ions will tend to have a nonuniform distribution in the grain boundary. As described in Chapter 5, the interaction between the boundary and the solute can arise from lattice strain energy due to ionic size mismatch between the solute and host ions, from electrostatic potential energy of interaction between charged grain boundaries and aliovalent solutes (i.e., solute atoms that have a different valence from the host atoms), or a combination of these two effects.

For a hypothetical stationary boundary, the concentration profile of the solute atoms (or ions) will be symmetrical on either side of the boundary (Figure 3.32a). The force of interaction due to the solute atoms to the right of the boundary balances that due to the solute atoms to the left of the boundary, so the net force of interaction is zero. If the boundary now starts to move, the solute concentration profile becomes asymmetric, since the diffusivity of the solute atoms across the boundary is expected to be different from that of the host (Figure 3.32b). This asymmetry results in a retarding force or drag on the boundary that reduces the driving force for migration. If the driving force for boundary migration is high enough, the boundary will break away from the high concentration of solute (sometimes called a solute cloud), and when this occurs (Figure 3.32c), its mobility will approach the intrinsic value, given by Equation 3.30.

Models for grain boundary migration controlled by solute drag have been developed by Cahn [70], Lücke and Stüwe [71], Hillert and Sundman [72], and others. We shall outline the model of Cahn, which is more quantitative and concise than the others and also has the advantage that the boundary mobility can be more directly related to the physical parameters of the process. The model analyzes the problem in one dimension and makes the following assumptions:

The solute concentration, C, expressed as a fraction of the host atoms, is fairly dilute, so the chemical potential of the solute can be expressed as $\mu = \mu_0 + kT \ln C$, where k is the Boltzmann constant and T is the absolute temperature (see Equation 1.94).

An interaction potential energy, U, exists between the solute atoms and the grain boundary, which is independent of the boundary velocity and the dopant concentration but is a function of the distance x from the grain boundary.

The Einstein relation between mobility, M, and diffusivity, D, is obeyed (i.e., $M = D/kT$). Steady-state conditions exist, so the boundary velocity v_b is constant.

The chemical potential of the solute atoms in the near grain boundary region can be written as

$$\mu = kT \ln C(x) + U(x) + U_o \tag{3.61}$$

where $C(x)$ and $U(x)$ are functions of x, and U_o is a constant chosen such that $U(\infty) = 0$. For steady-state conditions, the composition profile of the solute satisfies the equation

$$\frac{dC}{dt} = -v_b \frac{dC}{dx} \tag{3.62}$$

Following Equation 1.124, the flux of the solute atoms is given by

$$J = -\frac{D_b C}{\Omega kT} \frac{d\mu}{dx} \tag{3.63}$$

where D_b is the diffusion coefficient for the solute atoms across the boundary. Using Equation 3.61, we can also write Equation 3.63 as

$$J = -\frac{D_b}{\Omega} \frac{dC}{dx} - \frac{D_b C}{\Omega kT} \frac{dU}{dx} \tag{3.64}$$

The concentration profile of the solute can now be calculated from the continuity equation

$$\frac{dJ}{dx} + \frac{1}{\Omega} \frac{dC}{dt} = 0 \tag{3.65}$$

and the boundary conditions: $dC/dx = 0$, $dU/dx = 0$, and $C(x) = C_\infty$ at $x = \infty$. The concentration C_∞ can be taken as that in the bulk of the grain. Thus $C(x)$ must satisfy the equation

$$D_b \frac{dC}{dx} + \frac{D_b C}{kT} \frac{dU}{dx} + v_b(C - C_\infty) = 0 \tag{3.66}$$

The solute atom exerts a force $-(dU/dx)$ on the boundary, so the net force exerted by all the solute atoms is

$$F_s = -N_V \int_{-\infty}^{\infty} [C(x) - C_\infty] \frac{dU}{dx} dx \tag{3.67}$$

where N_V is the number of host atoms per unit volume. In the analysis, the $C(x)$ that satisfies Equation 3.66 is used to calculate F_s from Equation 3.67. An approximate solution that is valid for both low and high boundary velocities is

$$F_s = \frac{\alpha C_\infty v_b}{1 + \beta^2 v_b^2} \tag{3.68}$$

where the parameters α and β are given by

$$\alpha = 4 N_v kT \int_{-\infty}^{\infty} \frac{\sinh^2[U(x)/2kT]}{D_b(x)} dx \tag{3.69}$$

and

$$\frac{\alpha}{\beta^2} = \frac{N_v}{kT} \int_{-\infty}^{\infty} \left(\frac{dU}{dx} \right)^2 D_b(x) dx \tag{3.70}$$

Physically, α is the solute drag per unit velocity and per unit dopant concentration in the low boundary velocity limit, whereas $1/\beta$ is the drift velocity with which an impurity atom moves across the grain boundary. From the form of Equation 3.69, it is clear that solutes with either an attractive or repulsive interaction energy of the same magnitude will exert a similar drag force.

The total drag force on the boundary is the sum of the intrinsic drag F_b and the drag due to the solute atoms F_s, that is

$$F = F_b + F_s = \frac{v_b}{M_b} + \frac{\alpha C_\infty v_b}{1 + \beta^2 v_b^2} \tag{3.71}$$

where M_b is the intrinsic boundary mobility defined by Equation 3.30. In the low boundary velocity limit, the term $\beta^2 v_b^2$ in Equation 3.71 can be neglected, so

$$v_b = \frac{F}{1/M_b + \alpha C_\infty} \tag{3.72}$$

Initially the mobility due to solute drag is constant, but as the velocity increases the boundary continually sheds solute, so for sufficiently high velocity it is able to break away from the solute cloud and migrate at the intrinsic velocity. As Figure 3.33 shows for the theoretical relationship between velocity and driving force, transitions are predicted to occur from the solute drag-limited velocity to the intrinsic velocity over a range of driving forces. Direct observations indicate that the grain boundary motion is often not uniform but rather starts and stops as if making transitions between the solute drag and intrinsic regimes [73].

According to Equation 3.72, the boundary mobility M_b' equal to v_b/F can be written in terms of the intrinsic component M_b and the solute drag component M_s:

$$M_b' = \left(\frac{1}{M_b} + \frac{1}{M_s} \right)^{-1} \tag{3.73}$$

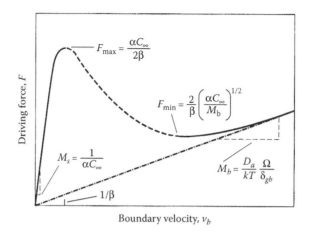

FIGURE 3.33 Driving force–velocity relationship for boundary migration controlled by solute drag and in the intrinsic regime.

where $M_s = 1/\alpha C_\infty$. For situations where the solute segregates to the grain boundary and the center of the boundary contributes most heavily to the drag effect, α can be approximated by

$$\alpha = \frac{4N_v kT\delta_{gb}Q}{D_b} \tag{3.74}$$

where Q is a partition coefficient (>1) for the dopant distribution between the boundary region and the bulk of the grain, such that the solute concentration in the boundary region is QC_∞. The mobility due to solute drag is, therefore,

$$M_s = \frac{D_b}{4N_v kT\delta_{gb}QC_\infty} \tag{3.75}$$

According to Equation 3.75, dopants are predicted to be most effective for reducing the boundary mobility when the diffusion coefficient of the rate-limiting species D_b is low and the segregated solute concentration QC_∞ is high. Since D_b decreases with increasing size of the diffusing atom, aliovalent cations with radii larger than that of the host cation are predicted to be effective for reducing the boundary mobility. Although some systems may show behavior consistent with this prediction, as for example CeO_2 doped with trivalent solutes [74,75], in general the selection of an effective dopant is complicated because of the multiplicity of the dopant role (see Chapter 5). Some examples of systems where the dopant approach has been used successfully for inhibiting grain growth are given in Table 3.1.

3.9 GRAIN GROWTH AND PORE EVOLUTION IN POROUS SOLIDS

Grain growth data for many porous ceramic systems during sintering follow trends similar to those shown in Figure 3.34 for the sintering of TiO_2 [82]. In the early stage of sintering, grain growth is limited, but some coarsening of the microstructure can occur by processes such as surface diffusion

TABLE 3.1
Examples of Dopants Used for Grain Growth
Control in Some Common Ceramics

Host	Dopant	Concentration (atomic %)	Ref.
Al_2O_3	Mg	0.025	30, 76
$BaTiO_3$	Nb; Co	0.5–1.0	77, 78
ZnO	Al	0.02	79
CeO_2	Y; Nd; Ca	3–5	74, 75
Y_2O_3	Th	5–10	80
SiC	(B + C)	0.3B + 0.5C	81

and vapor transport. The changes that occur in this early stage have a significant influence on the microstructural evolution in the later stages. Grain growth increases as the body densifies, but, frequently, grain growth does not become pronounced until the final stage of sintering when the pores pinch off and become isolated. In general, a region of significant densification with limited grain growth is followed by one of reduced densification but significant grain growth. Grain growth becomes prominent typically in the late intermediate and final stages of sintering. Gupta [83] found that the average grain size increases slowly and approximately linearly with density up to relative densities of 0.85 to 0.90, but then it increases much more rapidly above this density value (Figure 3.35). Migration of the boundaries leads to coalescence of the pores, so the average pore size increases (Figure 3.36a). It is believed that this process of grain growth and pore coalescence contributes to coarsening in the later stages of sintering. Figure 3.36b shows an example of coarsening by such a process in UO_2 [84].

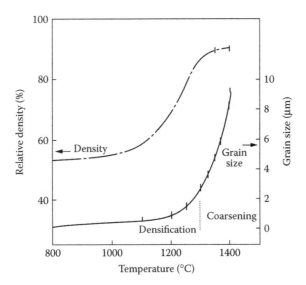

FIGURE 3.34 The density and grain size of a TiO_2 powder compact as a function of the sintering temperature. Rapid densification with limited grain growth occurs at lower temperatures and is followed by rapid grain growth with little densification at higher temperatures.

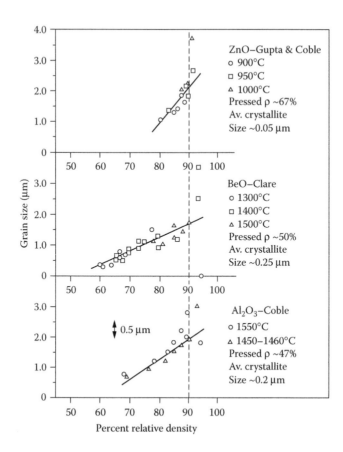

FIGURE 3.35 Simultaneous plots of grain size versus relative density for Al₂O₃, BeO, and ZnO. (From Gupta, T.K. Possible correlation between density and grain size during sintering, *J. Am. Ceram. Soc.,* 55, 276, 1972. With permission.)

3.9.1 PORE EVOLUTION DURING SINTERING

Microstructural coarsening during sintering is commonly considered in terms of how fast the grains grow. However, a more realistic view of the sintering compact may be that of a network of contacting grains interpenetrated by a network of porosity, so an understanding of the evolution of the porous network is also necessary for a broader understanding of microstructure evolution in porous systems. Figure 3.37 shows an example of a pore network in a partially sintered compact of ZnO with a relative density of 0.73. The pictures are a stereo pair of an epoxy resin replica of the network. The replica was obtained by forcing the epoxy resin, under high pressure, into the partially sintered compact, and, after curing of the resin, the ZnO was dissolved away in an acid solution. While the closed porosity cannot be viewed by this method, the complexity of the pore network in a real powder compact is readily apparent.

We may expect that for a compact formed from fully dense particles, the porosity is initially all connected. As sintering proceeds, more and more of the open porosity is converted to closed porosity. The conversion is, however, dependent on the packing uniformity of the structure. For a heterogeneously packed structure, the variation of the open and closed porosity may show trends similar to those of Figure 3.38 for a UO₂ powder compacted in a die [85]. Initially, the total porosity of the UO₂ compact was 0.37, made up of an open porosity of 0.36 and closed porosity of 0.01. During sintering at 1400°C, the open porosity decreased continuously, whereas the closed porosity

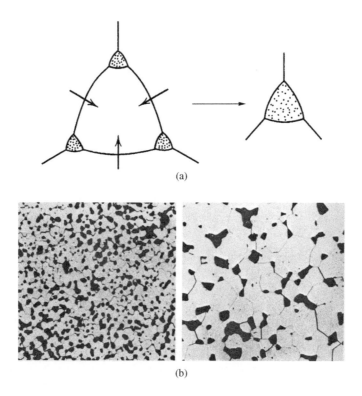

FIGURE 3.36 (a) Schematic illustration of grain growth accompanied by pore coalescence. (b) Grain growth and pore coalescence in a sample of UO_2 after 2 min, 91.5% dense, and 5 h, 91.9% dense, at 1600°C (magnification 400×). (From Kingery, W.D. and Francois, B., Grain growth in porous compacts, *J. Am. Ceram. Soc.,* 48, 546, 1965. With permission.).

increased initially to 0.04 but returned to the value of 0.01. This initial variation in the closed porosity may be due to the sintering of fairly dense agglomerates. When the open porosity had decreased to 0.15, the closed porosity began to increase again and reached a maximum value of 0.05, after which it decreased with further sintering. For a homogeneously packed powder, we may

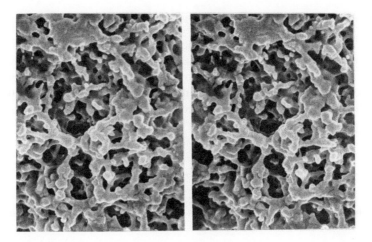

FIGURE 3.37 Stereographic pair of a replica of the pore space in a partially densified ZnO powder compact (density 73% of the theoretical). (Courtesy L.C. De Jonghe.)

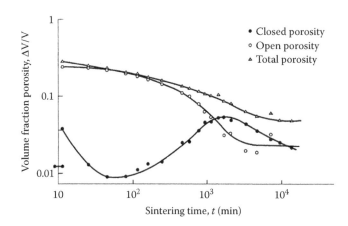

FIGURE 3.38 Change in porosity during the sintering of a UO_2 powder compact at 1400°C. (From Coleman, S.C., and Beeré, W.B., The sintering of open and closed porosity in UO_2, *Philos. Mag.*, 31, 1403, 1975.)

expect the conversion to closed porosity to occur later and more suddenly. Compared to the UO_2 results in Figure 3.38, a greater fraction of the total porosity may remain as an open network even after the compact has reached relative densities as high as 0.90.

The evolution of the pore network is difficult to analyze quantitatively. Experimentally, very few attempts have been made to produce a detailed stereological characterization of the evolution of the network. Such characterization is very time consuming, requiring detailed quantitative microscopy of a series of partially sintered samples. On the basis of such stereological observations, Rhines and DeHoff [86] concluded that the evolution of the pore network contained features that were comparable to those of other topological decay processes, such as grain growth in fully dense polycrystalline solids (see Section 3.5.3). The pore network, consisting of channels and junctions, apparently changes by the collapse of pore channels and the reforming of a new network of lower connectivity (Figure 3.39). Although it provides a more realistic picture of the evolution of the pore network, the topological model does not provide information on the kinetics of the coarsening process or how such coarsening can be controlled.

3.9.2 THERMODYNAMIC ANALYSIS OF PORE SHRINKAGE

Thermodynamically, whether a pore shrinks or not depends on the free energy change that accompanies the change in pore size. This can be easily explained by using the two-dimensional example shown in Figure 3.40 for a circular pore of radius r surrounded by N grains, where N is often referred to as the pore coordination number. If the pore shrinks slightly, as indicated by the dashed lines, the radius of the pore is reduced by δr, and the total change in free energy is given by

$$\delta E = N\delta r\gamma_{gb} - 2\pi\delta r\gamma_{sv} \qquad (3.76)$$

The pore will shrink only if $\delta E < 0$, which leads to the following condition for the pore to shrink:

$$N < 2\pi\frac{\gamma_{sv}}{\gamma_{gb}} \qquad (3.77)$$

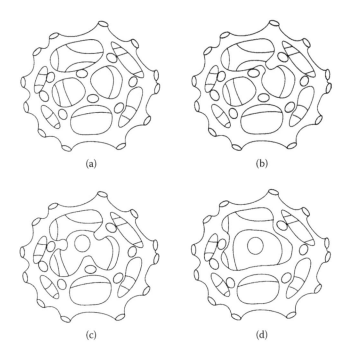

(a) (b)

(c) (d)

FIGURE 3.39 Sequence of pore channel collapse in a porous sintering solid. The sequential collapse can be envisioned as the coarsening process leaving a larger pore network as well as isolated pores. (From Rhines, F., and DeHoff, R., Channel network decay in sintering, in *Sintering and Heterogeneous Catalysis, Mater. Sci. Res.*, Vol. 16, Kuczynski, G.C., Miller, A.E., and Sargent, G.A., Eds., Plenum Press, New York, 1984.)

If the pore is surrounded by too many grains, then the increase in free energy due to the extension of the grain boundaries into the pores is greater than the decrease in the free energy due to the reduction of the pore surface area, so the pore should grow instead of shrink.

The argument can be extended to a pore with an equilibrium shape (Figure 2.4) dictated by the dihedral angle ψ, defined by

$$\cos\left(\frac{\psi}{2}\right) = \frac{\gamma_{gb}}{2\gamma_{sv}} \tag{3.78}$$

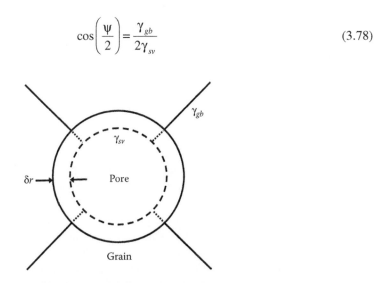

FIGURE 3.40 Circular pore surrounded by four identical grains.

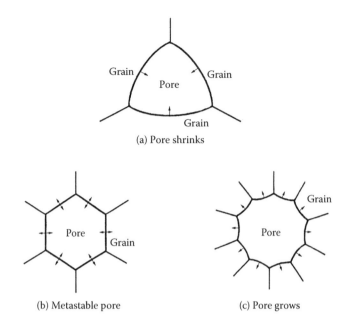

FIGURE 3.41 Pore stability in two dimensions for a dihedral angle of 120°.

Let us consider, in two dimensions, a pore with a dihedral angle $\psi = 120°$ that is surrounded by N grains. The pore has straight sides if $N = 6$, "convex" sides for $N < 6$, and "concave" sides for $N > 6$ (Figure 3.41). The surface of the pore will move toward its center of curvature, so the pore with $N < 6$ will shrink whereas the one with $N > 6$ will grow. The pore is metastable for $N = 6$, and this number is called the *critical pore coordination number*, denoted N_c. In fact, it can be shown that for the pore with "convex" sides ($N < 6$), if the pore shrinks slightly, the decrease in energy due to the reduction of pore surface area is greater than the increase in energy due to the growth of the grain boundaries. On the other hand, for the pore with "concave" sides ($N > 6$), the decrease in energy due to the reduction of the pore surface area is smaller than the increase in energy due to the growth of the grain boundaries. For the metastable situation ($N = 6$), the decrease in energy due to the reduction of the pore surface area just balances the increase in energy due to the growth of the grain boundaries. We can go on to consider other dihedral angles: for example, $N_c = 3$ for $\psi = 60°$. The general result is that N_c decreases with the dihedral angle. The two parameters are connected by a simple geometrical relationship given by:

$$\psi = \frac{180N_c - 360}{N_c} \tag{3.79}$$

The geometrical considerations can be extended to three dimensions, in which case the pore is a polyhedron. The analysis has been carried out by Kingery and Francois [87]. Taking r_s as the radius of curvature of the circumscribed sphere around a polyhedral pore surrounded by grains, the ratio of the radius of curvature of the pore r to r_s depends both on the dihedral angle and the pore coordination number, as shown in Figure 3.42. When the surfaces of the pore become flat ($r = \infty$), the pore is metastable, and there is no tendency for the pore to grow or shrink. The ratio r_s/r is zero, and this condition defines N_c. The value of N_c as a function of the dihedral angle is plotted in Figure 3.43. As an example, for a pore with a dihedral angle of 120°, we find that $N_c = 12$, so a pore with $N < 12$ will shrink whereas one with $N > 12$ will grow. A refinement of the theory by

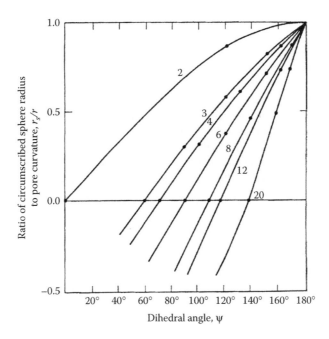

FIGURE 3.42 Change in the ratio r_s/r with dihedral angle for pores surrounded by different numbers of grains as indicated on the individual curves.

R. M. Cannon (unpublished work) indicates that the pore with $N > N_c$ may not grow continuously but may reach some limiting size, whereas Kellet and Lange [88] have extended the analysis to simple particle arrays and mass conservation.

From this discussion, we see that a poorly compacted powder, containing pores that are large compared to the grain size, would be difficult to densify, especially if the dihedral angle is low, because the pore coordination number is large. One practical solution to the difficulty of densification is to prepare compacts with reasonably high green density and a fairly uniform pore size distribution by cold isostatic pressing or by colloidal methods, so that the fraction of pores with $N > N_c$ is minimized.

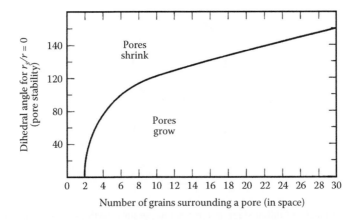

FIGURE 3.43 Conditions for pore stability in three dimensions as a function of pore coordination number.

The thermodynamic barrier to the closure of large pores ($N > N_c$) has long been recognized as a key principle in the fabrication of ceramics with high density and controlled (fine) grain size, but the theoretical framework has been questioned recently by Pan, Ch'ng, and Cocks [89]. Using computer simulations to study the sintering kinetics of a large pore in a dense polycrystalline matrix, these authors claim that the thermodynamic barrier predicted by the analysis of Kingery and Francois [87] results from the assumption that the large pore is surrounded by identical grains and from the imposed kinetic route that the grains move simultaneously into the pore as it shrinks. The simulations indicate that, in general, the elimination of a large pore from a dense polycrystalline solid does not necessarily have a free energy barrier and that a large pore will shrink even though N is greater than N_c. The sintering of large pores in a polycrystalline matrix will be discussed further in Chapter 5 when we consider the influence of inhomogeneities on sintering.

3.9.3 KINETICS AND MECHANISMS OF GRAIN GROWTH IN POROUS SOLIDS

Grain growth in porous ceramics is complex, but a few limiting cases can be discussed. Grain growth in very porous ceramics characteristic of the early stage of sintering is limited, but, because of the complexity of the microstructure, a detailed analysis of the coarsening process has not been performed. Instead, our understanding of the process is at a qualitative level. In less porous ceramics characteristic of the later stage of sintering, grain growth has been commonly analyzed in terms of an idealized microstructure consisting of a nearly spherical pore on an isolated grain boundary.

3.9.4 GRAIN GROWTH IN VERY POROUS SOLIDS

For a model consisting of two originally spherical particles separated by a grain boundary, which can be taken as a rough approximation to the microstructure in the initial stage of sintering, migration of the boundary is difficult. As sketched in Figure 3.44, the balance of interfacial tensions requires that the equilibrium dihedral angle of the grain boundary groove at the surface of the sphere be maintained. Migration of the boundary would involve a significant increase in the grain boundary area. The occurrence of grain growth is energetically unfavorable unless other processes that significantly reduce this energy barrier come into play.

One way in which grain growth can occur was suggested by Greskovich and Lay [90] on the basis of observations of the coarsening of Al_2O_3 powder compacts. Figure 3.45 illustrates the development of the porous Al_2O_3 microstructure during sintering at 1700°C. The green compact (Figure 3.45a) consists of individual grains with a large amount of open porosity. In going from

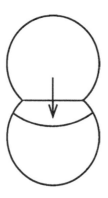

FIGURE 3.44 For an idealized initial stage microstructure, grain growth increases the total grain boundary area between two spheres. The dihedral angle constraint at the surface of the sphere creates a boundary curvature that opposes grain growth.

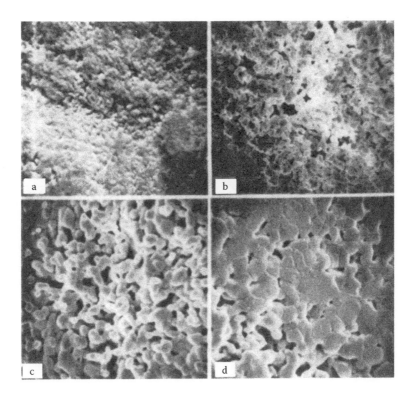

FIGURE 3.45 Development of the microstructure of an MgO-doped Al_2O_3 powder compact during sintering at 1700°C. (a) Green compact, (b) 1 min, (c) 2.5 min, and (d) 6 min (magnification 5000×). (From Greskovich, C., and Lay, K.W., Grain growth in very porous Al_2O_3 compacts, *J. Amer. Ceram. Soc.*, 55, 142, 1972. With permission.)

Figure 3.45a to Figure 3.45b, the relative density increased from 0.31 to 0.40, but even in such very porous structures the average grain size has nearly doubled.

The model put forward by Greskovich and Lay is shown in Figure 3.46 for two particles and for a cluster of particles. Greskovich and Lay assumed that surface diffusion assisted in the rounding of the particles and the growth of the necks between the particles. Whether the boundary can migrate depends on whether the structure permits it, i.e., there must be a decrease in the total energy of the system for an incremental movement of the boundary. As sketched for the two particles, after surface diffusion has produced the structural changes for the movement of the boundary to be energetically favorable, migration of the boundary occurs. The rate of boundary migration depends on the difference in initial size between the particles: the greater the difference in size, the greater the curvature of the boundary and the greater the driving force for the boundary to sweep through the smaller grain. The rate of the overall coarsening process is controlled by the slower of the two processes: neck growth by surface diffusion or boundary migration.

Although Greskovich and Lay considered surface diffusion to be the dominant neck growth mechanism in their model, other mechanisms may also be important in other systems. For example, grain boundary diffusion may also be important for fine particles, and vapor transport for systems with fairly high vapor pressure. Additional complications may be present in real powder compacts. For example, the coarsening of heterogeneously packed powder compacts may be controlled by more than one mechanism. As Figure 3.45d indicates, a heterogeneously packed powder compact will, after some sintering, contain highly porous regions as well as dense regions. For this type of structure, coarsening is likely to be controlled by two separate mechanisms: neck growth in the highly porous regions and curvature-driven boundary migration in the dense regions.

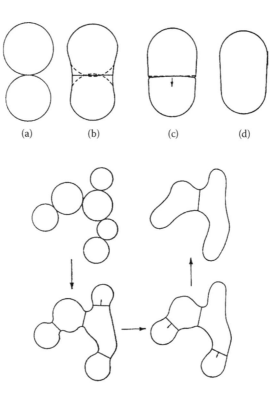

FIGURE 3.46 Qualitative mechanism for grain growth in porous powder compacts. *Top:* Two particles. (a) Particles of slightly different size in contact; (b) neck growth by surface diffusion between the particles; (c) grain boundary migrating from the contact plane; and (d) grain growth. *Bottom:* A cluster of particles. Arrows on the grain boundaries indicate the direction of boundary movement.

Recently, computer simulation techniques were used to model the coarsening of two cylindrical particles with different sizes by simultaneous neck growth and boundary migration [91]. It was assumed that neck growth occurred by grain boundary diffusion and that the matter deposited into the neck was transported over the free surfaces by surface diffusion. Figure 3.47 shows the results of the simulations for the cross-section at four different reduced times, assuming the following parameters: the ratio of the initial particle radii $a_1/a_2 = 0.5$, the ratio of the surface and grain boundary diffusion coefficients $D_s/D_{gb} = 10$, the normalized boundary mobility $M_b a_2^2 kT/D_{gb} = 100$, and $\gamma_{sv}/\gamma_{gb} = 3$. An interesting observation is that under the assumed conditions, the time taken for the neck size to become approximately equal to the size of the smaller particle (Figure 3.47b) is much smaller than the entire coarsening process, indicating that under the assumed conditions, neck growth and coarsening are almost two separate processes.

Another way in which grain growth can occur in porous compacts was described by Edelson and Glaeser [92]. As sketched in Figure 3.48a, if large grain size differences exist at the necks between the particles, elimination of grain boundaries by the advancing boundary releases enough energy to make the overall process favorable. In this way, the finer neighboring grains can be consumed in an incremental growth process. Figure 3.48b shows that the growing grain can also entrap porosity, a process that can limit the density of the final article. However, this process is less likely to occur in the earlier stages of sintering because the large continuous pores that are closely spaced provide a significant drag on the boundary and limit its mobility.

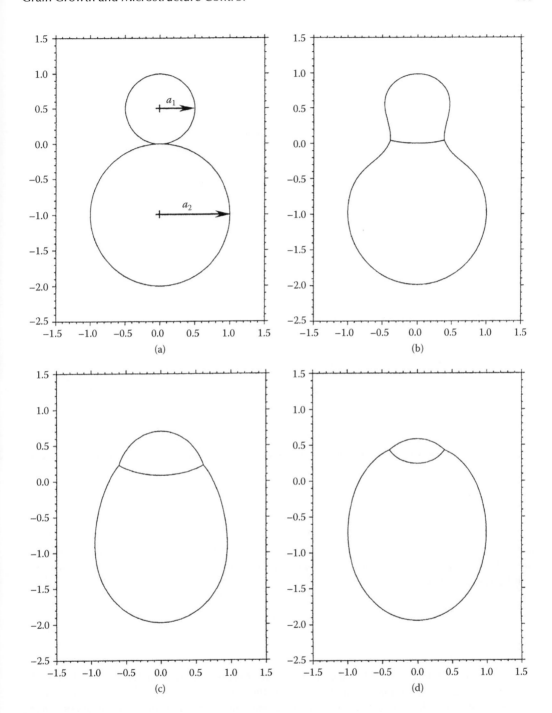

FIGURE 3.47 Computer simulation of sintering of two cylindrical particles with different sizes by grain boundary diffusion, surface diffusion, and grain boundary migration at four reduced times: (a) $t_o = 0.0$, (b) $t_1 = 3.425 \times 10^{-5}$, (c) $t_2 = 1.558 \times 10^{-3}$, (d) $t_3 = 3.109 \times 10^{-3}$. (From Ch'ng, H.N., and Pan, J., Cubic spline elements for modeling microstructural evolution of materials controlled by solid-state diffusion and grain boundary migration, *J. Comput. Phys.*, 196, 724, 2004. With permission.)

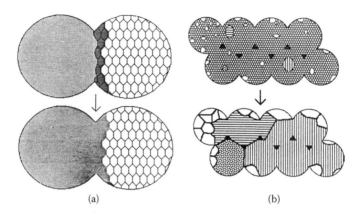

(a)   (b)

FIGURE 3.48 Growth of intra-agglomerate grains through polycrystalline necks. (a) Growth of a large grain proceeds across a neck by incremental consumption of smaller grains. (b) Exaggerated intra-agglomerate crystalline growth results in the development of large grains in the early stage of sintering. The resulting grain structure may contain some large grains that grow abnormally during inter-agglomerate sintering. (Courtesy of A.M. Glaeser.)

3.9.5 Grain Growth in Less Porous Solids

Grain growth in the later stages of sintering is often discussed in terms of an idealized final stage sintering model consisting of small, isolated spherical pores situated at the grain boundaries. In the analysis of the grain growth, the approach is to consider an isolated region of the grain boundary containing a single pore and to assume that the kinetic equations derived for the model represent the average behavior of the whole system. By adopting this approach, we are assuming that the microstructure is homogeneous. The approach may be viewed as being equivalent to that of Burke and Turnbull for dense solids but with the added complication of the interaction of a pore and the grain boundary.

Two cases may be distinguished, depending on whether the pores are immobile or mobile. The case of immobile pores on the grain boundary can be discussed in terms of the Zener-type analysis discussed earlier for pinning particles. In principle, if the drag exerted by the pores is sufficiently high, the boundary will be pinned, and boundary migration will stop. A limiting grain size, G_L, will be reached, given by

$$G_L = \left(\frac{2\alpha}{3} \right) \frac{r}{P} \tag{3.80}$$

where r is now the pore radius, P is the porosity, and α is a geometrical factor that depends on the pore shape. If the drag exerted by the pores is insufficient, the boundary will break away, leaving the pores trapped in the grains. Observations of ceramic microstructures do not provide convincing evidence for Equation 3.80. Instead, much of the evidence is in favor of mobile pores.

3.9.5.1 Pore Mobility

Small, isolated pores situated at the grain boundary can be dragged along by the moving grain boundary. The reason is that the boundary, moving under the influence of its curvature, applies a force on the pore that causes the pore to change its shape, as illustrated in Figure 3.49. The leading

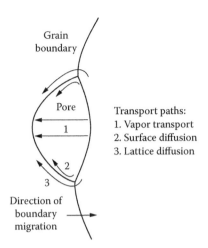

Transport paths:
1. Vapor transport
2. Surface diffusion
3. Lattice diffusion

FIGURE 3.49 Possible transport paths for a pore moving with a grain boundary: 1, vapor transport (evaporation/condensation); 2, surface diffusion; 3, lattice diffusion.

surface of the pore becomes less strongly curved than the trailing surface. The difference in curvature leads to a chemical potential difference that drives matter flux from the leading surface to the trailing surface. The result is that the pore moves forward in the direction of the boundary motion. Matter transport from the leading surface to the trailing surface can occur by three separate mechanisms: vapor transport, also referred to as evaporation/condensation, surface diffusion, and lattice diffusion.

The flux of matter from the leading surface to the trailing surface of the pore can be analyzed to derive an equation for the pore mobility M_p. By analogy with the case of a moving boundary, the pore velocity v_p is defined by a force-mobility relationship,

$$v_p = M_p F_p \tag{3.81}$$

where F_p is the driving force acting on the pore. Let us consider a pore of average radius r in which matter transport occurs by surface diffusion from the leading surface to the trailing surface [93]. The net atomic flux is

$$J_s A_s = \left(\frac{D_s}{\Omega kT} F_a \right) 2\pi r \delta_s \tag{3.82}$$

where J_s is the flux of atoms, A_s is the area over which surface diffusion occurs, D_s is the surface diffusion coefficient, Ω is the atomic volume, F_a is the driving force on an atom, δ_s is the thickness for surface diffusion, k is the Boltzmann constant, and T is the absolute temperature. If the pore moves forward a distance dx in a time dt, the volume of matter that must be moved per unit time is $\pi r^2 (dx/dt)$. Equating the number of atoms moved to the net flux gives

$$\frac{\pi r^2}{\Omega} \frac{dx}{dt} = -\frac{D_s 2\pi r \delta_s}{\Omega kT} F_a \tag{3.83}$$

where the negative sign is inserted because the flux is opposite to the direction of motion. The work done in moving the pore a distance dx is equal to that required to move $\pi r^2 dx/\Omega$ atoms a distance $2r$, so

$$F_p dx = - F_a \frac{\pi r^2 dx}{\Omega} 2r \qquad (3.84)$$

Substituting for F_a from Equation 3.83 and rearranging gives

$$F_p = \frac{\pi r^4 kT}{D_s \delta_s \Omega} \frac{dx}{dt} \qquad (3.85)$$

Putting the velocity of the pore v_p equal to dx/dt, the pore mobility is given by

$$M_p = \frac{D_s \delta_s \Omega}{\pi kT r^4} \qquad (3.86)$$

The pore mobility for matter transport by vapor transport or by lattice diffusion can be derived by a similar procedure, and the formulae are summarized in Table 3.2. For all three mechanisms, M_p is found to have a strong dependence on r, decreasing rapidly with increasing r. Fine pores are beneficial for sintering because they are highly mobile and, therefore, better able to remain attached to the moving boundary during grain growth. In practice, however, strategies for avoiding

---

TABLE 3.2
Pore and Boundary Parameters

A. Mobilities[a]

M_p	Mobility of spherical pore; migration by faster of	
	(a) Surface diffusion	$D_s \delta_s \Omega / kT \pi r^4$
	(b) Lattice diffusion	$D_l \Omega / f kT \pi r^3$
	(c) Vapor transport	$D_g d_g \Omega / 2kT d_s \pi r^3$
M_b	Mobility of boundary	
	Pure system	$D_a \Omega / kT \gamma_{gb}$
	Solute drag	$\Omega / kT \delta_{gb} (1/D_a + 4N_V \Omega Q C_\infty / D_b)^{-1}$

B. Forces[b]

F_p	Maximum drag force of pore	$\pi r \gamma_{gb}$
F_b	Force per unit area of pore-free boundary	$\alpha \gamma_{gb} / G$

[a] f = correlation factor; d_g = density in the gas phase of the rate-controlling species; d_s = density in the solid phase of the rate-controlling species.

[b] α = geometrical constant depending on the grain shape (e.g., $\alpha = 2$ for spherical grains).

Source: Brook, R.J., Controlled grain growth, in *Ceramic Fabrication Processes*, Treatise on Materials Science and Technology, Vol. 9, Wang, F.F.Y., Ed., Academic Press, New York, 1976, p. 331. With permission.

breakaway of the boundary from the pore (abnormal grain growth) are based mainly on suppressing M_b by the solute drag mechanism or the use of fine second-phase particles to pin the boundary.

3.9.5.2 Kinetics of Pore–Boundary Interactions

The definition of the pore mobility allows us to go on to analyze how the interactions between the pores and the grain boundaries influence the kinetics of grain growth. There are two cases that can be considered: (1) the pore becomes separated from the boundary and (2) the pore remains attached to the boundary. Pore separation will occur when $v_p < v_b$, and this condition can also be written as

$$F_p M_p < FM_b \qquad (3.87)$$

where F is the effective driving force on the boundary. If F_d is the drag force exerted by a pore, then a balance of forces requires that F_d is equal and opposite to F_p. Considering unit area of the boundary in which there are N_A pores, Equation 3.87 can be written

$$F_p M_p < (F_b - N_A F_p) M_b \qquad (3.88)$$

where F_b is the driving force on the pore-free boundary due to its curvature. Rearranging Equation 3.88, the condition for pore separation can be expressed as

$$F_b > N_A F_p + \frac{M_p F_p}{M_b} \qquad (3.89)$$

The condition for pore attachment to the boundary is $v_p = v_b$, which can also be written as

$$F_p M_p = (F_b - N_A F_p) M_b \qquad (3.90)$$

Putting $v_p = F_p M_p = v_b$ and rearranging gives

$$v_b = F_b \frac{M_p M_b}{N_A M_b + M_p} \qquad (3.91)$$

Two limiting conditions can be defined. When $N_A M_b \gg M_p$, then

$$v_b = F_b M_p / N_A \qquad (3.92)$$

The effective driving force on the boundary is $F = F_b - N_A F_p$, and using Equation 3.90 gives $F = v_p / M_b$. Putting $v_p = v_b = F_b M_p / N_A$ gives $F = F_b M_p / N_A M_b \ll F_b$. The driving force on the boundary is nearly balanced by the drag of the pores, and the boundary motion is limited by the pore mobility. This condition is referred to as *pore control*. The other limiting condition is when $N_A M_b \ll M_p$, in which case

$$v_b = F_b M_b \tag{3.93}$$

The drag exerted by the pores is $N_A F_p = N_A(F_b M_b / M_p) \ll F_b$. The presence of the pores has almost no effect on the boundary velocity, a condition referred to as *boundary control*.

3.9.5.3 Grain Growth Kinetics

For the simplified model consisting of nearly spherical isolated pores on the grain boundary, we can derive equations for the grain growth kinetics. Consider the situation of grain growth controlled by the pore mobility (pore control). If pore migration occurs by surface diffusion, then Equation 3.86 and Equation 3.92 give

$$v_b \approx \frac{dG}{dt} = \frac{F_b}{N_A} \left(\frac{D_s \delta_s \Omega}{\pi k T r^4} \right) \tag{3.94}$$

According to Equation 3.26, $F_b \sim 1/G$, and the interpore distance X is approximately equal to G, so $N_A \sim 1/X^2 \sim 1/G^2$. Assuming coarsening by grain growth and pore coalescence, then $r \sim G$. Substituting into Equation 3.94, after some rearrangement we have

$$\frac{dG}{dt} = \frac{K_1}{G^3} \tag{3.95}$$

where K_1 is a constant at a given temperature. After integration we obtain

$$G^4 = G_0^4 + K_2 t \tag{3.96}$$

where G_0 is the grain size at $t = 0$ and K_2 is a constant. Grain growth equations can be derived by a similar procedure for the other mechanisms.

Using the general form of the grain growth equation as $G^m = G_0^m + Kt$, the exponent m for the various mechanisms is summarized in Table 3.3. Except for the mechanism involving solution of second-phase particles ($m = 1$), the m values lie in the range 2 to 4. In many ceramics, the value $m = 3$ has often been reported, and it is seen that this value can correspond to at least five different mechanisms. Attaching physical significance to the m values determined from grain growth data can, therefore, be somewhat dubious. Furthermore, the fitting of experimental data to produce m values that are exact whole numbers may not be realistic because the occurrence of simultaneous mechanisms is likely to give m values that are not integers. We must also remember the many approximations used in the derivation of the equations, which include a structurally homogeneous compact, isotropic grain boundary energy, and isolated spherical pores at the grain boundary interfaces. These conditions are rarely achieved in real powder compacts.

3.10 SIMULTANEOUS DENSIFICATION AND GRAIN GROWTH

In our consideration of solid-state sintering, we have so far treated the processes of densification and grain growth (coarsening) separately. However, as sketched in Figure 3.50 for an idealized final-stage microstructure, the two processes occur simultaneously, so the occurrence of one process

TABLE 3.3
Grain Growth Exponent *m* in the Equation $G^m = G_0^m + Kt$
for Various Mechanisms

Mechanism	Exponent *m*
Pore control	
Surface diffusion	4
Lattice diffusion	3
Vapor transport (vapor pressure p = constant)	3
Vapor transport ($p = 2\gamma_{sv}/r$)	2
Boundary control	
Pure system	2
System containing second-phase particles	
Coalescence of second phase by lattice diffusion	3
Coalescence of second phase by grain boundary diffusion	4
Solution of second phase	1
Diffusion through continuous second phase	3
Doped system	
Solute drag (low solubility)	3
Solute drag (high solubility)	2

Note: The pore control kinetics are given for the situation where the pore separation is related to the grain size, i.e., the number of pores per unit area of the boundary $N_A \sim 1/G^2$. Changes in distribution during growth would change the kinetics.

Source: Brook, R.J., Controlled grain growth, in *Ceramic Fabrication Processes*, Treatise on Materials Science and Technology, Vol. 9, Wang, F.F.Y., Ed., Academic Press, New York, 1976, p. 331. With permission.

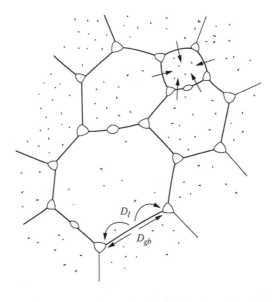

FIGURE 3.50 Late-stage sintering of a powder compact. The two contributions to free energy reduction are (1) densification (lower part of figure), where the arrows show the direction of atom flow, and (2) coarsening or grain growth (top right), where the arrows show the direction of boundary movement.

will influence the rate at which the other can proceed. To be able to understand and control the sintering process, one must consider both densification and coarsening and address the interaction between them. In addressing this interaction, we should remember that each process by itself is fairly complex and that the models provide only a qualitative understanding of each process in practical systems. Faced with this situation, it appears unlikely that a detailed theory of simultaneous densification and coarsening would provide any significant advantage. Most models that have been developed so far provide only a qualitative understanding of the interaction, and the results are commonly plotted in maps showing the evolution of the microstructural parameters. More recently, computer simulations have started to address some of the complexities of microstructural evolution in simple model systems.

3.10.1 MICROSTRUCTURAL MAPS

3.10.1.1 Brook Model

A treatment of the simultaneous densification and grain growth processes and the transition to abnormal grain growth was developed by Brook [94]. Assuming an idealized final-stage microstructure consisting of a nearly spherical pore on an isolated grain boundary, it involves the determination of the transition from pore drag-controlled boundary migration to intrinsic or solute drag boundary migration, with the pores either attached or separated from the boundary, as functions of grain size G and pore size $2r$. As discussed earlier, the conditions for attachment of the pore to the boundary fall into two limiting cases: pore control and boundary control. The conditions that separate these two cases are represented by a curve, called the *equal mobility curve,* defined by the condition that the pore mobility is equal to the boundary mobility, that is,

$$N_A M_b = M_p \tag{3.97}$$

Assuming that pore migration occurs by surface diffusion, the appropriate relations for M_b and M_p can be found in Table 3.2. Using the approximation that $N_A \approx 1/X^2$, where X is the interpore distance, and taking $X \approx G$, Equation 3.97 gives

$$G_{em} = \left(\frac{D_a \pi}{D_s \delta_s \delta_{gb}} \right)^{1/2} r^2 \tag{3.98}$$

where G_{em} is the grain size defined by the equal mobility condition. Using logarithmic axes for a plot of G versus $2r$ (Figure 3.51), the equal mobility condition is represented by a straight line with a slope of 2.

In determining the conditions for separation of the boundary from the pore, we recall that the maximum force exerted by the grain boundary on a pore to drag it along is given by Equation 3.53, that is,

$$F_p^{max} = \pi r \gamma_{gb} \tag{3.99}$$

The maximum velocity that the pore can attain is therefore

$$v_p^{max} = M_p F_p^{max} \tag{3.100}$$

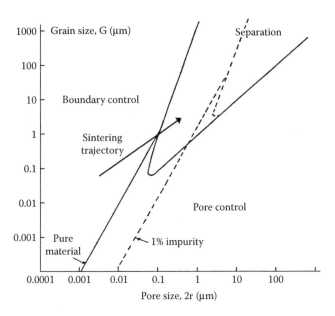

FIGURE 3.51 The dependence of the type of pore-boundary interaction on microstructural parameters when pores migrate by surface diffusion. The interpore spacing is assumed to be equal to the grain size. (From Brook, R.J., Pore–grain boundary interactions and grain growth, *J. Amer. Ceram. Soc.*, 52, 56, 1969. With permission.)

If the velocity of the boundary with the attached pore were to exceed v_p^{max}, then separation will occur. The limiting condition for separation can, therefore, be written as

$$v_b = v_p^{max} \qquad (3.101)$$

Substituting for v_b from Equation 3.91, we obtain

$$\frac{M_b M_p}{N_A M_b + M_p} F_b = M_p F_p^{max} \qquad (3.102)$$

Putting $N_A \approx 1/X^2$ and substituting for the other parameters from Table 3.2, after some rearranging, Equation 3.102 gives

$$G_{sep} = \left(\frac{\pi r}{X^2} + \frac{D_s \delta_s \delta_{gb}}{D_a r^3} \right)^{-1} \qquad (3.103)$$

where G_{sep} is the grain size when the boundary separates from the pore. Assuming that $X \approx G$, Equation 3.103 can be written

$$\left(\frac{D_s \delta_s \delta_{gb}}{D_a r^3} \right) G_{sep}^2 - G_{sep} + \pi r = 0 \qquad (3.104)$$

The solution to this quadratic equation determines the *separation curve* shown in Figure 3.51.

To allow for the use of quantitative axes for the G versus $2r$ diagram in Figure 3.51, Brook used realistic values, taken from the literature, for the parameters in Equation 3.98 and Equation 3.104. Qualitatively, three basic types of pore-boundary interactions are found in the diagram. For larger, less mobile pores that are closely separated, the pores remain attached to the boundary and control the motion. In another region of the diagram, smaller, more mobile pores remain attached to the boundary, and they do not exert a significant drag on the boundary, so the boundary migration is controlled by the boundary mobility. The third region occurs for larger pores that are widely separated, where boundary-controlled migration occurs with pores separated. For the achievement of high density with controlled grain size, this separation region must be avoided. Dopants, as discussed earlier, have the effect of reducing M_b. In the G versus $2r$ diagram, this has the effect of extending the region of pore attachment (Figure 3.51). According to this diagram, a possible explanation for the effectiveness of dopants is that they delay the onset of abnormal grain growth (pore separation) beyond the grain size at which final densification is achieved.

Possible trajectories for sintering can be identified on the basis of the G versus $2r$ diagram. For example, grain growth and pore coalescence often contribute to the coarsening of porous ceramics, particularly in the later stages of sintering, resulting in an increase in both the average grain size and the average pore size during sintering. For this situation, a possible trajectory for the microstructural evolution runs diagonally upward from left to right (Figure 3.51). According to this representation, it is likely that the interaction changes from boundary control, with possibly a small region of pore control, to separation.

The kinetic analysis of the pore-boundary interactions and its representation in terms of the G versus $2r$ diagram provide a qualitative basis for understanding microstructural evolution during the later stages of sintering, but because of the considerable simplifying assumptions in the geometry and microstructure, the treatment may be inadequate for describing the microstructural evolution of real systems. It is to be expected that factors such as grain-size distribution, the number of pores at the grain boundaries, and the dihedral angle will affect the simple relationships between grain boundary curvature and pore size assumed in the analysis for the separation condition [95].

A refinement to Brook's basic treatment was introduced by Carpay [96], who observed that when pore separation occurs, the assumption $X \approx G$ is no longer valid. By solving Equation 3.104 for fixed values of X (in the range of 1 to 4 μm), the most significant result of this refinement is that after a region of separation, the boundary may again become attached to the pores. Hsueh, Evans, and Coble [97] and Sakarcan, Hsueh, and Evans [98] considered the effect of dihedral angle. As shown in Figure 3.52, the effective area of the boundary intersected by a pore of constant volume increases as the dihedral angle decreases, resulting in greater pore drag and less likelihood of the boundary separating from the pore. The effects of a grain size distribution and solute drag were incorporated into the basic Brook model by Yan, Cannon, and Bowen [99]. Compared to a powder with a narrow size distribution, the critical density at which pore separation occurs is found to be

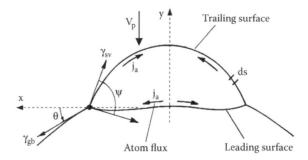

FIGURE 3.52 Schematic of the distortion that accompanies the motion of pores attached to grain boundaries and the atomic flux from the leading surface to the trailing surface.

significantly lower for powder compacts with a wide distribution of particle size. The statistical nature of the separation process must also be recognized. Separation of only a fraction of the pores from the boundary is sufficient to cause abnormal grain growth, as opposed to the assumption in the Brook model that all pores separate from the boundaries. In this case, the separation region in the G versus $2r$ diagram becomes larger as the fraction of pores required to separate from the boundary decreases.

3.10.1.2 Yan, Cannon, and Chowdhry Model

Another treatment of simultaneous densification and grain growth was developed by Yan, Cannon, and Chowdhry [82]. Assuming an idealized final stage microstructure consisting of tetrakaideca-hedral grains with spherical pores at the grain corners, the treatment determines how the achievable final density and the extent of coarsening depend on the ratio of the coarsening rate to the densification rate. The instantaneous rate of change of the pore radius r is taken as

$$\frac{dr}{dt} = \left(\frac{dr}{dt}\right)_P + \left(\frac{dr}{dt}\right)_G \tag{3.105}$$

where the first term on the right-hand side gives the rate of change of the pore size at constant porosity due to the coarsening process (a positive value) and the second term gives the rate of change at constant grain size due to the densification process (a negative value). For a coarsening process involving grain growth and pore coalescence (where r is proportional to G), we can write

$$\frac{dr}{dt} = \frac{r}{G}\left(\frac{dG}{dt}\right)_P + \left(\frac{dr}{dt}\right)_G \tag{3.106}$$

Since $(dG/dr) = (dG/dt)/(dr/dt)$, then

$$\frac{dG}{dr} = \frac{(dG/dt)}{(r/G)(dG/dt) + (dr/dt)_G} \tag{3.107}$$

Equation 3.107 can be written as

$$\frac{d\ln G}{d\ln r} = \frac{\Gamma}{\Gamma - 1} \tag{3.108}$$

where Γ is the ratio of the coarsening rate to the densification rate given by

$$\Gamma = -\frac{(r/G)(dG/dt)}{(dr/dt)_G} \tag{3.109}$$

The term Γ is calculated by assuming specific models for densification and grain growth, and it is necessary to consider simple cases where only one densification mechanism and one coarsening mechanism are dominant.

    As an example, let us assume (1) the Coble model for final-stage densification by grain boundary diffusion (see Chapter 2) to calculate the change in pore size with time, and (2) that the change in

TABLE 3.4
Simultaneous Densification-Coarsening Equations for Various Rate-Controlling Mechanisms

Mechanisms		Ratio of Coarsening Rate/Densification Rate
Densification	Coarsening	Γ
Grain boundary diffusion	Surface diffusion	$(3/176)(D_s\delta_s/D_{gb}\delta_{gb})(\gamma_{gb}/\gamma_{sv})\xi$
Lattice diffusion	Surface diffusion	$(1/32)(D_s\delta_s/D_l r)(\gamma_{gb}/\gamma_{sv})\xi$
Grain boundary diffusion	Boundary control	$(3/11)(M_b kTr^4/D_{gb}\delta_{gb}\,\Omega G^2)(\gamma_{gb}/\gamma_{sv})\xi$

grain size with time is controlled by grain boundary migration limited by surface diffusion-controlled pore drag (see Equation 3.92 and Table 3.2), then

$$\Gamma = \frac{3}{176}\frac{D_s\delta_s\gamma_{gb}}{D_{gb}\delta_{gb}\gamma_{sv}}\xi \qquad (3.110)$$

where D_s and D_{gb} are the surface and grain boundary diffusion coefficients, respectively; δ_s and δ_{gb} are the effective widths for surface and grain boundary diffusion, respectively; γ_{sv} and γ_{gb} are the surface and grain boundary energies, respectively; and ξ is function of the porosity. The equations for Γ for different combinations of densification and coarsening mechanisms are given in Table 3.4.

Selection of the initial values for G and r, followed by the use of Equation 3.108 and Equation 3.110, and iteration allow the calculation of the grain size versus density trajectory. Figure 3.53 shows the calculated results for different values of Γ assuming densification by grain boundary diffusion and coarsening by surface diffusion. If $\Gamma > 1$, grain growth dominates and the approach to the final density is asymptotic, so a high density may be unachievable in a realistic time. On the other hand, little grain growth occurs during densification if $\Gamma < 1$, and theoretical density is achievable.

A refinement of the treatment of simultaneous densification and grain growth was introduced by Bennison and Harmer [100], who modified the analysis of Yan, Cannon, and Chowdhry to include the conditions under which separation of the boundary from the pore occurs. The calculated results for Al_2O_3, where densification is controlled by lattice diffusion and grain growth (coarsening) is controlled by surface diffusion-controlled pore drag, is shown in Figure 3.54. The combined features of grain size G versus density ρ trajectory and the conditions for separation make the treatment particularly effective for discussing the influence of dopants, temperature, and other variables on microstructural evolution. Figure 3.54 indicates that the separation region is shifted to higher grain sizes by the use of MgO as a dopant, which lowers the grain boundary mobility, thereby making it easier for the G versus ρ trajectory to bypass the separation region. Flattening of the trajectory, that is, increasing the densification rate relative to the grain growth rate, is also beneficial and can be achieved, for example, by the use of a dopant that increases the rate of densification or reduces the boundary mobility. As described later, the trajectory can also be flattened by controlling the heating cycle during sintering, as, for example, by rapid heating to a temperature where densification dominates over grain growth.

The treatment by Yan, Cannon, and Chowdhry and the refined treatment by Bennison and Harmer provide significant benefits for predicting microstructural evolution during sintering, but their limitations for application to real systems must also be recognized. The treatments assume an idealized final-stage microstructure and the sintering and grain growth equations derived for these idealized geometries. They also rely on the use of data for material parameters, such as

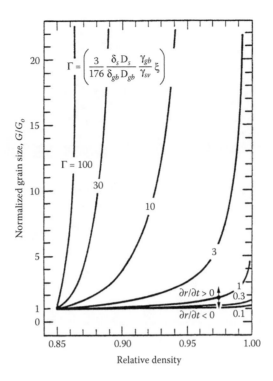

FIGURE 3.53 Grain size–density evolution during final-stage sintering for different values of the ratio of the coarsening rate to the densification rate Γ, assuming coarsening by surface diffusion and densification by grain boundary diffusion.

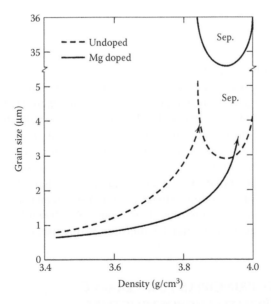

FIGURE 3.54 Grain size–density map for A_2O_3, illustrating the effect of raising the surface diffusion coefficient by a factor of 4, reducing the lattice diffusion coefficient by a factor of 2, and reducing the grain boundary mobility by a factor of 34. This has the effect of flattening the grain size–density trajectory and raising the separation region to larger grain sizes, thereby making it possible to sinter to full density. (Courtesy of M.P. Harmer.)

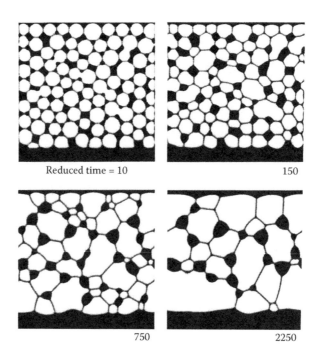

Reduced time = 10   150

750   2250

FIGURE 3.55 Simulated microstructural evolution during sintering of a sample in two dimensions using a continuum phase-field method. (From Wang, Y., Liu, Y., Ciobanu, C., and Patton, B.R., Simulating microstructural evolution and electrical transport in ceramic gas sensors, *J. Am. Ceram. Soc.* 83, 2219, 2000. With permission.)

diffusion coefficients, that may have a doubtful reliability or may have widely different values in the literature.

3.10.2 COMPUTER SIMULATIONS OF MICROSTRUCTURAL EVOLUTION

Computer simulation techniques are beginning to address some of the complexities of micro-structural evolution, but because of the computing power required to analyze three-dimensional powder compacts, the analyses consider simple models or two-dimensional models that have limited applicability to real systems. Pan, Cocks, and Kucherenko [101] have developed finite element formulations based on variational calculus to model microstructural evolution by a combination of grain boundary diffusion, surface diffusion, and grain boundary migration. The numerical techniques have been applied only to simple examples, such as thermal grooving and the morphological evolution of a two-dimensional array of equal-sized spheres adherent on a substrate. Wang and coworkers [102] used a continuum phase-field method [103] to model the microstructural evolution of a two-dimensional system of spherical particles in which the starting microstructure was simulated by a particle flow model. As shown in Figure 3.55, the images of the simulated microstructure can often be almost indistinguishable from those of real micro-structures.

3.11 FABRICATION PRINCIPLES FOR CERAMICS WITH CONTROLLED MICROSTRUCTURE

Most applications of ceramics require products with high density and controlled (small) grain size [104]. The discussion in the previous section indicates that when suitable processing procedures are employed, such an endpoint microstructure is achievable by fabrication routes that have the effect of increasing the ratio of the densification rate to the grain growth (coarsening) rate and avoiding the separation region (abnormal grain growth). The principles governing these fabrication routes are discussed here; the technology will be considered in Chapter 6.

Sintering with an external pressure — When compared to sintering, hot pressing produces an increase in the driving force for densification. For an equivalent microstructure, the dependence of the densification rate on the driving force can be written

$$\dot{\rho}_{hp} \sim (\Sigma + p_a); \qquad \dot{\rho}_s \sim \Sigma \tag{3.111}$$

where the subscripts *hp* and *s* refer to hot pressing and sintering, respectively; Σ is the sintering stress; and p_a is the applied pressure. Since the grain boundary mobility is unaltered,

$$\dot{G}_{hp} \approx \dot{G}_s \tag{3.112}$$

For $p_a \gg \Sigma$, we find that $(\dot{\rho}/\dot{G})_{hp} \gg (\dot{\rho}/\dot{G})_s$. According to this discussion, the effectiveness of hot pressing for the production of high density coupled with small grain size arises from the ability to increase $\dot{\rho}/\dot{G}$. For MgO-doped Al$_2$O$_3$, the measured trajectories for the grain size versus density are shown in Figure 3.56 for hot pressing and for sintering. The flatter trajectory for the case of hot pressing is consistent with the present discussion.

Use of dopants—The role of dopants in sintering is fairly complex, and we shall consider this topic further in Chapter 5. We found that dopants that segregate to the grain boundary can reduce the boundary mobility by the solute drag effect, so in this case $\dot{G}_{doped} < \dot{G}_{undoped}$. Dopants may also influence the densification process, but this effect is often small compared to the effect on the grain

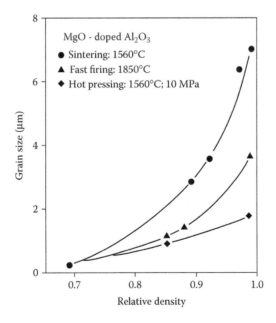

FIGURE 3.56 Experimental results for microstructural development in Al$_2$O$_3$ doped with 200 ppm MgO, showing the grain size-density trajectories for fabrication by hot pressing, conventional sintering, and fast firing. (Courtesy of M.P. Harmer.)

growth rate. Under these conditions, the effectiveness of dopants arises from the ability to reduce $\dot{G}$ or, equivalently, to increase $\dot{\rho}/\dot{G}$.

Use of fine inclusions — Fine, inert inclusions, as we observed earlier, exert a drag on the grain boundary and, if the drag is large enough, may pin the boundary. They often appear not to have any significant influence on the diffusional transport of matter leading to densification. As in the case of dopants, the effectiveness of inclusions can be interpreted in terms of a decrease in $\dot{G}$ or, equivalently, an increase in $\dot{\rho}/\dot{G}$.

Uniform packing of fine particles — Uniformly compacted fine powders have small pores with low coordination numbers (i.e., $N < N_c$). The densification rate for such a system is higher than for a similar system with heterogeneous packing [1]. Furthermore, if the particle size distribution is narrow, the driving force for grain growth due to the curvature of the boundary is small. The effectiveness of this route can therefore be interpreted in terms of an increase in $\dot{\rho}/\dot{G}$.

Controlling the temperature schedule — For powder systems in which the activation energy for densification is significantly greater than for grain growth (Figure 3.57), fast heat-up to a high enough temperature where $\dot{\rho} > \dot{G}$ can provide an effective fabrication route, often referred to as *fast firing*. For MgO-doped Al_2O_3, the grain size versus density trajectory during fast firing is compared with those for hot pressing and sintering in Figure 3.56. In this case, the benefit of fast firing over the more conventional sintering route is clear.

Liquid-phase sintering — A second phase that forms a liquid at the sintering temperature can provide a fast diffusion path for densification, but grain growth by the Ostwald ripening process may also be enhanced. In this case, high density is normally accompanied by appreciable grain growth. This commonly used fabrication approach is the subject of Chapter 4.

The separation region (abnormal grain growth) that we are also trying to avoid is a moving target that depends on the porosity. Another complication is that the models used for predicting the separation region assume an idealized geometry of uniform pore size and grain size. In practice, a range of pore sizes and grain sizes exist in any given microstructure, and abnormal grain growth can occur with only local separation of the pore from the boundary (i.e., it is not required for all

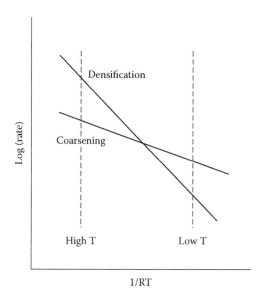

FIGURE 3.57 Under conditions where the densification mechanism has a higher activation energy than the coarsening mechanism, a fast heating rate to high firing temperatures (fast firing) can be beneficial for the achievement of high density.

the pores to separate, as assumed in many models). Generally, the sintering techniques that produce a reduction of the grain growth rate or an enhancement of the densification rate, coupled with forming methods that produce a homogeneous microstructure of fine pores and small interpore separation, are beneficial for achieving high density and fine grain size.

3.12 CONCLUDING REMARKS

In this chapter, we outlined the basic principles that govern microstructural development during solid-state sintering of polycrystalline ceramics. Simple models allow the derivation of equations for the kinetics of normal grain growth, but they commonly analyze a system consisting of an isolated grain boundary or a single grain and neglect the topological requirements of space filling. Computer simulations are playing an increasing role in exploring the interplay between the kinetics of grain growth and the topological requirements of space filling. They provide additional insight into the grain growth process. Abnormal grain growth cannot be explained in terms of the reduction of a uniform grain boundary energy. True abnormal grain growth is believed to occur when the local driving force is higher than that resulting from the geometry, or when the boundary mobility of the growing grain is higher than that of ordinary boundaries. A considerable gap exists between the theoretical understanding of microstructural evolution and the practical application to real powder systems. The analytical models provide useful guidelines for the control of the microstructure during sintering, whereas computer simulation techniques have started to address some of the complexities of microstructural evolution in simple models. High density coupled with controlled grain size is achievable during sintering by using techniques that decrease the rate of grain growth (coarsening) or increase the densification rate, coupled with the use of forming methods that produce a homogeneous green microstructure of fine pores with small interpore distances. Finally, if solid-state sintering is inadequate for achieving the desired endpoint density, then liquid-phase sintering, the subject of the next chapter, may provide an effective fabrication route.

PROBLEMS

3.1 Distinguish between normal and abnormal grain growth. Why is it important to control normal grain growth and avoid abnormal grain growth for the achievement of high density during sintering?

3.2 For a dense, pure polycrystalline ZnO in which the grain growth follows normal, parabolic kinetics, the average grain size after annealing for 120 min at 1200°C is found to be 5 μm. Annealing for 60 min at 1400°C gives an average grain size of 11 μm. If the average grain size at $t = 0$ is 2 μm, estimate what the average grain size will be after annealing for 30 min at 1600°C.

3.3 A dense, pure polycrystalline MgO has an average grain size of 1 μm, and the grain growth follows normal, parabolic kinetics. Estimate what the average grain size will be after 5 h annealing at 1700°C, given that the grain boundary diffusion coefficient $D_{gb} = 10^{-8}$ cm²/s, the grain boundary width $\delta_{gb} = 0.5$ nm, and the grain boundary energy $\gamma_{gb} = 0.5$ J/m².

If some SiO_2 is present, leading to the formation of a continuous, liquid film with a thickness of 50 nm at the grain boundaries, estimate the change in the boundary mobility if the diffusion coefficient through the liquid D_L is (i) equal to D_{gb} and (ii) equal to $100 D_{gb}$.

3.4 List and briefly explain the possible causes of abnormal grain growth.

3.5 Compare the grain growth phenomena that can occur in a dense polycrystalline solid and in a dense thin adherent polycrystalline film.

3.6 Consider the grain boundary between two spherical particles with approximately the same radius. In one case, both particles are single crystalline. In the other case, one particle is polycrystalline and the second particle single crystalline. Compare the grain growth phenomena that may be expected to occur in the two cases.

3.7   Compare the densification of a homogeneously packed powder compact of 5 μm single crystalline particles with the densification of a compact of 5 μm agglomerates consisting of 0.5 μm single crystalline particles. Assume the particles have the same chemical composition.

3.8   Compare the conditions that control the stability of a pore in a polycrystalline ceramic with those for a pore in a glass.

3.9   A dense Al_2O_3 ceramic (grain size of 1 μm) contains 10 vol% of fine ZrO_2 particles that are uniformly distributed at the grain boundaries. If the particle size of the ZrO_2 is 0.2 μm, and the grain boundary energy is 0.5 J/m^2, estimate the maximum pinning force that the ZrO_2 particles can exert on unit area of the grain boundary. If the ceramic is annealed at a high enough temperature (e.g., 1600°C), discuss the main features of the grain growth phenomena that may be expected to occur.

3.10 Derive the equation in Table 3.2 for the pore mobility when the pore migration is controlled by vapor transport.

3.11 According to the sintering equations, grain growth during sintering leads to a reduction of the densification rate. Can grain growth actually lead to densification of ceramic powder compacts? Explain your answer.

3.12 For a similar packing arrangement of the grains, will an increase in the dihedral angle enhance or retard pore elimination? Explain.

3.13 Consider a polycrystalline ceramic with pores of various sizes situated at the grain boundaries and within the grains, as sketched schematically below:

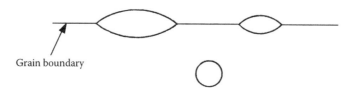

Grain boundary

(a) Use arrows to sketch the possible atomic transport paths from the grain boundaries to the pores.

(b) Which pore will disappear first? Which pore will disappear last? Explain your answer.

(c) If the three pores are close enough so that the diffusion distances between the pores are small, will any matter be transported between the pores? Use arrows to show the direction of matter transport and explain your answer.

REFERENCES

1. Rahaman, M.N., *Ceramic Processing*, CRC Press, Boca Raton, FL, 2006.
2. Brook, R.J., Controlled grain growth, in *Ceramic Fabrication Processes*, Treatise on Materials Science and Technology, Vol. 9, Wang, F.F.Y., Ed., Academic Press, New York, 1976, p. 331.
3. Kingery, W.D., Bowen, H.K., and Uhlmann, D.R., *Introduction to Ceramics*, 2nd ed., Wiley, New York, 1976.
4. Greenwood, G.W., The growth of dispersed precipitates in solutions, *Acta Metall.*, 4, 243, 1956.
5. Wagner, C., Theory of Ostwald ripening, *Z. Electrochem.*, 65, 581, 1961.
6. Lifshitz, I.M., and Slyozov, V.V., The kinetics of precipitation from supersaturated solutions, *Phys. Chem. Solids*, 19, 35, 1961.

7. Fischmeister, H., and Grimvall, G., Ostwald ripening — a survey, in *Sintering and Related Phenomena, Mater. Sci. Res.* Vol. 6, Kuczynski, G.C., Ed., Plenum Press, New York, 1973, p. 119.

8. Ardell, A.J., The effect of volume fraction on particle coarsening: theoretical considerations, *Acta Metall.*, 20, 61, 1972.

9. Davies, C.K.L., Nash, P., and Stevens, R.N., The effect of volume fraction of precipitate on Ostwald ripening, *Acta Metall.*, 28, 179, 1980.

10. Brailsford, A.D., and Wynblatt, P., The dependence of Ostwald ripening kinetics on particle volume fraction, *Acta Metall.*, 27, 489, 1979.

11. Enomoto, Y., Kawasaki, K., and Tokuyama, M., Computer modeling of Ostwald ripening, *Acta Metall.*, 35, 904, 1987.

12. Enomoto, Y., Kawasaki, K., and Tokuyama, M., The time dependent behavior of the Ostwald ripening for the finite volume fraction, *Acta Metall.*, 35, 915, 1987.

13. Smith, C.S., Grain shapes and other metallurgical applications of topology, in *Metal Interfaces*, ASM, Cleveland, OH, 1952, p. 65.

14. Atkinson, H.V., Theories of normal grain growth in pure single phase systems, *Acta Metall.*, 36, 469, 1988.

15. Burke, J.E., and Turnbull, D., Recrystallization and grain growth, *Progr. Metal Phys.*, 3, 220, 1952.

16. Hillert, M., On the theory of normal and abnormal grain growth, *Acta Metall.*, 13, 227, 1965.

17. Feltham, P., Grain growth in metals, *Acta metall.*, 5, 97, 1957.

18. Louat, N.P., On the theory of normal grain growth, *Acta Metall.*, 22, 721, 1974.

19. Rhines, F.N., and Craig, K.R., Mechanism of steady-state grain growth in aluminum, *Metall. Trans.* A, 5, 413, 1974.

20. Doherty, R.D., Discussion of "Mechanism of steady-state grain growth in aluminum," *Metall. Trans.* 6A, 588, 1975.

21. Weaire, D., and Kermode, J.P., Computer simulation of a two-dimensional soap froth, I. Method and motivation, *Phil. Mag.*, B48, 245, 1983.

22. Weaire, D., and Kermode, J.P., Computer simulation of a two-dimensional soap froth, II. Analysis of results, *Phil. Mag.*, B50, 379, 1984.

23. Anderson, M.P., Srolovitz, D.J., Grest, G.S., and Sahni, P.S., Computer simulation of grain growth — I. Kinetics, *Acta Metall.*, 32, 783, 1984.

24. Srolovitz, D.J., Anderson, M.P., Sahni, P.S., and Grest, G.S., Computer simulation of grain growth — II. Grain size distribution, topology, and local dynamics, *Acta Metall.*, 32, 793, 1984.

25. Srolovitz, D.J., Grest, G.S., and Anderson, M.P., Computer simulation of grain growth — V. Abnormal grain growth, *Acta Metall.*, 33, 2233, 1985.

26. Thompson, C.V., Frost, H.J., and Spaepen, F., The relative rates of secondary and normal grain growth, *Acta Metall.*, 35, 887, 1987.

27. Rollet, A.D., Srolovitz, D.J., and Anderson, M.P., Simulation and theory of abnormal grain growth — anisotropic grain boundary energies and mobilities, *Acta Metall.*, 37, 1227, 1989.

28. Yang, W., Chen, L., and Messing, G.L., Computer simulation of anisotropic grain growth, *Mater. Sci. Eng. A* 195, 179, 1995.

29. Kunaver, U., and Kolar, D., Three-dimensional computer simulation of anisotropic grain growth in ceramics, *Acta Mater.*, 46, 4629, 1998.

30. Harmer, M.P., A history of the role of MgO in the sintering of α-Al_2O_3, *Ceram. Trans.*, 7, 13, 1990.

31. Berry, K.A., and Harmer, M.P., Effect of MgO solute on microstructure development in Al_2O_3, *J. Amer. Ceram. Soc.*, 69, 143, 1986.

32. Handwerker, C.A., Dynys, J.M., Cannon, R.M., and Coble, R.L., Dihedral angles in magnesia and alumina: distributions from surface thermal grooves, *J. Am. Ceram. Soc.*, 73, 1371, 1990.

33. Bennison, S.J., and Harmer, M.P., Effect of MgO solute on the kinetics of grain growth in Al_2O_3, *J. Am. Ceram. Soc.,* 66, C90, 1983.

34. Horn, D.S., and Messing, G.L., Anisotropic grain growth in TiO_2-doped alumina, *Mater. Sci. Eng., A* 195, 169, 1995.

35. Powers, J.D., and Glaeser, A.M., Titanium effects on sintering and grain growth of alumina, in *Sintering Technology*, German, R.M., Messing, G.L., and Cornwall, R.G., Eds., Marcel Dekker, New York, 1996, p. 333.

36. Kaysser, W.A., Sprissler, M., Handwerker, C.A., and Blendell, J.E., Effect of liquid phase on the morphology of grain growth in alumina, *J. Am. Ceram. Soc.*, 70, 339, 1987.

37. Song, H., and Coble, R.L., Origin and growth kinetics of platelike abnormal grains in liquid-phase-sintered alumina, *J. Am. Ceram. Soc.*, 73, 2077, 1990.

38. Bateman, C.A., Benison, S.J., and Harmer, M.P., Mechanism for the role of MgO in the sintering of Al_2O_3 containing small amounts of a liquid phase, *J. Am. Ceram. Soc.*, 72, 1241, 1989.

39. Matsuzawa, S., and Mase, S., Method for producing a single crystal of ferrite, U.S. Patent No. 4,339,301 (1981).

40. Yamamoto, T., and Sakuma, T., Fabrication of barium titanate single crystals by solid-state grain growth, *J. Am. Ceram. Soc.*, 77, 1107, 1994.

41. Scott, C., Strok, J., and Levinson, L., Solid-state thermal conversion of polycrystalline alumina to sapphire using a seed crystal, U.S. Patent No. 5,549,746 (1996).

42. Li, T., Scotch, A.M., Chan, H.M., Harmer, M.P., Park, S.E., Shrout, T.E., and Michael, J.R., Single crystals of $Pb(Mg_{1/3}Nb_{2/3})O_3$–35 mol% $PbTiO_3$ from polycrystalline precursors, *J. Am. Ceram. Soc.*, 81, 244, 1998.

43. Sajgalik, P., Dusza, J., and Hoffmann, M.J., Relationship between microstructure, toughening mechanisms, and fracture toughness of reinforced silicon nitride ceramics, *J. Am. Ceram. Soc.*, 78, 2619, 1995.

44. Cao, J.J., Moberly-Chan, W.J., De Jonghe, L.C., Gilbert, C.J., and Ritchie, R.O., *In situ* toughened silicon carbide with Al-B-C additions, *J. Am. Ceram. Soc.*, 79, 461, 1996.

45. Huang, T., Rahaman, M.N., Mah, T.I., and Parthasarathay, T.A., Effect of SiO_2 and Y_2O_3 additives on the anisotropic grain growth of dense mullite, *J. Mater. Res.*, 15, 718, 2000.

46. Patwardhan, J.S., and Rahaman, M.N., Compositional effects on densification and microstructural evolution of bismuth titanate, *J. Mater. Sci.*, 39, 133, 2004.

47. Kwon, S., Sabolsky, E.M., and Messing, G.L., Control of ceramic microstructure by templated grain growth, in *Handbook of Advanced Ceramics*, Vol. 1. Materials Science, Smiya, S., Aldinger, F., Claussen, N., Spriggs, R.M., Uchino, K., Koumoto, K., and Kaneno, M., Eds., Elsevier, New York, 2003, p. 459.

48. Horn, J.A., Zhang, S.C., Selvaraj, U., Messing, G.L., and Trolier-McKinstry, S., Templated grain growth of textured bismuth titanate, *J. Am. Ceram. Soc.*, 82, 921, 1999.

49. Beck, P.A., Holtzworth, M.L., and Sperry, P.R., Effect of a dispersed phase on grain growth in Al–Mn alloys, *Trans. AIME*, 180, 163, 1949.

50. Palmer, J.E., Thompson, C.V., and Smith, H.I., Grain growth and grain size distributions in thin germanium films, *J. Appl. Phys.*, 62, 2492, 1987.

51. Mullins, W.W., The effect of thermal grooving on grain boundary motion, *Acta Metall.*, 6, 414, 1958.

52. Frost, H.J., Thompson, C.C., and Walton, D.T., Simulation of thin film grain structures — I. Grain growth stagnation, *Acta Metall. Mater.*, 38, 1455, 1990.

53. Frost, H.J., Thompson, C.C., and Walton, D.T., Simulation of thin film grain structures — II. Abnormal grain growth, *Acta Metall. Mater.*, 40, 779, 1992.

54. Carel, R., Thompson, C.V., and Frost, H.J., Computer simulation of strain energy effects vs. surface and interface effects on grain growth in thin films, *Acta Mater.*, 44, 2479, 1996.

55. Yan, M.F., Cannon, R.M., and Bowen, H.K., Grain boundary migration in ceramics, in *Ceramic Microstructures '76*, Fulrath, R.M., and Pask, J.A., Eds., Westview Press, Boulder, CO, 1977, p. 276.

56. Smith, C.S., Grains, phases, and interfaces: an interpretation of microstructure, *Trans. AIME*, 175, 15, 1948.

57. Gladman, T., On the theory of the effect of precipitate particles on grain growth in metals, *Proc. Roy. Soc. London A*, 294, 298, 1966.

58. Haroun, N.A., Theory of inclusion-controlled grain growth, *J. Mater. Sci. Lett.*, 15, 2816, 1980.

59. Louat, N., The inhibition of grain boundary motion by a dispersion of particles, *Philos. Mag. A*, 47, 903, 1983.

60. Nes, E., Ryum, N., and Hunderi, O., On the Zener drag, *Acta Metall.*, 33, 11, 1985.

61. Srolovitz, D.J., Anderson, M.P., Grest, G.S., and Sahni, P.S., Computer simulation of grain growth — III. Influence of a particle dispersion, *Acta Metall.*, 32, 1429, 1984.

62. Doherty, R.D., Srolovitz, D.J., Rollett, A.D., and Anderson, M.P., On the volume fraction dependence of particle-inhibited grain growth, *Scripta Metall.*, 21, 675, 1987.

63. Anderson, M.P., Grest, G.S., Doherty, R.D., Li, K., and Srolovitz, D.J., Inhibition of grain growth by second-phase particles: three-dimensional Monte Carlo simulations, *Scripta Metall.*, 23, 753, 1989.

64. Lange, F.F., and Hirlinger, M., Hindrance of grain growth in Al_2O_3 by ZrO_2 inclusions, *J. Am. Ceram. Soc.*, 67, 164, 1984.

65. Hori, S., Kurita, R., Yoshimura, M., and Somiya, S., Suppressed grain growth in final-stage sintering of Al_2O_3 with dispersed ZrO_2 particles, *J. Mater. Sci. Lett.*, 4, 1067, 1985.

66. Lange, F.F., Yamaguchi, T., Davis, B.I., and Morgan, P.E.D., Effect of ZrO_2 inclusions on the sinterability of Al_2O_3, *J. Am. Ceram. Soc.*, 71, 446, 1988.

67. Lange, F.F., and Hirlinger, M., Grain growth in two-phase ceramics: Al_2O_3 inclusions in ZrO_2, *J. Am. Ceram. Soc.*, 70, 827, 1987.

68. Stearns, L.C., and Harmer, M.P., Particle-inhibited grain growth in Al_2O_3-SiC: I, experimental results, *J. Am. Ceram. Soc.*, 79, 3013, 1996.

69. Stearns, L.C., and Harmer, M.P., Particle-inhibited grain growth in Al_2O_3-SiC: II, equilibrium and kinetic analyses, *J. Am. Ceram. Soc.*, 79, 3020, 1996.

70. Cahn, J.W., The impurity drag effect in grain boundary motion, *Acta Metall.*, 10, 789, 1962.

71. Lücke, K., and Stüwe, H.P., On the theory of impurity controlled grain boundary migration, *Acta Metall.*, 19, 1087, 1971.

72. Hillert, M., and Sundman, B., The treatment of the solute drag on moving grain boundaries and phase interfaces in binary alloys, *Acta Metall.*, 24, 731, 1976.

73. Glaeser, A.M., Bowen, H.K., and Cannon, R.M., Grain-boundary migration in LiF: I, mobility measurements, *J. Am. Ceram. Soc.*, 69, 119, 1986.

74. Chen, P.-L., and Chen, I.-W., Role of defect interaction in boundary mobility and cation diffusivity of CeO_2, *J. Am. Ceram. Soc.*, 77, 2289, 1994.

75. Rahaman, M.N., and Zhou, Y.-C., Effect of dopants on the sintering of ultra-fine CeO_2 powder, *J. Europ. Ceram. Soc.*, 15, 939, 1995.

76. Coble, R.L., Transparent alumina and method of preparation, U.S. Patent No. 3,026,210 (1962).

77. Rahaman, M.N., and Manalert, R., Grain boundary mobility of $BaTiO_3$ doped with aliovalent cations, *J. Europ. Ceram. Soc.*, 18, 1063, 1998.

78. Brook, R.J., Tuan, W.H., and Xue, L.A., Critical issues and future directions in sintering science, *Ceramic Trans.*, 1, 811, 1988.

79. Rahaman, M.N., De Jonghe, L.C., Voight, J.A., and Tuttle, B.A., Low temperature sintering of zinc oxide varistors, *J. Mater. Sci.*, 25, 737, 1990.

80. Jorgensen, P.J., and Anderson, R.C., Grain boundary segregation and final stage sintering of Y_2O_3, *J. Am. Ceram. Soc.*, 50, 553, 1967.

81. Prochazka, S., and Scanlan, R.M., Effect of boron and carbon on sintering of SiC, *J. Am. Ceram. Soc.*, 58, 72, 1975.

82. Yan, M.F., Microstructural control in the processing of electronic ceramics, *Mater. Sci. Eng.*, 48, 53, 1981.

83. Gupta, T.K., Possible correlation between density and grain size during sintering, *J. Am. Ceram. Soc.*, 55, 276, 1972.

84. Kingery, W.D., and Francois, B., Grain growth in porous compacts, *J. Am. Ceram. Soc.*, 48, 546, 1965.

85. Coleman, S.C., and Beeré, W.B., The sintering of open and closed porosity in UO_2, *Philos. Mag.*, 31, 1403, 1975.

86. Rhines, F., and DeHoff, R., Channel network decay in sintering, in *Sintering and Heterogeneous Catalysis*, *Mater. Sci. Res.*, Vol. 16, Kuczynski, G.C., Miller, A.E., and Sargent, G.A., Eds., Plenum Press, New York, 1984, p. 49.

87. Kingery, W.D., and Francois, B., Sintering of crystalline oxides, I. Interactions between grain boundaries and pores, in *Sintering and Related Phenomena*, Kuczynski, G.C., Hooton, N.A., and Gibbon, G.F., Eds., Gordon & Breach, New York, 1967, p. 471.

88. Kellet, B.J., and Lange, F.F., Thermodynamics of densification: I, sintering of simple particle arrays, equilibrium configurations, pore stability, and shrinkage, *J. Am. Ceram. Soc.*, 72, 725, 1989.

89. Pan, J., Ch'ng, H.N., and Cocks, A.C.F., Sintering kinetics of large pores, *Mech. Mater.*, 37, 705, 2004.

90. Greskovich, C., and Lay, K.W., Grain growth in very porous Al_2O_3 compacts, *J. Am. Ceram. Soc.*, 55, 142, 1972.

91. Ch'ng, H.N., and Pan, J., Cubic spline elements for modeling microstructural evolution of materials controlled by solid-state diffusion and grain boundary migration, *J. Comput. Phys.*, 196, 724, 2004.

92. Edelson, L.H., and Glaeser, A.M., Role of particle substructure in the sintering of monosized titania, *J. Am. Ceram. Soc.*, 71, 225, 1988.

93. Shewmon, P.G., Movement of small inclusions in solids by a temperature gradient, *Trans. AIME*, 230, 1134, 1964.

94. Brook, R.J., Pore–grain boundary interactions and grain growth, *J. Am. Ceram. Soc.*, 52, 56, 1969.

95. Handwerker, C.A., Cannon, R.M., and Coble, R.L., Final stage sintering of MgO, *Adv. Ceram.*, 10, 619, 1984.

96. Carpay, F.M.A., Discontinuous grain growth and pore drag, *J. Am. Ceram. Soc.*, 60, 82, 1977.

97. Hsueh, C.H., Evans, A.G., and Coble, R.L., Microstructure development during final/initial stage sintering: I, pore/grain boundary separation, *Acta Metall.*, 30, 1269, 1982.

98. Sakarcan, M., Hsueh, C.H., and Evans, A.G., Experimental assessment of pore breakaway during sintering, *J. Am. Ceram. Soc.*, 66, 456, 1983.

99. Yan, M.F., Cannon, R.M., and Bowen, H.K., Effect of grain size distribution on sintered density, *Mater. Sci. Eng.*, 60, 275, 1983.

100. Bennison, S.J., and Harmer, M.P., Effect of magnesia solute on surface diffusion in sapphire and the role of magnesia in the sintering of alumina, *J. Am. Ceram. Soc.*, 73, 833, 1990.

101. Pan, J., Cocks, A.C.F., and Kucherenko, S., Finite element formulation of grain boundary and surface diffusion with grain boundary migration, *Proc. Roy. Soc. London A*, 453, 2161, 1997.

102. Wang, Y., Liu, Y., Ciobanu, C., and Patton, B.R., Simulating microstructural evolution and electrical transport in ceramic gas sensors, *J. Am. Ceram. Soc.*, 83, 2219, 2000.

103. Chen, L.Q., and Wang, Y., The continuum field approach to modeling microstructural evolution, *JOM*, 48, 13, 1996.

104. Brook, R.J., Fabrication principles for the production of ceramics with superior mechanical properties, *Proc. Brit. Ceram. Soc.*, 32, 7, 1982.

4 Liquid-Phase Sintering

4.1 INTRODUCTION

Our discussion of the sintering process has so far been concerned with solid-state sintering, in which the material remains entirely in the solid state. In many ceramic systems, the formation of a liquid phase is commonly used to assist in the sintering and microstructural evolution. Usually the purpose of *liquid-phase sintering* is to enhance densification rates, to achieve accelerated grain growth, or to produce specific grain boundary properties. The distribution of the liquid phase and of the resulting solidified phases produced on cooling after densification is critical to achieving the required properties of the sintered material. Commonly, the amount of liquid formed during sintering is small, typically less than a few volume percent (vol%), which can make precise control of the liquid composition difficult. In some systems, such as Al_2O_3, the amount of liquid phase can be very small indeed and so difficult to detect that many studies that were believed to involve solid-state sintering actually involved liquid silicate phases, as later revealed by careful high-resolution transmission electron microscopy.

Liquid-phase sintering is particularly effective for ceramics such as Si_3N_4 and SiC that have a high degree of covalent bonding and are, therefore, difficult to densify by solid-state sintering. The process is also important when the use of solid-state sintering is too expensive or requires too high a fabrication temperature. However, the enhanced densification rates achieved by liquid-forming additives are of interest only if the properties of the fabricated ceramic remain within the required limits. A disadvantage of liquid-phase sintering is that the liquid phase used to promote sintering commonly remains as a glassy intergranular phase that may degrade high-temperature mechanical properties such as creep and fatigue resistance.

Some examples of ceramic liquid-phase sintering systems and their applications are given in Table 4.1. For consistency, we use the following nomenclature to describe the systems in liquid-phase sintering: the particulate solid forming the major component is written first, and the liquid-producing component is written in parentheses. In this nomenclature, Si_3N_4 sintered in the presence of a liquid phase produced by the addition of MgO is written Si_3N_4(MgO).

A related process is *activated sintering,* in which minor amounts of additives that segregate strongly to the grain boundaries can significantly enhance mass transport rates along the grain boundary, giving rise to accelerated densification, even at temperatures well below that for liquid formation in the system. In many systems, there is no clear difference in principles between activated sintering and liquid-phase sintering, except that for the activated system, the amount of additive is small, so that the presence of a liquid grain boundary film can be difficult to detect.

If sufficient liquid is present (on the order of 25–30 vol%), rearrangement of the solid phase coupled with liquid flow can lead to a fully dense material, without the need for contributions from other processes. Such large volume fractions of liquid are commonly used in traditional, clay-based ceramics such as porcelains and in cemented carbides. In the traditional ceramics, the liquid phases are molten silicates that remain as a glassy phase after cooling, giving the fabricated materials a glassy appearance. The ceramics are referred to as *vitrified*, and the sintering process is referred to as *vitrification.*

TABLE 4.1
Examples of Common Ceramic Liquid-Phase Sintering Systems

Ceramic System	Additive Content (wt%)	Application	Ref.
Al_2O_3(talc)	~5	Electrical insulators	1
$MgO(CaO-SiO_2)$	<5	Refractories	2
MgO(LiF)	<3	Refractories	3
$ZnO(Bi_2O_3)$	2-3	Electrical varistors	4, 5
$BaTiO_3(TiO_2)$	<1	Dielectrics	6
$BaTiO_3$(LiF)	<3	Dielectrics	7, 8
$UO_2(TiO_2)$	~1	Nuclear ceramics	9
$ZrO_2(CaO-SiO_2)$	<1	Ionic conductors	10
Si_3N_4(MgO)	5–10	Structural ceramics	11, 12
$Si_3N_4(Y_2O_3-Al_2O_3)$	5–10	Structural ceramics	12, 13
$SiC(Y_2O_3-Al_2O_3)$	5–10	Structural ceramics	14
WC(Ni)	~10	Cutting tools	15

4.2 ELEMENTARY FEATURES OF LIQUID-PHASE SINTERING

4.2.1 ENHANCEMENT OF DENSIFICATION

Compared to solid-state sintering, the presence of the liquid phase leads to enhanced densification through enhanced rearrangement of the particulate solid and enhanced matter transport through the liquid. Figure 4.1 shows a sketch of an idealized two-sphere model in which the microstructural aspects of liquid-phase sintering are compared with those of solid-state sintering. In liquid-phase sintering, if, as we assume, the liquid wets and spreads to cover the solid surfaces, the particles will be separated by a liquid bridge. The friction between the particles is significantly reduced, so they can rearrange more easily under the action of the compressive capillary stress exerted by the liquid. Once a quasi-steady-state grain boundary film is established, densification proceeds similar to solid-state sintering at a comparatively enhanced rate. In solid-state sintering by, for example, grain boundary diffusion, a key parameter that controls the rate of diffusion is the product of the grain boundary diffusion coefficient D_{gb} and the grain boundary thickness δ_{gb}. In liquid-phase sintering, the corresponding parameter is the product of the diffusion coefficient D_L of the solute atoms in the liquid and the thickness of the liquid bridge, δ_L. Since δ_L is typically many times greater than δ_{gb}, and diffusion through a liquid is much faster than in solids, the liquid therefore provides a path for enhanced matter transport. Rapid transport in the liquid also makes grain growth more prominent in liquid-phase sintering. Inclusion of a dispersion of fine, inert particles can assist in limiting grain growth.

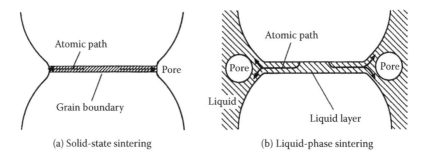

FIGURE 4.1 Sketch of an idealized two-sphere model comparing the microstructural aspects of (a) solid-state sintering with (b) liquid-phase sintering.

4.2.2 Driving Force for Densification

Assuming as above that the liquid wets and spreads over the solid surfaces, the solid–vapor interface of the particulate system will be eliminated and pores will form in the liquid. The reduction of the liquid–vapor interfacial area provides a driving force for shrinkage (densification) of the system. For a spherical pore of radius r in a liquid, the pressure difference across the curved surface is given by the equation of Young and Laplace:

$$p = -\frac{2\gamma_{lv}}{r} \tag{4.1}$$

where γ_{lv} is the specific surface energy of the liquid–vapor interface. The pressure in the liquid is lower than that in the pore, and this generates a compressive *capillary stress* on the particles. This compressive stress due to the liquid is equivalent to placing the system under an external hydrostatic pressure, the magnitude of which is given by Equation 4.1. Taking $\gamma_{lv} \approx 1$ J/m^2 and $r \approx 0.5$ μm gives $p \approx 4$ MPa. Pressures of this magnitude can provide an appreciable driving force for sintering.

4.2.3 Formation of the Liquid Phase

For liquid-phase sintering, the green body is commonly formed from a mixture of two powders: a major component and an additive phase. On heating, the additive either melts or reacts with a small part of the major component to form a *eutectic liquid*. The formation of the liquid phase by melting of the additive is common in metallic systems, e.g., Fe(Cu) and W(Ni), whereas in ceramic systems the formation of a eutectic liquid is more common, e.g., MgO(CaO-SiO$_2$) and ZnO(Bi$_2$O$_3$). For systems that rely on the formation of a eutectic liquid, *phase diagrams* play a key role in the selection of the additive and in the choice of the sintering conditions. Despite the presence of a viscous liquid between the particles, the structure does not collapse unless the volume of liquid is very large. The relatively large capillary stress exerted by the liquid holds the solid particles together. The effective viscosity of the system is, however, much lower than that of a similar system without the liquid phase.

In most systems, the liquid persists throughout the sintering process, and its volume does not change appreciably. This situation is sometimes referred to as *persistent liquid-phase sintering*. On cooling, the liquid commonly forms a glassy grain boundary phase, which, as outlined earlier, may degrade the high-temperature mechanical properties. In a small number of systems, the liquid may be present over a major portion of the sintering process but then essentially disappears by incorporation into the solid phase to produce a *solid solution*, e.g., Si$_3$N$_4$(Al$_2$O$_3$-AlN); crystallization of the liquid, e.g., Si$_3$N$_4$(Al$_2$O$_3$-Y$_2$O$_3$); or evaporation, e.g., BaTiO$_3$(LiF). The term *transient liquid-phase sintering* is used to describe the sintering in which the liquid phase disappears prior to the completion of sintering. The interest in ceramics for mechanical engineering applications at high temperatures has led to the investigation of transient liquid-phase sintering in a few Si$_3$N$_4$ systems. However, the enhanced sensitivity of the system to the process variables makes microstructural control difficult in most cases. In this book, the term liquid-phase sintering will refer most generally to the case of a persistent liquid. A distinction between persistent and transient liquid-phase sintering will be made only when it is required.

4.2.4 Microstructures

In addition to any porosity that may be present, the microstructures of the ceramics produced by liquid-phase sintering consist of two phases: (1) the crystalline grains and (2) the grain boundary phase resulting from the solidified liquid. Unless it is crystallizable, the grain boundary phase is commonly amorphous. As discussed later, depending on the interfacial tensions, liquid phase may penetrate the grain boundaries completely, in which case the grains will be separated from one another by a thin layer (≈ 1 nm to several microns), or the liquid phase may only partially penetrate the grain boundaries, in which case solid–solid contacts will exist between neighboring grains.

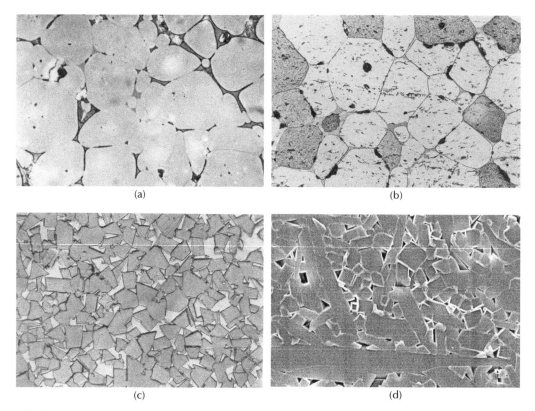

FIGURE 4.2 Commonly observed microstructures of ceramics produced by liquid-phase sintering. (a) Rounded grains in a moderate amount of liquid (> 5 vol%); (b) grains with flat contact surfaces in a low volume fraction of liquid (<2–5 vol%); (c) prismatic grains dictated by anisotropic interfacial energy with a moderate-to-high liquid volume (>10 vol%); (d) elongated grains with flat sides resulting from anisotropic interfacial energy with a low liquid content (<2–5 vol%).

Depending on the composition of the particulate solid and the liquid phase, a variety of grain shapes, ranging from nearly equiaxial grains to elongated grains with curved sides or straight (faceted) sides, are observed. Here we describe only a few typical examples. For systems with isotropic interfacial energies, when the amount of liquid is moderate (above ≈5 vol%), grains with fairly rounded shapes are observed (Figure 4.2a), whereas for higher liquid content, the grain shape becomes almost spheroidal. For low volume fraction of liquid (<2–5 vol%), the grains undergo considerable changes in shape and develop a morphology in which the contact regions between neighboring grains are relatively flat (Figure 4.2b). The shape changes allow the grains to pack more efficiently, a phenomenon usually described as *grain shape accommodation*. Anisotropic abnormal grain growth can occur in systems with nonisotropic interfacial energies (see Chapter 3). The grains may take up a prismatic shape when the liquid content is high (Figure 4.2c), whereas at lower liquid content elongated grains with curved sides or plate-like grains with straight (faceted) sides may be observed (Figure 4.2d).

4.3 STAGES OF LIQUID-PHASE SINTERING

In most liquid-phase sintering systems, chemical reactions between the particulate solid and the liquid are relatively weak, so that the interfacial energies have a dominant effect on the rate of sintering. Under these conditions, as illustrated in Figure 4.3, liquid-phase sintering is generally regarded as proceeding in a sequence of three dominant stages [16,17]:

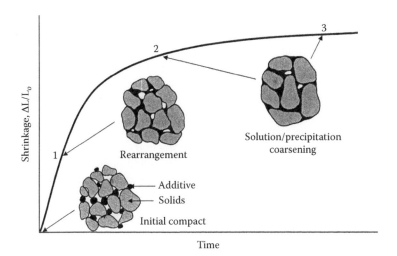

FIGURE 4.3 Schematic evolution of a powder compact during liquid-phase sintering. The three dominant stages overlap significantly.

Redistribution of the liquid and rearrangement of the particulate solid under the influence of capillary stress gradients
Densification and grain growth by solution-precipitation
Final-stage sintering dominated by Ostwald ripening

As the temperature of the powder compact is raised, solid-state sintering may occur prior to the formation of the liquid, producing significant densification in some systems, for example, those in which the major phase has a fine particle size. Assuming good wetting between the liquid and the particulate solid, further densification occurs as a result of the capillary force exerted by the liquid on the particles. The particles shrink as solid dissolves in the liquid, and they rapidly rearrange to produce a higher packing density, releasing liquid to fill pores between the particles. Capillary stresses will cause the liquid to redistribute itself between the particles and into the small pores, leading to further rearrangement. Contact points between agglomerates will be dissolved due to their higher solubility in the liquid, and the fragments will also undergo rearrangement. Throughout the process, dissolution of sharp edges will make the particle surfaces smoother, reducing the interfacial area and aiding the rearrangement of the system. Initially, rearrangement occurs rapidly, but, as densification occurs, the viscosity of the system increases, causing the densification rate to decrease continuously.

As densification by rearrangement slows, effects dependent on the solid solubility in the liquid and the diffusivity in the liquid dominate, giving the second stage, termed *solution-precipitation*. The solid dissolves at the solid–liquid interfaces with a higher chemical potential, diffuses through the liquid, and precipitates on the particles at other sites with a lower chemical potential. One type of dissolution site is the wetted contact area between the particles, where the capillary stress due to the liquid, or an externally applied stress, leads to a higher chemical potential. Precipitation occurs at sites away from the contact area. For systems with a distribution of particle sizes, matter can also be transported from the small particles to the large particles by diffusion through the liquid, a process described as Ostwald ripening (see Chapter 3). The net result is a coarsening of the microstructure. Densification by the solution-precipitation mechanism is accompanied by changes in the shape of the grains. When the amount of liquid is fairly large, the grains normally take a rounded shape (Figure 4.2a). For a small amount of liquid, the grains develop flat faces and assume the shape of a polyhedron to achieve more efficient packing (grain shape accommodation; Figure 4.2b).

The final stage of liquid-phase sintering is controlled by the densification of the solid particulate skeletal network. The process is slow because of the large diffusion distances in the coarsened structure and the rigid skeleton of contacting solid grains. Ostwald ripening dominates the final stage, and the residual pores become larger if they contain trapped gas, leading to compact swelling. Coarsening is accompanied by grain shape accommodation. Liquid may be released from the more efficiently packed regions, and it may flow into the isolated pores, leading to densification.

In the stages of liquid-phase sintering summarized in Figure 4.3, the extent to which each stage influences densification is dependent on the volume fraction of liquid, so there are many variants in this conceptual picture. When the volume fraction of liquid is high, complete densification can be achieved through the rearrangement process alone. On the other hand, at the low liquid contents common for many systems, the solid skeleton inhibits densification, so that solution–precipitation and final-stage sintering are required to achieve further densification.

4.4 THERMODYNAMIC AND KINETIC FACTORS

Production of the required microstructure in liquid-phase sintering depends on several kinetic and thermodynamic factors, as well as several processing parameters to be described later.

4.4.1 WETTING AND SPREADING OF THE LIQUID

Good wetting of the solid by the liquid is a fundamental requirement for liquid-phase sintering. It is generally found that liquids with a low surface tension readily wet most solids, giving a low contact angle, whereas liquids with a high surface tension show poor wetting with a large contact angle (Figure 4.4). At a molecular level, if the cohesion between the liquid molecules is greater than the adhesion between the liquid and the solid, the liquid will not tend to wet the solid.

The degree of wetting is characterized by the contact angle θ, which depends on the various interfacial energies for the solid–liquid–vapor system; it is usually discussed in terms of a droplet

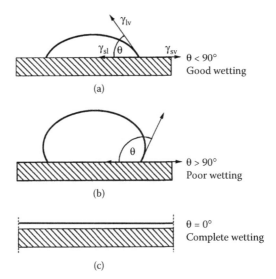

FIGURE 4.4 Wetting behavior between a liquid and a solid showing (a) good wetting, (b) poor wetting, and (c) complete wetting, for a liquid with a contact angle of θ.

of liquid on a flat solid surface (Figure 4.4). If the specific energies of the liquid–vapor, solid–vapor, and solid–liquid interfaces are γ_{lv}, γ_{sv}, and γ_{sl}, respectively, then by the principle of virtual work

$$\gamma_{sv} = \gamma_{sl} + \gamma_{lv} \cos \theta \qquad (4.2)$$

This equation, derived by Young and by Dupré, is what we would obtain by taking the horizontal components of the interfacial tensions. The surface energies of many inorganic melts, such as silicates, are often in the range of 0.1–0.5 J/m^2, with a value of ~0.3 J/m^2 commonly cited for molten silicates, and the surface energies of liquid metals and metal oxides are as high as 2 J/m^2. The change in surface energy with temperature for melts such as silicates is not well documented. Careful measurements in the range of 1300°C–1500°C indicate that the surface energy of a calcium aluminosilicate melt (62 wt% SiO_2, 15 wt% Al_2O_3, 23 wt% CaO) increases slowly with temperature and can be represented by

$$\gamma_{lv} = 0.293 + 0.67 \times 10^{-4}(T - 273.2) \qquad (4.3)$$

where γ_{lv} is in J/m^2 and T is the absolute temperature [18]. Compositional changes also modify the surface energy. Table 4.2 gives the interfacial energies of several materials.

TABLE 4.2
Measured Solid–Vapor, Liquid–Vapor, and Solid–Liquid Interfacial Energies for Various Materials

Material	Temperature (°C)	Interfacial Energy (J/m^2)
Solids		
Al_2O_3	1850	0.905
MgO	25	1.000
TiC	1100	1.190
NaCl (100)	25	0.300
0.20 Na_2O–0.80 SiO_2	1350	0.380
Cu	1080	1.430
Ag	750	1.140
Fe (γ phase)	1350	2.100
Liquids		
Water	25	0.072
Al_2O_3	2080	0.700
MgO	2800	0.660
B_2O_3	900	0.080
Bi_2O_3	825	0.213
FeO	1420	0.585
0.13 Na_2O–0.13 CaO–0.74 SiO_2	1350	0.350
0.15 Al_2O_3–0.23 CaO–0.62 SiO_2	1400	0.387
0.30 FeO–0.21 CaO–0.49 SiO_2	1400	0.396
Cu	1120	1.270
Ag	1000	0.920
Fe	1535	1.880
Solid–Liquid		
$Al_2O_3(s)$–silicate glaze(*l*)	1000	<0.700
$Al_2O_3(s)$–Ag(*l*)	1000	1.770
$Al_2O_3(s)$–Fe(*l*)	1570	2.300
MgO(*s*)–Ag(*l*)	1300	0.850
MgO(*s*)–Fe(*l*)	1725	1.600
SiO_2(*glass*)–Cu(*l*)	1120	1.370

Source: Kingery, W.D., Bowen, H.K., and Uhlmann, D.R., *Introduction to Ceramics*, 2nd ed., Wiley, New York, 1976.

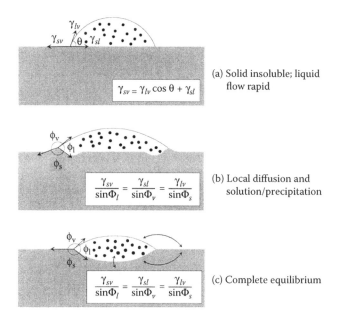

$$\gamma_{sv} = \gamma_{lv} \cos\theta + \gamma_{sl}$$

(a) Solid insoluble; liquid
flow rapid

$$\frac{\gamma_{sv}}{\sin\Phi_l} = \frac{\gamma_{sl}}{\sin\Phi_v} = \frac{\gamma_{lv}}{\sin\Phi_s}$$

(b) Local diffusion and
solution/precipitation

$$\frac{\gamma_{sv}}{\sin\Phi_l} = \frac{\gamma_{sl}}{\sin\Phi_v} = \frac{\gamma_{lv}}{\sin\Phi_s}$$

(c) Complete equilibrium

FIGURE 4.5 Shape changes associated with partial solubility of the solid in the liquid. (From Cannon, R.M., Saiz, E., Tomsia, A.P., and Carter, W.C., Reactive wetting taxonomy, *Mater. Res. Soc. Symp. Proc.*, 357, 279, 1995. With permission.)

The geometries shown in Figure 4.4 ignore the complication that for effective liquid-phase sintering, the solid must have some solubility in the liquid. This solubility effect leads to detailed wetting geometries or to liquid distribution that requires modification of the Young and Dupré equation. Complete equilibrium must include a force-balance in both dimensions as well as the attainment of constant curvature surfaces. These effects have been examined by Cannon et al. [19], and Figure 4.5 shows some of the configurational changes when the solubility of the solid in the liquid is taken into account.

Spreading refers to the kinetic process in which the liquid distributes itself to cover the surfaces of the particulate solid. It is important in the rearrangement stage soon after the formation of the liquid. For spreading to occur, the total interfacial energy must be reduced. For an infinitesimal change in the contact area between the solid and the liquid (Figure 4.4), spreading will occur if

$$\gamma_{lv} + \gamma_{sl} - \gamma_{sv} = 0 \tag{4.4}$$

Thus a spreading liquid has contact angle of zero.

4.4.2 DIHEDRAL ANGLE

Consider a liquid in contact with the corners of the grains. As sketched in Figure 4.6 for the two-dimensional situation, grooves are formed where the grain boundaries intersect the surface of the liquid. The dihedral angle is defined as the angle between the solid–liquid interfacial tensions. Applying a force balance and rearranging, we obtain

$$\cos\left(\frac{\psi}{2}\right) = \frac{\gamma_{ss}}{2\gamma_{sl}} \tag{4.5}$$

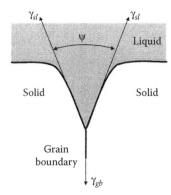

FIGURE 4.6 Dihedral angle Ψ for a liquid at a grain boundary.

The solid–solid interfacial tension, γ_{ss}, is the same as the interfacial tension in the grain boundary, γ_{gb}, defined earlier in the discussion of solid-state sintering.

4.4.2.1 Liquid Penetration of the Grain Boundary

The variation of the dihedral angle, ψ, as a function of the ratio γ_{ss}/γ_{sl} is sketched in Figure 4.7. For $\gamma_{ss}/\gamma_{sl} < 2$, the dihedral angle has values between 0° and 180°, and the liquid does not completely penetrate the grain boundary. In particular, when the ratio γ_{ss}/γ_{sl} is small, liquid penetration of the grain boundary is limited. In this case, rearrangement of the solid network is difficult, and solid-state processes can be significant. The condition $\gamma_{ss}/\gamma_{sl} = 2$ represents the limiting condition for complete penetration of the grain boundary by the liquid. For $\gamma_{ss}/\gamma_{sl} > 2$, no value of ψ satisfies Equation 4.5, and this represents the condition for complete penetration of the grain boundary. A consequence of this complete penetration of the grain boundary is the lack of rigidity of the system. Physically, the condition $\gamma_{ss}/\gamma_{sl} > 2$ means that the sum of the specific energy associated with the two solid–liquid interfaces is less than that for the solid-solid interface, so penetration of the boundary leads to an overall decrease in energy.

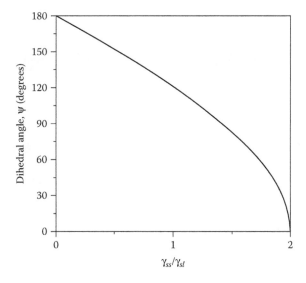

FIGURE 4.7 Variation of the dihedral angle Ψ with the ratio of the solid–solid interfacial tension γ_{ss} to the solid–liquid interfacial tension γ_{sl}.

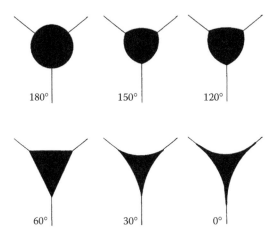

FIGURE 4.8 Effect of the dihedral angle on the idealized shape (in two dimensions) of the liquid at the corners of three grains.

4.4.2.2 Shape of the Liquid and the Grains

The dihedral angle affects the shapes of the grains and the liquid. If it is assumed that no porosity exists in the structure, then the equilibrium shape of the liquid phase can be calculated [20,21]. Much earlier, Smith [22] showed that the equilibrium distribution of second phases in a granular structure can be explained on the assumption that at three-grain junctions, the interfacial tensions should be in a state of balance. In two dimensions, Figure 4.8 shows in an idealized form the shapes a small volume of liquid phase must have if it appears at the corners of three grains.

To obtain a more realistic idea of the shape of the liquid phase in the structure, we must consider the three-dimensional situation (Figure 4.9). For $\psi = 0$, the liquid completely penetrates

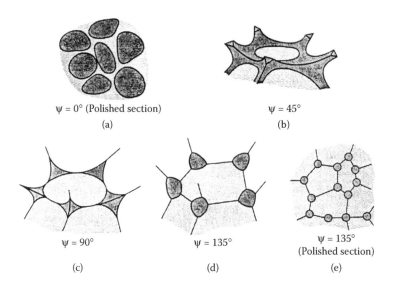

FIGURE 4.9 Idealized liquid-phase distribution (in three dimensions) for selected values of the dihedral angle. (From Kingery, W.D., Bowen, H.K., and Uhlmann, D.R., *Introduction to Ceramics*, 2nd ed., Wiley, New York, 1976. With permission.)

TABLE 4.3
Microstructural Features for Two-Phase Ceramics Produced by Liquid-Phase Sintering for Various Ratios of the Interfacial Energies γ_{ss}/γ_{sl} and the Corresponding Dihedral Angles ψ

γ_{ss}/γ_{sl}	ψ	Microstructure
≥ 2	$0°$	All grains separated by liquid phase
$\sqrt{3} - 2$	$0°–60°$	Continuous liquid phase penetrating all three-grain junctions; partial penetration of the grain boundaries by the liquid
$1 - \sqrt{3}$	$60°–120°$	Isolated liquid phase partially penetrating the three-grain junctions
≤ 1	$\geq 120°$	Isolated liquid phase at four-grain junctions

the grain boundary, and no solid–solid contact exists. As ψ increases, the penetration of the liquid phase between the grains should decrease, whereas the amount of solid–solid contacts (i.e., grain boundary area) should increase. However, for values of ψ up to 60°, the liquid should still be capable of penetrating indefinitely along the three-grain edges, and the structure should, therefore, consist of two continuous interpenetrating phases. When ψ is greater than 60°, the liquid should form isolated pockets at the corners of four grains. The microstructural features obtained for various values of the dihedral angle are summarized in Table 4.3. It should be remembered that these equilibrium liquid shapes represent idealizations; local departures from them will occur in real systems.

4.4.3 Effect of Solubility

The two solubilities to be considered are the solid solubility in the liquid and the liquid solubility in the solid. Good solid solubility in the liquid leads to densification and is essential for liquid-phase sintering. On the other hand, a high liquid solubility in the solid should be avoided, because it leads to a transient liquid phase and swelling of the compact. Transient liquid-phase sintering, as outlined earlier, has been used successfully for a few systems, but, in general, the process is difficult to control. The effect of solubility on densification and compact swelling is illustrated in Figure 4.10.

Particle size has an effect on solubility. Following Equation 3.3 for the relationship between solute concentration and particle radius, and taking the solubility as equivalent to the concentration, we have

$$\ln\left(\frac{S}{S_o}\right) = \frac{2\gamma_{sl}\Omega}{kTa} \tag{4.6}$$

where S is the solubility of a particle with a radius a in the liquid, S_o is the equilibrium solubility of the solid in the liquid at a planar interface, γ_{sl} is the specific energy of the solid–liquid interface, Ω is the atomic volume, k is the Boltzmann constant, and T is the absolute temperature. According to Equation 4.6, the solubility increases with decreasing particle radius, so matter transport will occur from small particles to large particles, leading to Ostwald ripening. In addition, asperities have a small radius of curvature, and they tend to dissolve. Pits, crevices, and necks between the particles have negative radii of curvature, so solubility is diminished and precipitation is enhanced in those regions.

4.4.4 Capillary Forces

For a liquid that completely wets the solid, the pressure deficit in the liquid, given by Equation 4.1, leads to a compressive stress on the particles. In general, the magnitude and nature of the stress will depend on several factors, such as the contact angle, the volume of liquid, the separation of

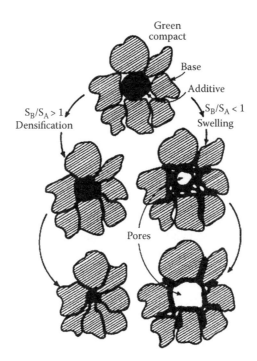

FIGURE 4.10 Schematic diagram comparing the effects of solubility on densification or swelling during liquid-phase sintering. (From German, R.M., *Liquid Phase Sintering*, Plenum Press, New York, 1985. With permission.)

the particles, and the particle size. To estimate how these variables influence the capillary force exerted by the liquid, let us consider an idealized model consisting of two spheres of the same radius, a, separated by a distance h by a liquid with a contact angle θ (Figure 4.11). The shape of the liquid meniscus is called a *nodoid,* and an analytical solution for the shape is known in terms of elliptical integrals. However, the calculations of the meniscus shape and the capillary force are complicated. As an approximation, we shall use the circular shape approximation in which the liquid surface is assumed to be part of a circle of radius r. With this approximation, the pressure difference across the liquid–vapor meniscus is given by

$$p = \gamma_{lv}\left(\frac{1}{Y} - \frac{1}{r}\right) \qquad (4.7)$$

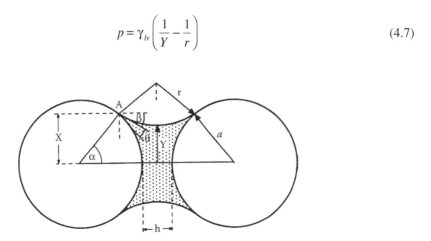

FIGURE 4.11 Geometrical parameters for an idealized model of two spheres separated by a liquid bridge.

where Y and r are the principal radii of curvature of the meniscus. The force acting on the two spheres is the sum of the contributions from the pressure difference, p, across the liquid–vapor meniscus and the surface tension of the liquid [23,24]. It is commonly evaluated at the point labeled A in Figure 4.11. In this case, the force equation is

$$F = -\pi X^2 p + 2\pi X \gamma_{lv} \cos \beta \tag{4.8}$$

where F is taken to be positive when the force is compressive. Some of the earlier papers in the literature include only the term with p in Equation 4.8, but Heady and Cahn [23] showed that Equation 4.8 is the correct expression for F. Substituting for p and putting $X = a \sin \alpha$, Equation 4.8 becomes

$$F = -\pi a^2 \gamma_{lv} \left(\frac{1}{Y} - \frac{1}{r} \right) \sin^2 \alpha + 2\pi a \gamma_{lv} \sin \alpha \cos \alpha \tag{4.9}$$

The distance of separation between the spheres is

$$h = 2[r \sin \beta - a(1 - \cos \alpha)] \tag{4.10}$$

and the angles are related by

$$\alpha + \beta + \theta = \pi/2 \tag{4.11}$$

Substituting for β in Equation 4.10 and rearranging gives

$$r = \frac{h + 2a(1 - \cos \alpha)}{2 \cos(\theta + \alpha)} \tag{4.12}$$

The positive radius of curvature of the meniscus is given by

$$Y = a \sin \alpha - r[1 - \sin(\theta + \alpha)] \tag{4.13}$$

Substituting for r from Equation 4.12 gives

$$Y = a \sin \alpha - \left[\frac{h + 2a(1 - \cos \alpha)}{2 \cos(\theta + \alpha)} \right] [1 - \sin(\theta + \alpha)] \tag{4.14}$$

The volume of the liquid bridge is

$$V = 2\pi(r^3 + r^2 Y) \left[\cos(\theta + \alpha) - \left[\frac{\pi}{2} - (\theta + \alpha) \right] \right] + \pi Y^2 r \cos(\theta + \alpha) \tag{4.15}$$

A situation of interest is the solution of Equation 4.8 for F as a function of the interparticle distance h for a given volume fraction of liquid. Figure 4.12 shows the calculated values for F between two tungsten spheres separated by a liquid copper bridge as a function of h for several

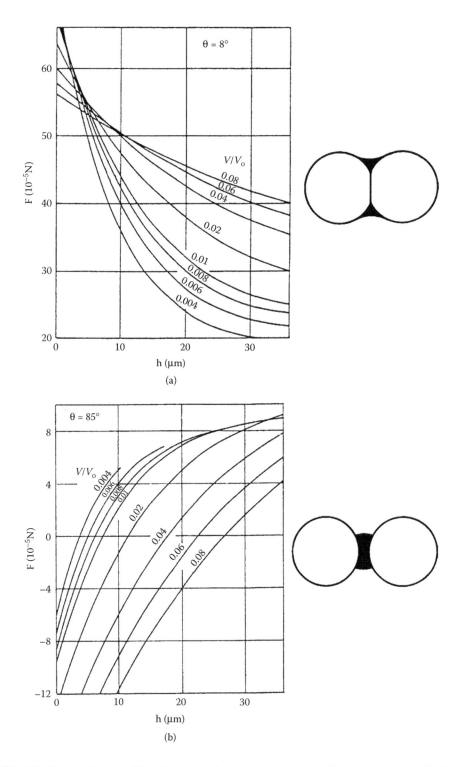

FIGURE 4.12 The calculated capillary force acting between two tungsten spheres separated by a liquid copper bridge, as a function of the interparticle distance, for several values of the liquid volume V, normalized to the volume of a tungsten sphere V_o at two extreme values of the contact angle. (a) $\theta = 8°$; (b) $\theta = 85°$. (From Huppmann, W.J., and Riegger, H., Modeling rearrangement processes in liquid-phase sintering, *Acta Metall.*, 23, 965, 1975. With permission.)

values of V/V_o, where V_o is the volume of a tungsten sphere [25]. The results are shown for two extreme values of the contact angle: $\theta = 8°$ and $\theta = 85°$. In the calculations, the surface energy of liquid copper was taken as 1.28 J/m². For $\theta = 8°$, the capillary force is large and compressive for all amounts of liquid and decreases strongly with increasing values of h (Figure 4.12a). The dependence of F on the volume of liquid is also reversed for large values of h. For $\theta = 85°$, F is negative (repulsive) for small values of h but compressive for large separations. An equilibrium separation defined by the condition $F = 0$ will be attained (Figure 4.12b).

An important result arising from these calculations is that very different rearrangement processes can be expected for low and for high contact angles. For efficient rearrangement and the achievement of high density, the contact angle must be kept low.

4.4.5 EFFECT OF GRAVITY

Because of the relatively low effective viscosity of powder compacts containing a liquid phase, the weight of the system itself can cause significant distortion, particularly when a significant amount of liquid is present (>5–10 vol%). The effect of kinetic and microstructural factors on distortion during liquid-phase sintering has been studied, and macroscopic models to predict the role of parameters such as compact mass, green density, particulate solid content, and dihedral angle on distortion of liquid-phase-sintered heavy metal alloys have been developed [26–29]. In practice, judicious support for large objects is necessary [30].

An additional effect of gravity is the possibility of liquid redistribution, leading to particle settling or liquid drainage. Solid–liquid separation or segregation is particularly important for systems where the amount of liquid is high and where a large density difference exists between the particulate solid and the liquid phase, as for example in the case of cemented carbides and heavy metal alloys [31]. The heavier particulate solid settles toward the bottom, leaving the lighter liquid at the top. The solid–liquid separation leads to distortion of the object and to large differences in the microstructure from the top to the bottom of the object. These effects have been examined under conditions of microgravity [32]. The particulate solid and liquid do not separate as they do under gravity. However, for high liquid content, even in microgravity the solid is still not dispersed throughout the liquid. Instead, the solid particles agglomerate, as shown in Figure 4.13 for a tungsten heavy metal alloy. Since this is a metallic system, surface charge effects are not expected to be present. The particles are propelled as a result of uneven dissolution and precipitation rates along their surfaces, leading to collision and joining. It is possible that a liquid film may be present at the contact areas between the particles, but this has not yet been determined.

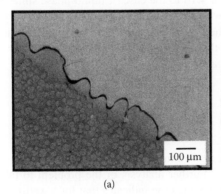
(a)

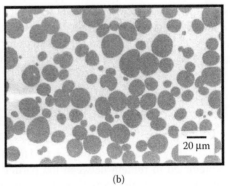
(b)

FIGURE 4.13 Micrographs of W(15.4 wt% Ni + 6.6 wt% Fe) after liquid-phase sintering for 1 min in microgravity at 1507°C: (a) low-magnification view of sample edge and wavy solidified liquid; (b) high-magnification view of sample interior showing grain agglomeration. (Courtesy of R.M. German.)

4.5 GRAIN BOUNDARY FILMS

It is often assumed in the theoretical analysis of liquid-phase sintering that the composition and structure of the grain boundary liquid phase is constant. However, this is not always the case. A wetting liquid layer, as described earlier, leads to the development of a compressive capillary force, which is equivalent to placing the system under a fairly large hydrostatic compression. The evolution of the intergranular liquid layer under the influence of the capillary force, as well as other forces, has been the subject of recent investigation.

After the rearrangement stage, we may envisage a situation where densification occurs by the solution-precipitation mechanism and the liquid layer separating the grains becomes progressively thinner with time. Assuming that there is no repulsion between the grain surfaces, we would eventually reach a stage where the liquid capillary becomes so narrow that liquid flow is very difficult. The solution-precipitation can still continue, but at a reduced rate. Kinetically, a competition is set up between the rate at which the liquid layer becomes thinner by viscous flow and the rate of densification. This type of argument appears to point to the development of a finite thickness of the liquid layer when densification is completed, because liquid flow through the capillary becomes very slow.

Observations by high-resolution transmission electron microscopy often reveal an amorphous film, commonly a silicate glass, with a thickness of 0.5–2 nm between the grains in many ceramics that have been fabricated by liquid-phase sintering. Examples include Si_3N_4, SiC, ZnO, and Al_2O_3 (Figure 4.14). There is evidence that the thickness of the intergranular film has a constant *equilibrium* value from one boundary to another in any given material regardless of the volume fraction of the silicate glass (the excess glass is located elsewhere, such as at three- and four-grain junctions).

Lange [33] applied a theory developed for flat plates separated by a liquid layer to calculate the rate of thinning by viscous flow of a liquid layer separating two spherical particles undergoing liquid-phase sintering. The rate of approach of two plates separated by a Newtonian viscous liquid is [34]

$$\frac{dh}{dt} = -\frac{2\pi h^5}{3\eta A^2 h_o^2} F \qquad (4.16)$$

where h is the thickness of the liquid layer at time t, h_0 is the initial thickness of the layer, F is the compressive force exerted on the plates, η is the viscosity of the liquid, and A is the contact area between the liquid and the plate. To apply Equation 4.16 to the case of liquid-phase sintering, let us consider the model shown in Figure 4.15 for two spheres held together by the capillary force due to a liquid bridge. Instead of asking whether the liquid will be squeezed out or not, it is more convenient to determine how long it will take for the liquid to be squeezed out. Putting $y = h/h_0$ in Equation 4.16 and integrating, the time for y to approach zero is

$$t_f = \frac{3\eta A^2}{8\pi h_o^2 F}\left(\frac{1}{y^4} - 1\right) \qquad (4.17)$$

This equation shows that $t_f = \infty$ for $y = 0$. Therefore, within any experimental time frame, some liquid is predicted to remain between the particles. Further analysis of the problem to take into account the approach of the particle centers due to densification indicates that for a wetting liquid, although the thickness decreases progressively with time, a layer of finite thickness always remains between the particles within any experimental time frame. However, when the liquid layer becomes very thin (in the nanometer range), effects other than viscous flow become dominant. These effects include structural and chemical forces, as well as charge interactions.

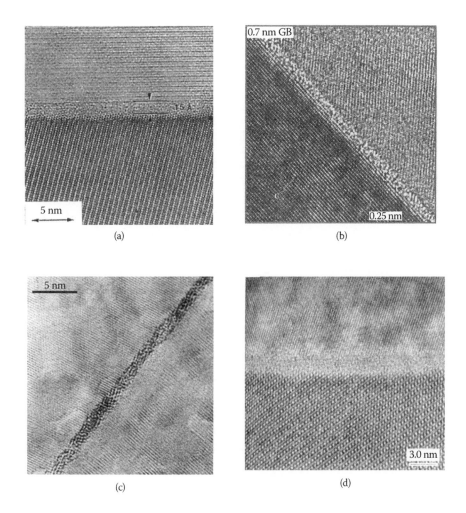

FIGURE 4.14 Amorphous grain boundary film with a constant equilibrium thickness in (a) $Si_3N_4(Y_2O_3)$, film thickness of 1.5 nm; (b) $SiC(Al + B + C)$, 0.7 nm; (c) $ZnO(Bi_2O_3)$, 1.5 nm; (d) $Al_2O_3(CaO-SiO_2)$, 1.5 nm.

Clarke [35] proposed an explanation for the equilibrium thickness of the intergranular film in terms of a balance between the attractive van der Waals forces between the grains and short-range repulsive forces due to the resistance to deformation of the silicate liquid. The repulsive forces, referred to as *disjoining* or *steric* forces, are attributed to the structural ordering of the silica tetrahedra of the intergranular film at the crystalline grain surfaces. The idea of a structural disjoining pressure is not new, having been introduced by Derjaguin and Churaev [36].

The assumption in the present system is that for microscopic thicknesses of the silicate phase, the structure of the liquid is not random, as might be expected for bulk volumes of the liquid, but instead exhibits a form of spatially varying orientation order close to each grain. If the correlation length associated with orientational fluctuations is ξ, then it is assumed that the orientational order imposed on the liquid-phase structure by the grains will extend a distance ξ into the liquid. Assuming that the grains have flat sides and are separated by a distance h, the disjoining force (per unit area of the boundary) is given by the expression [35]

$$F_S = \frac{\alpha \eta_o^2}{\sinh^2(h/2\xi)} \tag{4.18}$$

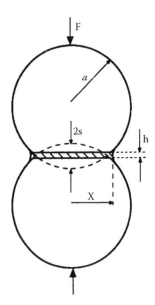

FIGURE 4.15 Parameters of an idealized two-sphere model separated by a liquid layer of thickness h used to analyze the change in the thickness of the layer during sintering.

where $\alpha \eta_o^2$ corresponds to the free energy difference between the intergranular film with and without ordering.

Clarke et al. [37] later explored the possibility of an electrical double-layer contributing to the repulsion, thereby assisting in the stabilization of the film thickness. This double-layer repulsion is identical in origin to that in the DLVO theory of colloid stability [38]. Assuming that the grains have flat, parallel faces separated by a distance h, the expression for the double-layer repulsion is

$$F_R = \frac{8kT}{\pi z^2 b_L h^2} \left(\tanh \frac{ze\phi_o}{4kT} \right)^2 (Kh)^2 \exp(-Kh) \tag{4.19}$$

where k is the Boltzmann constant, T is the absolute temperature, z is the valence of the charge-determining ions, e is the electronic charge, ϕ_o is the surface potential at the grain faces, K is the inverse Debye length, and b_L, is the *Bjerrum length*, given by

$$b_L = \frac{e^2}{4\pi\varepsilon\varepsilon_o kT} \tag{4.20}$$

where ε is the dielectric constant of the liquid medium and ε_o is the permittivity of vacuum.

The attractive van der Waals force between the grains (assumed to have flat sides and separated by a distance h) is given by [38]

$$F_A = -\frac{A}{6\pi h^3} \tag{4.21}$$

where A is the Hamaker constant. At equilibrium, the overall force balance is satisfied:

$$F_R + F_S = F_A + p_C \tag{4.22}$$

where p_C is the compressive capillary pressure that acts to draw the grains together.

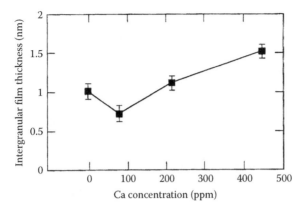

FIGURE 4.16 Equilibrium grain boundary thickness measured by HREM plotted versus the bulk calcium doping level in silicon nitride. (Courtesy of D.R. Clarke.)

Much of the data required to calculate the contributions of the van der Waals, disjoining, and electrical double-layer forces are not available, but equilibrium separation distances estimated from the theory are in qualitative agreement with observations. For Si_3N_4 doped with Ca [39], measurements appear to indicate that the equilibrium film thickness initially decreases with Ca concentration but then increases (Figure 4.16). This variation in the film thickness has been qualitatively explained in terms of the balance of the van der Waals attraction, the repulsive disjoining force, and the electrical double-layer repulsion, as illustrated schematically in Figure 4.17. In the undoped Si_3N_4, it is assumed that there is no net electrical charge, so the equilibrium thickness is controlled by the van der Waals and steric forces only, and the grain boundary film is idealized as a pure silica network structure (Figure 4.17a). The addition of Ca^{2+} ions to the intergranular phase can have a dual effect. It may disrupt the silica network and the ordering imposed by the Si_3N_4 grains, reducing the magnitude of the disjoining force (represented schematically in Figure 4.17b). The addition of Ca^{2+} ions also provides a charged species to the system, and, provided the ions adsorb to the grain surfaces, they may provide a double-layer repulsion (Figure 4.17c). Since Ca^{2+} is known to be a potent silica network modifier, it may be expected that the strongest effect is to reduce the disjoining force. Further additions of Ca^{2+} may be expected to enhance the double-layer repulsion, leading to an increase in the film thickness.

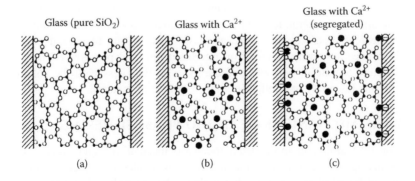

FIGURE 4.17 Model for force balance in Ca-doped silicon nitride. (Courtesy of D.R. Clarke.)

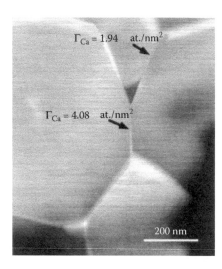

FIGURE 4.18 STEM image showing two neighboring grain boundary films with remarkably different calcium excess in calcia-doped silicon nitride.

A problem with the force-balance explanation for the equilibrium film thickness in Ca-doped Si_3N_4 is that later studies revealed no clear evidence for Ca^{2+} segregation at the grain boundaries [40]. The Ca^{2+} concentration appears to be constant along the length and thickness of the film. Based on these studies, it was suggested that the increase in the film thickness with Ca^{2+} concentration (Figure 4.16) could be due to a reduction of the attractive van der Waals force rather than an increase in the electrical double-layer repulsion.

For the Ca-doped Si_3N_4 system, it is also observed that for a given material (with a fixed Ca^{2+} concentration), the composition of the grain boundary film is different from that of the liquid in the three- or four-grain junction, or even from boundary to boundary (Figure 4.18). This compositional variation is also observed in several other systems [41], and it may be common rather than an exception. Since the film composition should dictate the balance of forces, and, hence, the film thickness, the nearly constant film thickness in a given material indicates that the origins of the compositional variation from boundary to boundary may lie elsewhere.

Another explanation, based on thermodynamics, has been proposed for the $ZnO(Bi_2O_3)$ system, in which a thin amorphous film (1.0–1.5 nm) is also observed at the grain boundaries [42]. It was argued that the thin amorphous film might represent the minimum free energy configuration for the interface. This would be the case if the free energy of the grain boundary containing the amorphous film were lower than the sum of the crystal–crystal grain boundary energy and the free energy for accommodating the grain boundary glass as a solid solution or nonwetting secondary phase. In this case, it would not be energetically favorable to remove the intergranular amorphous film by crystallization or dewetting.

Grain boundary films in ceramics are not just restricted to the thin equilibrium films discussed so far. In a variety of other ceramics, thicker glass films (10 nm to several microns) are also observed, which have a significant effect on microstructure and properties. These thicker films represent a different regime of behavior and may vary in thickness with the volume fraction of liquid and from one boundary to another in a given sample of material.

4.6 THE BASIC MECHANISMS OF LIQUID-PHASE SINTERING

We now go on to examine in more detail the basic mechanisms and processes occurring in liquid-phase sintering. For convenience, the discussion is divided into three main parts, related to the three stages of liquid-phase sintering, but it should be remembered that there is some degree of overlap between the successive stages.

4.6.1 STAGE 1: REARRANGEMENT AND LIQUID REDISTRIBUTION

The fairly large capillary pressures developed for liquid-phase sintering of fine particles cause rapid particle rearrangement when the viscosity of the system is still very low. However, other processes may accompany rearrangement, and these may have important consequences for liquid-phase sintering. Capillary pressure gradients will cause liquid to flow from regions with large pores to regions with smaller pores, leading to redistribution of the liquid.

4.6.1.1 Liquid Redistribution

Experimental observations indicate that considerable redistribution of the liquid may occur during liquid-phase sintering. Perhaps the most dramatic evidence for such redistribution comes from the experiments of Kwon and Yoon [43], who investigated the sintering of fine tungsten powder containing coarse nickel particles that melt to form the liquid phase. As shown in Figure 4.19, sequential filling of the pores occurs in such a way that the small pores are filled first and the larger pores later. A difficulty with modeling the process is that the local liquid surface curvature depends sensitively on the local geometry, so quantitative statements can be made only for highly idealized microstructures.

Liquid redistribution was analyzed by Shaw [44] for a two-dimensional model of circular particles. The approach involves a determination of the equilibrium distribution of the liquid in different packing arrangements of particles under the condition that the chemical potential of the liquid in all the pores in a particle array is the same at equilibrium. The chemical potential of an atom under the surface of a liquid–vapor meniscus with an average radius of curvature r is

$$\mu = \mu_o + \frac{\gamma_{lv}\Omega}{r} \tag{4.23}$$

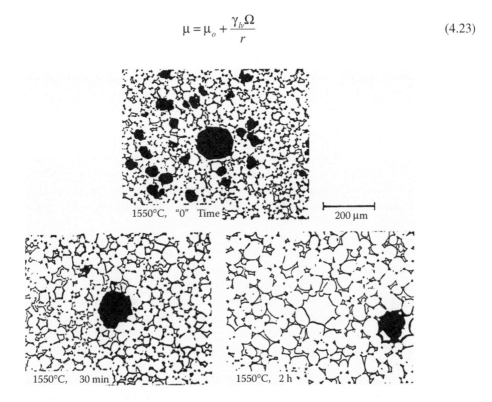

FIGURE 4.19 The change in microstructure during liquid-phase sintering of a mixture of fine (10 μm) W powder, 2 wt% of 30 μm Ni spheres, and 2 wt% 125 μm Ni spheres, showing sequential filling of the pores. (From Kwon, O.-J., and Yoon, D.N., Closure of isolated pores in liquid-phase sintering of W-Ni, *Internat. J. Powder Metall. & Powder Technol.*, 17, 127, 1981.)

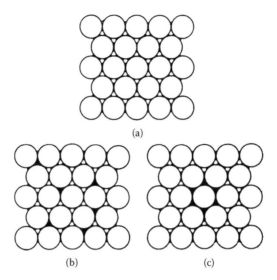

(a)

(b)   (c)

FIGURE 4.20 Possible equilibrium configurations that can be adopted by a liquid in a close-packed array of particles. (a) Isolated necks filled with liquid; (b) fraction of pores completely filled with liquid; (c) same as (b) but with inhomogeneous liquid distribution.

where μ_o is the chemical potential of an atom under a flat surface, γ_{lv} is the liquid–vapor surface energy, and Ω is the atomic volume. The condition that the chemical potential must be the same is, therefore, equivalent to the radius of the liquid menisci being the same.

For a regular array of circles, with threefold coordinated pores, in which no shrinkage occurs, Figure 4.20 shows two possible ways in which the liquid will distribute itself. At small volume fractions, the liquid will be distributed evenly in the necks between the particles (Figure 4.20a). Increasing the volume of liquid will cause the necks to fill with liquid until some critical volume fraction is reached when, instead of filling each pore evenly, the liquid adopts a distribution, driven by minimization of surface area, in which a certain fraction of the pores is completely filled and the remainder of the liquid is in isolated necks (Figure 4.20b). Altering the amount of liquid in this regime has no effect on the amount of liquid situated at the necks but simply alters the fraction of pores that are filled. Increasing the contact angle decreases the range of liquid volume over which the liquid is evenly distributed in the necks.

An important finding of the analysis is the way in which a liquid that is initially inhomogeneously distributed will redistribute itself in the two-dimensional array. For low liquid content where the liquid is situated at isolated necks, there is always a force driving the liquid to redistribute itself homogeneously. In the second regime, where some of the pores are filled, it is immaterial which pores fill with liquid, as long as the correct fraction of pores is filled. Consequently, once an inhomogeneous distribution, such as that pictured in Figure 4.20c, forms, there is no driving force to redistribute the liquid homogeneously.

A model consisting of an array of circles containing pores with three-fold and six-fold coordination (Figure 4.21), while simple in geometry, can provide some insight into how a liquid will redistribute itself in the practical situation of an inhomogeneously packed powder system. The free energy calculations can be used to construct a diagram such as that given in Figure 4.22, showing the distribution of the liquid in the array. The diagram shows the fraction of particles surrounding six-fold coordinated pores as a function of the volume fraction of the liquid. Four regions can be distinguished in addition to the shaded region to the right of the diagram corresponding to the case where the pores are completely filled.

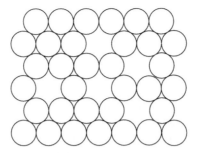

FIGURE 4.21 Example of a two-dimensional arrangement of pores that contains three- and six-fold coordinated pores.

To illustrate the significance of Figure 4.22, let us consider a structure containing a fixed fraction of particles at six-fold coordinated pores, and consider how the pores will fill as the volume of liquid is increased. For low volume fraction, the liquid is situated only at isolated necks between the particles. Each point in this region lies on a tie line representing liquid-filled necks at three-fold coordinated pores at the bottom of the diagram and liquid-filled necks at six-fold coordinated pores at the top of the diagram. The tie lines connect systems in which the chemical potential of the liquid is the same in the two types of pores. As the volume fraction of liquid is increased, we enter the second region, in which it becomes energetically more favorable to fill three-fold coordinated pores than to continue filling the necks between the particles. In this region, the liquid content between the necks remains constant, and the increase in the liquid volume goes into filling the three-fold coordinated pores. When the third region is reached, all of the three-fold coordinated pores are filled with liquid, and an increase in liquid volume goes into filling the necks between particles surrounding the six-fold coordinated pores. Finally, in the fourth region, the six-fold coordinated pores become completely filled. When the analysis is extended to include the effect of shrinkage on the liquid distribution, the results show the general sequence of behavior described for the simple model.

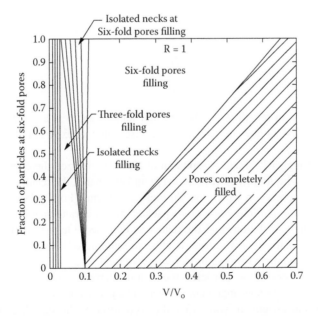

FIGURE 4.22 Liquid distribution diagram for a particle array that contains three- and six-fold coordinated pores.

The analysis shows that the pores in the simple model fill sequentially. If we extend the results to the more complex situation of a powder compact with a distribution of pore sizes, the same behavior of sequential filling of the pores is expected to occur. The pores with the smallest coordination number will be the first to fill, because such pores have high surface-to-volume ratio, so a given volume of liquid eliminates more solid–vapor interfacial area. If there is sufficient liquid, the pores with higher coordination number start to fill. However, the pore filling leads to a percolation problem, and the liquid might not have access to all of the small pores, so some may remain empty while large pores start to fill.

The concerns about the homogeneity of the starting powder compact for solid-state sintering also apply to liquid-phase sintering [38]. It is desirable to start with a compact that is packed homogeneously to produce pores with a narrow distribution of sizes, in which the major component and the additive are mixed homogeneously to provide a homogeneous distribution of the liquid phase. Heterogeneous packing leads to sequential filling of the pores, so the larger pores are filled later in the sintering process, producing regions that are enriched with the liquid composition. Inhomogeneous mixing leads to an inhomogeneous liquid distribution such that there is no driving force for redistribution of the liquid. Also, using large particles to create the liquid phase (as in Figure 4.19) leaves huge voids when the particles melt and the liquid invades the smaller pores. The optimum situation is to start with particles that are precoated with the liquid-forming additive. Methods by which particles can be coated include fluidized bed vapor deposition and precipitation from solution [38].

4.6.1.2 Particle Rearrangement

After formation of the liquid, as the liquid wets and spreads over the solid surfaces, particle rearrangement of the initial network is rapid, occurring in as little as a few minutes. Rearrangement leads to initial densification for a wetting liquid and also determines the initial microstructure of the sintering compact, which affects further densification and microstructural development. Although several theoretical and experimental approaches have been used to investigate the process [25,45–48], the analysis of rearrangement in a randomly packed array of particles is a difficult problem, and our understanding of the process in real systems is limited. Computational approaches to rearrangement have been made recently [49]. Kingery [45] used an empirical approach in which the surface tension forces driving densification are balanced by the viscous forces resisting rearrangement, deriving a simple kinetic relationship for the variation of the shrinkage with time t:

$$\frac{\Delta L}{L_o} \sim t^{1+y} \tag{4.24}$$

where ΔL is the change in length, L_o is the original length, and y is a positive number less than one. Although Equation 4.24 does not appear to be unreasonable, experimental verification has not been convincing.

Huppmann and Riegger [25] and Huppmann et al. [47] analyzed the capillary forces between particles separated by a liquid bridge (Figure 4.12) and compared the results of their analysis with data for Cu-coated W spheres in which the Cu melted to form a homogeneous distribution of the liquid phase between the particles. For close-packed planar arrangements of the spheres, good agreement between the calculations and the data was obtained. On the other hand, for randomly packed arrays, local densification occurred, leading to opening of large pores (Figure 4.23). Less densification than that predicted by the model was observed.

For a system of polycrystalline particles, the overall rearrangement process can consist of two stages, referred to as primary rearrangement and secondary rearrangement. *Primary rearrangement* describes the rapid rearrangement, soon after the formation of the liquid, of the polycrystalline particles under the surface tension forces of the liquid bridge. As described earlier, if $\gamma_{ss}/\gamma_{sl} > 2$, the liquid can penetrate the grain boundaries between the particles in the polycrystals, and

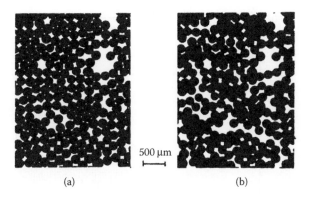

(a)   (b)

FIGURE 4.23 Planar array of copper-coated tungsten spheres before (*left*) and after (*right*) liquid-phase sintering. (From Huppmann, W.J., and Riegger, H., Modeling rearrangement processes in liquid-phase sintering, *Acta Metall.*, 23, 965, 1975. With permission.)

fragmentation of the polycrystalline particles will occur. *Secondary rearrangement* describes the rearrangement of these fragmented particles. Since it depends on the rate at which the grain boundaries are dissolved away, secondary rearrangement occurs more slowly than primary rearrangement. The two types of rearrangement are sketched in Figure 4.24.

If sufficient liquid is present, rearrangement alone may lead to a fully dense material. The relative amounts of solid and liquid required for this to occur depend on the rearranged density of the particulate solid. As an example, let us consider a powder compact with a relative density of 60% (porosity of 40%), consisting of a major component and a liquid-producing additive. After

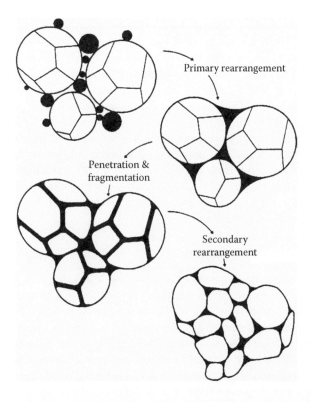

FIGURE 4.24 Schematic diagram illustrating fragmentation and rearrangement of polycrystalline particles.

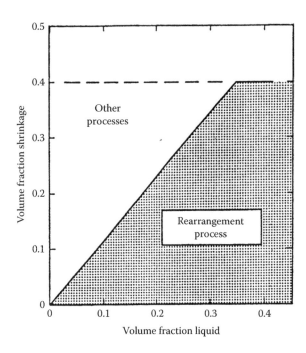

FIGURE 4.25 Fractional shrinkage due to the rearrangement process for different liquid contents. (From Kingery, W.D., Densification during sintering in the presence of a liquid phase, I. Theory, *J. Appl. Phys.*, 30, 301, 1959. With permission.)

formation of the liquid, suppose that the solid particles rearrange to form 64% of the final product (i.e., a packing density close to that for dense random packing of monosize spheres). For a liquid volume fraction greater than 36%, full densification is achieved by rearrangement alone. When the liquid content is less than 36%, full densification requires the occurrence of additional processes, such as solution-precipitation. The volumetric shrinkage due to rearrangement as a function of liquid volume is shown in Figure 4.25.

The results in the figure imply that rearrangement to a denser packing is possible with an arbitrarily small amount of liquid, but in practice it is commonly observed that rearrangement is difficult when the liquid volume is small (less than 2–3 vol%), particularly if the solid particles have irregular shapes. Rearrangement is also difficult for systems in which the dihedral angle is greater than 0°, if significant solid-state sintering occurs prior to the formation of the liquid phase, leading to a skeleton of interconnected grains.

4.6.2 Stage 2: Solution-Precipitation

In Stage 2, rearrangement decreases considerably and the solution-precipitation mechanism becomes dominant. The major processes that occur by solution-precipitation are densification and coarsening. They occur concurrently and may be accompanied by grain shape accommodation if the liquid volume is small. Coalescence of small grains with contacting large grains also contributes to coarsening and grain shape accommodation. There are two models for densification, which we will refer to as densification by contact flattening and densification accompanied by Ostwald ripening. It has been suggested that for some systems, pore filling by the liquid phase, to be described in Stage 3, can make a significant contribution to densification from the early stages of liquid-phase sintering, and not just in the later stages [50].

4.6.2.1 Densification by Contact Flattening

Densification by contact flattening was described by Kingery [45]. As a result of the compressive capillary force of a wetting liquid, the solubility at the contact points between the particles is higher than that at other solid surfaces. This difference in solubility (or chemical potential) leads to matter transport away from the contact points, allowing center-to-center approach under the action of the surface tension forces and the formation of a flat contact zone (Figure 4.26). As the radius of the contact zone increases, the stress along the interface decreases and densification slows. The rate of matter transport is controlled by the slower of the two mechanisms: (1) diffusion through the liquid and (2) the interface reaction: dissolution into the liquid or precipitation onto the particle surfaces.

Kingery assumed a model consisting of two spherical particles of the same radius, a, for which the geometrical parameters are similar to those outlined in Figure 2.11. If each sphere is dissolved away along the center-to-center line by a distance h to give a circular contact area of radius X, then

$$h \approx X^2/2a \tag{4.25}$$

The volume of material removed from each sphere is given by $V \approx \pi X^2 h/2$. Using Equation 4.25, this becomes

$$V \approx \pi a h^2 \tag{4.26}$$

4.6.2.1.1 Rate Control by Diffusion through the Liquid

Kingery adopted a diffusional flux equation similar to that assumed by Coble for the intermediate stage of solid-state sintering (see Chapter 2). In this case, the flux from the boundary per unit thickness is given by

$$J = 4\pi D_L \Delta C \tag{4.27}$$

where D_L is the diffusion coefficient for the solute atom in the liquid and ΔC is the solute concentration difference between that at the contact area, C, and that at a flat, stress-free surface, C_o. If the thickness of the liquid bridge is δ_L, the rate of removal of the solid is

$$\frac{dV}{dt} = \delta_L J = 4\pi D_L \delta_L \Delta C \tag{4.28}$$

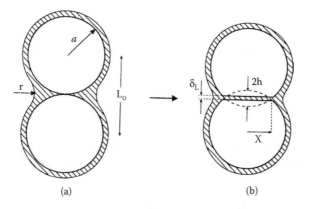

FIGURE 4.26 Idealized two-sphere model for densification by contact flattening.

As outlined in Chapter 1, if ΔC is small, we can write

$$\frac{\Delta C}{C_o} = \frac{p_e \Omega}{kT} \tag{4.29}$$

where p_e is the local stress on an atom, Ω is the atomic volume, k is the Boltzmann constant, and T is the absolute temperature. The capillary pressure p due to a spherical pore in the liquid is given by Equation 4.1, which is equivalent to the externally applied hydrostatic pressure that would have to be applied to the system in order that the resulting interparticle force be equal to that produced by the liquid bridge. Because the contact area is less than that of the external area, the local pressure is magnified at the contact area. Assuming a simple force balance, as described in Chapter 2 for the stress intensification factor in solid-state sintering, the local pressure p_e at the contact area is given by

$$p_e X^2 = k_1 p a^2 \tag{4.30}$$

where k_1 is a geometrical constant. Substituting for p (Equation 4.1) and X (Equation 4.25) gives

$$p_e = k_1 \frac{\gamma_{lv} a}{rh} \tag{4.31}$$

Assuming that the radius of the pore is proportional to the sphere radius, i.e., $r \approx k_2 a$, where k_2 is assumed to remain constant during sintering, Equation 4.31 becomes

$$p_e = \frac{k_1}{k_2}\left(\frac{\gamma_{lv}}{h}\right) \tag{4.32}$$

From Equation 4.28, Equation 4.29, and Equation 4.32, we obtain

$$\frac{dV}{dt} = \frac{4\pi k_1 D_L \delta_L C_o \Omega \gamma_{lv}}{k_2 hkT} \tag{4.33}$$

Equation 4.26 gives $dV/dt = 2\pi ah(dh/dt)$, so Equation 4.33 can be written as

$$h^2 dh = \frac{2k_1 D_L \delta_L C_o \Omega \gamma_{lv}}{k_2 akT} dt \tag{4.34}$$

Integrating and applying the boundary condition that $h = 0$ at $t = 0$, Equation 4.34 gives

$$h = \left(\frac{6k_1 D_L \delta_L C_o \Omega \gamma_{lv}}{k_2 akT}\right)^{1/3} t^{1/3} \tag{4.35}$$

Since $h/a = -\Delta L/L_o = -(1/3)\Delta V/V_o$ for small $\Delta L/L_o$, where $\Delta L/L_o$ and $\Delta V/V_o$ are the linear shrinkage and the volumetric shrinkage of the powder compact, respectively, we can write

$$-\frac{\Delta L}{L_o} = -\frac{1}{3}\frac{\Delta V}{V_o} = \left(\frac{6k_1 D_L \delta_L C_o \Omega \gamma_{lv}}{k_2 a^4 kT}\right)^{1/3} t^{1/3} \tag{4.36}$$

When diffusion through the liquid is the rate-controlling mechanism, the shrinkage is predicted to be proportional to the one-third power of the time and inversely proportional to the four-thirds power of the initial particle size. It is also predicted to increase with the one-third power of the thickness of the intergranular liquid layer.

4.6.2.1.2 Control by Phase Boundary Reaction

When the phase boundary reaction leading to dissolution of the solid into the liquid is the rate-controlling mechanism, the volumetric rate of material transfer is assumed to be directly proportional to the contact area, times a rate constant for the phase boundary reaction, times the increase of the activity of the solid at the contact area due to the capillary pressure; that is,

$$\frac{dV}{dt} = k_3 \pi X^2 (a' - a'_o) = 2\pi k_3 ha(C - C_o) \tag{4.37}$$

where k_3 is the reaction rate constant, and the activities a' and a'_o are taken to be equal to the concentrations. Following the steps outlined above for the case of diffusion control, the shrinkage is given by

$$-\frac{\Delta L}{L_o} = -\frac{1}{3}\frac{\Delta V}{V_o} = \left(\frac{2k_1 k_3 C_o \Omega \gamma_{lv}}{k_2 a^2 kT} \right)^{1/2} t^{1/2} \tag{4.38}$$

In this case, the shrinkage is predicted to be proportional to the square root of the time and inversely proportional to the initial particle radius.

4.6.2.2 Densification Accompanied by Ostwald Ripening

The second mechanism of densification is based on the observations by Yoon and Huppmann [51] for liquid-phase sintering of W(Ni) powder mixtures. In an idealized model (Figure 4.27), the small grains dissolve and the material is precipitated on the large grains away from the contact points, in a way such that grain shape accommodation occurs, so that the large-growing grains fill the space more efficiently. This process, coupled with rearrangement of the resulting polyhedral structure, leads to center-to-center approach of the large grains, and, therefore, to shrinkage. In this way, densification is accompanied by Ostwald ripening. Densification by dissolution of the small grains and precipitation on the large grains depends on the particle size distribution and other factors.

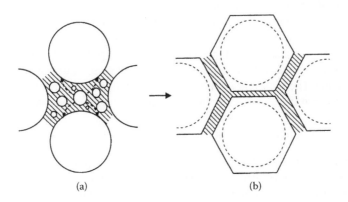

(a)   (b)

FIGURE 4.27 Schematic diagram illustrating densification accompanied by Ostwald ripening. Grain shape accommodation can also occur when the liquid volume fraction is low.

Theoretical analysis of the shrinkage rate is difficult [52,53], but one estimate, based on diffusion-controlled Ostwald ripening, leads to a form similar to that predicted by the contact-flattening model [17]:

$$-\frac{\Delta L}{L_o} = -\frac{1}{3}\frac{\Delta V}{V_o} = \left(\frac{48D_L C_o \Omega \gamma_{lv}}{a^3 kT}\right)^{1/3} t^{1/3} \tag{4.39}$$

4.6.2.3 Assessment of the Densification Models

Experimental data often show reasonable agreement with the prediction that the shrinkage depends on the one-third power of time, which is taken to indicate a diffusion-controlled solution-precipitation mechanism [54,55]. In several studies, however, fitting of the data to determine the time dependence of the shrinkage appears somewhat arbitrary. The time at which solution-precipitation is assumed to begin has a significant effect on the exponent, and this time is often difficult to determine because of the overlap with the previous rearrangement stage. Often, the shrinkage data can be fitted with a curve having a smoothly varying slope instead of a line having a fixed slope equal to the predicted value.

In view of the initially high capillary forces exerted on small grain contacts, it appears reasonable that Kingery's contact-flattening model is initially important. The model can also account for the phenomenon of grain shape accommodation. On the other hand, observations with real powder systems, which normally have a distribution of particle sizes, show that densification correlates with the onset of coarsening in the solution-precipitation stage and the dissolution of the small grains. Kingery's model assumes no grain growth, so it cannot account for this observed coarsening. Densification accompanied by coarsening is clearly observed in the experiments by Yoon and Huppmann [51], who studied the sintering of a mixture of coarse, single crystal, spherical tungsten particles (200–250 μm in diameter), fine tungsten particles (average size of ≈10 μm), and fine nickel powder that melts to form the liquid phase. When the tungsten dissolves in the molten nickel and precipitates out, the precipitated material is not pure tungsten but a solid solution containing a small amount of nickel (~0.15 wt%). By etching in Murakami's solution, the precipitated tungsten can be distinguished from the pure tungsten. Figure 4.28 shows the microstructure, after sintering at 1670°C for 3 min, 20 min, 120 min, and 360 min, of a powder mixture that contained 48 wt% large tungsten spheres, 48 wt% fine tungsten particles, and 4 wt% nickel. The microstructures show that the fine tungsten particles dissolve and precipitate on the coarse tungsten spheres. These observations, coupled with the measured decrease in porosity of the system with time, clearly indicate that Ostwald ripening accompanies densification.

Figure 4.28 also indicates that the precipitated tungsten does not grow uniformly around the large tungsten spheres but rather occurs preferentially in the regions where no neighboring large grains impede it. The coarsened large tungsten spheres develop flat faces and assume the shape of a polyhedron, so densification and coarsening for this powder composition are accompanied by grain shape accommodation. The occurrence of accommodation indicates that Kingery's contact-flattening mechanism is active, but there is no clear indication as to which mechanism causes shrinkage.

4.6.2.4 Grain Shape Accommodation

Grain shape accommodation is clearly observed in the micrographs of Figure 4.28 for the W(Ni) system containing 4 wt% Ni. However, many microstructures of materials produced by liquid-phase sintering also show rounded (spheroidal) grains. As an example, Figure 4.29 shows the microstructure for the same W(Ni) system, but with a higher amount of liquid (14 wt% Ni). It is clear from Figure 4.28 and Figure 4.29 that the volume fraction of liquid is a key factor in determining whether or not grain shape accommodation will occur.

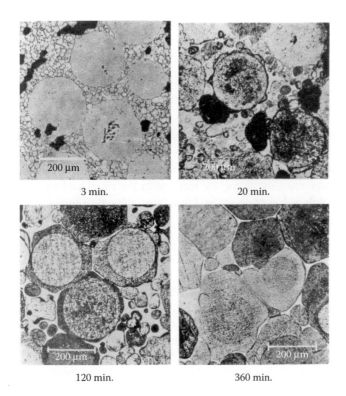

FIGURE 4.28 The microstructures of a mixture of 48 wt% large W spheres, 48 wt% fine W powder, and 4 wt% Ni, after sintering for various times at 1670°C. (From Yoon, D.N., and Huppmann, W.J., Grain growth and densification during liquid-phase sintering of W-Ni, *Acta Metall.*, 27, 693, 1979. With permission.)

The occurrence of grain shape accommodation is normally favored when the amount of liquid is small. In this case, the liquid is insufficient to completely fill the voids between the grains as long as the grains take up a rounded or spheroidal shape. Grain shape accommodation produces a polyhedral grain shape with flat contact surfaces, and this leads to a more efficient packing of the grains. Liquid released from the well-packed regions can flow into the pores.

Grain shape accommodation will occur if it leads to a decrease in the energy of the system. Compared to a sphere with the same volume, the polyhedral grain shape has a higher surface area.

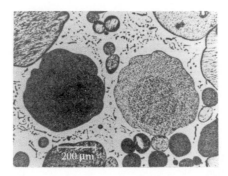

FIGURE 4.29 The microstructure of a mixture of 43 wt% large W spheres, 43 wt% fine W powder, and 14 wt% Ni, after sintering for 60 min at 1670°C. (From Yoon, D.N., and Huppmann, W.J., Grain growth and densification during liquid-phase sintering of W-Ni, *Acta Metall.*, 27, 693, 1979. With permission.)

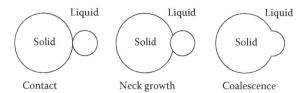

FIGURE 4.30 Schematic diagram illustrating grain growth by coalescence of small and large grains.

If grain shape accommodation is to occur, the decrease in the interfacial energy associated with the filling of the pores must overcome the increase in the interfacial energy associated with the development of the polyhedral grain shape. For a large amount of liquid, the capillary pressure is low, so the driving force for contact flattening and shape accommodation is reduced, and the spheroidal grain shape is maintained.

4.6.2.5 Coalescence

The pulling of grains into contact by a wetting liquid can lead to coarsening by coalescence [56]. As sketched in Figure 4.30, a possible mechanism for coalescence involves contact formation between the grains, neck growth, and migration of the grain boundary. Coalescence can occur by several mechanisms, such as solid-state grain boundary migration, liquid-film migration, and solution-precipitation through the liquid (Figure 4.31). When the dihedral angle is low, the liquid partially penetrates the grain boundary, and movement of the boundary will involve initially an increase in the grain boundary energy. Coalescence is, therefore, impeded, but solution-precipitation from the small grain to the large grain can act to eliminate the energy barrier. For larger dihedral angles, where the liquid penetration of the boundary is significantly reduced, coalescence may become energetically favorable. The process is enhanced when the difference between the particle sizes becomes greater, so the conditions are most favorable for coalescence in the early portion of the solution-precipitation stage. Although coalescence by solid-state migration of the boundary has been reported in a few metallic systems [57], the occurrence of the process is probably low.

In systems where no solid–solid contact exists, coalescence may be possible by migration of the liquid film separating the grains. This mechanism has been found to occur in a few metallic systems and is normally referred to as *directional grain growth*. The coalescence process is not driven by a reduction in interfacial energy but by a reduction in chemical energy. Yoon and Huppmann [58] observed that when single crystal W spheres are sintered in the presence of liquid Ni, one W sphere grows at the expense of its neighbor (Figure 4.32a). An electron microprobe

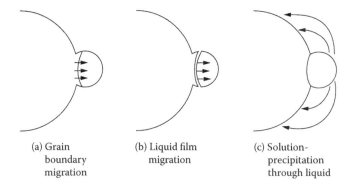

FIGURE 4.31 Three possible mechanisms of coalescence between contacting grains: (a) solid-state grain boundary migration; (b) liquid-film migration; (c) solution-precipitation through the liquid.

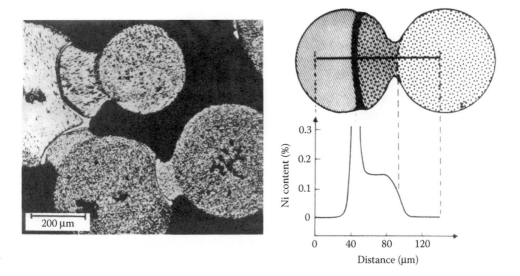

FIGURE 4.32 Directional grain growth during liquid-phase sintering of single-crystal W spheres with nickel at 1640°C, showing (a) the microstructure and (b) the microprobe analysis. (From Yoon, D.N., and Huppmann, W.J., Chemically driven growth of tungsten grains during sintering in liquid nickel, *Acta Metall.*, 27, 973, 1979. With permission.)

analysis shows that the shrinking grain consists of pure W, whereas the precipitated material on the growing grain is a solid solution of W containing 0.15 wt% Ni (Figure 4.32b). This compositional difference between the pure W and the solid solution provides a large decrease in chemical energy, which more than offsets the increase in the interfacial energy.

4.6.3 Stage 3: Ostwald Ripening

In Stage 3, densification slows considerably and microstructural coarsening clearly becomes the dominant process. For $\gamma_{ss}/\gamma_{sl} > 2$, the grains are completely separated by a liquid layer, the thickness of which, as discussed earlier, may decrease asymptotically. When the dihedral angle is greater than zero, solid–solid contacts lead increasingly to the formation of a rigid skeleton that inhibits the elimination of the isolated pores present in the liquid phase. With the development of the solid–solid contacts, solid-state sintering and coarsening take place, but matter transport through the liquid is much faster than through the solid, so solution-precipitation can still dominate over solid-state transport processes.

4.6.3.1 Densification by Pore Filling

When the volume of the liquid phase is low, solution-precipitation and grain shape accommodation lead to a slow, continuous elimination of the isolated porosity. In some cases, continued densification of the solid skeletal network can lead to partial expulsion of the liquid phase. For a larger volume of liquid, Kang, Kim, and Yoon [59] have shown that the filling of the isolated pores may occur in a discontinuous manner. The filling of the pores appears to be determined by grain growth rather than by grain shape accommodation. In their experiments, Kang et al. studied the shape change of grains surrounding large, isolated pores in the Mo(Ni) system. The sample was heated for 30 min at 1460°C and cooled. When the sintering process was repeated three times, the shape of the growing grains after each heating cycle was revealed by ghost boundaries formed within the grains as a result of strong etching. As shown in Figure 4.33a, the grains surrounding the pore grow laterally along the pore surface (e.g., the grains labeled A and B), so it is clear that the pore is not

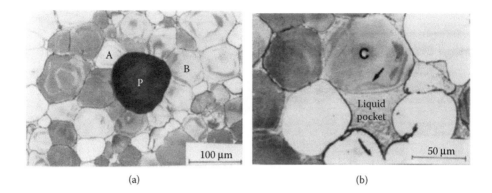

(a)   (b)

FIGURE 4.33 Microstructures of a Mo (4 wt% Ni) system showing (a) lateral growth of grains (such as A and B) around a pore (P) for a specimen sintered in three cycles (for 30 min in each cycle) at 1460°C; (b) after pore filling, preferential growth of a grain (C) into a liquid pocket formed at a pore site, resulting in a more homogeneous microstructure in a specimen sintered in three cycles (60 min, 30 min, and 30 min) at 1460°C. (From Lee, S.M., and Kang, S.J.L., Theoretical analysis of liquid-phase sintering: pore filling theory, *Acta Mater.*, 46, 3691, 1998. With permission.)

filled continuously by matter deposited on its surface. The pore can remain roughly unchanged for an extended period of sintering, but, at some point, when the size of the grains surrounding the pore reaches a critical value, the pore is filled rapidly. Upon further sintering (Figure. 4.33b), the etch boundaries of the grain labeled C reveal that the grains grow preferentially into the pocket of liquid, leading to a more rounded grain shape. Therefore, the microstructures in Figure 4.33 indicate that as long as the large pore is not filled with liquid, the grains grow around the pore, but once the pore is filled, the grains grow into the liquid pocket. The growth of the grains into the liquid pocket has the effect of improving the homogeneity of the microstructure, because the size of the heterogeneity (the liquid pocket) is reduced.

As sketched in Figure 4.34, a large pore remains unfilled because of preferential wetting of the necks between the particles. As the grains grow, the liquid reaches a favorable condition for filling the pore as determined by the curvature of the liquid–vapor meniscus [60,61]. In Figure 4.35, we see that the radius of curvature of the meniscus r_m increases with the radius a of the grain (assumed to be spherical) according to

$$r_m = a \frac{(1 - \cos \alpha)}{\cos \alpha}$$ (4.40)

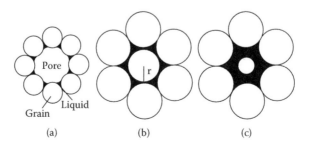

(a)   (b)   (c)

FIGURE 4.34 Pore filling during grain growth. A large pore is stable until grain growth increases the liquid meniscus radius sufficiently for capillary refilling of the pore.

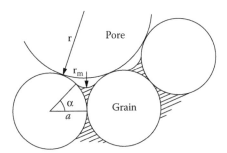

FIGURE 4.35 Calculation model for pore filling based on spherical grains surrounding the pore. Pore filling depends on the liquid meniscus radius exceeding the pore radius.

For a contact angle of zero, the critical point for pore filling occurs when r_m becomes equal to the pore radius, r, because after this point r_m can only decrease, leading to a reduction in the capillary pressure surrounding the pore. Liquid is drawn from the numerous menisci at the surface of the sample, whose radii change only very slightly because the volume of liquid required to fill the pore is small. If the contact angle is greater than zero, r_m must exceed r for pore filling to occur.

Pore filling driven by grain growth is an important densification mechanism for large isolated pores in Stage 3 of liquid-phase sintering, but a more recent theory of pore filling indicates that for some systems, the mechanism can make a significant contribution to densification from the early stages of liquid-phase sintering, and not just in the later stages [50]. Soon after rearrangement when a skeleton of solid grains forms, it is claimed that pore filling is the dominant densification mechanism when compared to contact flattening. The recent model appears to provide a good description of the microstructural evolution when a small amount of liquid phase is present and the dihedral angle is greater than zero, such as in the W(Ni-Fe) system.

Gases trapped in the isolated pores during final-stage densification can prevent complete filling of the pores. Sintering in a gaseous atmosphere that is not soluble in the liquid phase leads to trapped gas in the isolated pores, so densification stops when the increasing gas pressure in the shrinking pores equals the sintering stress, resulting in a limiting final density (see Chapter 6). Prolonged sintering, leading to pore growth by coalescence or by Ostwald ripening, results in a reduction of the sintered density. These problems can be alleviated by sintering in a vacuum or in a gaseous atmosphere that is soluble in the liquid.

4.6.3.2 Microstructural Coarsening

Coarsening by the Ostwald ripening mechanism was suggested as early as the 1930s to account for the significant grain growth observed in the liquid-phase sintering of heavy metals. Following the theoretical analysis of the process by Lifshitz, Slyozov, and Wagner, referred to as the LSW theory (see Chapter 3), several studies have been made to compare coarsening data obtained in liquid-phase sintering with the theory. According to the LSW theory, as well as modified analyses that take into account varying volume fraction of precipitates, the growth of the average grain size G with time t is given by

$$G^m = G_o^m + Kt \qquad (4.41)$$

where G_o is the initial value grain size, K is a temperature-dependent parameter, and the exponent m is dependent on the rate-controlling mechanism: $m = 3$ when diffusion through the liquid is rate controlling, and $m = 2$ for rate control by the interface reaction.

The grain growth exponent m is found to be close to 3 for many ceramic and metallic systems, indicating that diffusion-controlled coarsening is commonly active. An example is the work of Buist et al. [2], who investigated the growth of periclase (MgO), lime (CaO), and corundum (α-Al$_2$O$_3$) grains in a variety of liquid-phase compositions for a liquid content of 10–15 vol%. Typical microstructures of the sintered samples are shown in Figure 4.36. In all cases, a grain growth exponent very close to 3 was observed. Figure 4.37 shows the data for the growth of periclase grains in four different liquids corresponding to systems with the compositions given in Table 4.4. Despite the difference in the composition of the liquids, the results show good agreement with the cubic growth law.

The solubility of the solid in the liquid and the diffusion of the solute atoms through the liquid are expected to increase with temperature in accordance with the Arrhenius relationship, so the rate of coarsening increases with temperature, as can be observed in Figure 4.37. An increase in the dihedral angle produces a reduction in the area of the solid in contact with the liquid and a consequent increase in the grain boundary area. Because matter transport through the liquid is faster than by solid-state diffusion, a reduction in the solid–liquid contact area is expected to lower the rate of matter transport by solution-precipitation, leading to a reduction

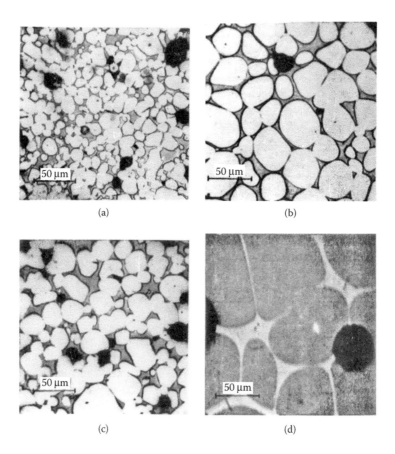

(a)

(b)

(c)

(d)

FIGURE 4.36 Microstructures of (a) composition 3 (see Table 4.4) after sintering for 0.5 h at 1550°C, showing rounded grains in a silicate matrix; (b) composition 3 after sintering for 8 h at 1550°C; (c) composition 2 after sintering for 8 h at 1550°C; and (d) a composition of 85 wt% CaO(15 wt% Ca$_2$Fe$_2$O$_5$) after firing for 8 h at 1550°C, showing rounded grains in a ferrite matrix. (The very dark areas are pores, some of which contain Araldite.) (From Buist, D.S., Jackson, C., Stephenson, I.M., Ford, W.F., and White, J., The kinetics of grain growth in two-phase (solid–liquid) systems, *Trans. Br. Ceram. Soc.*, 64, 173, 1965.)

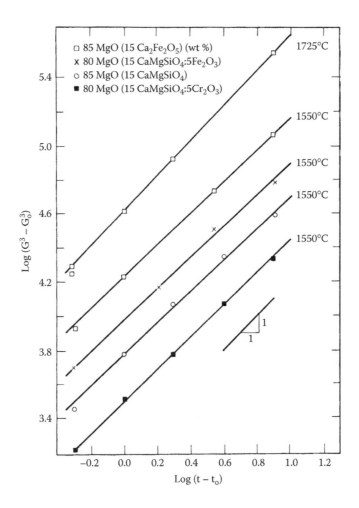

FIGURE 4.37 Plots of log $(G^3 - G_o^3)$ against log $(t - t_o)$ for systems containing periclase grains and liquid (see Table 4.4) at the sintering temperatures shown. G and G_o are the mean grain diameters (in μm) at times t and t_o (in hours).

TABLE 4.4

Parameters for the Liquid-Phase Sintering of MgO* in Four Different Liquids at 1550°C and 1725°C

		Dihedral Angle		Vol% Liquid	
Number	Composition (wt%)	1550°C	1725°C	1550°C	1725°C
1	85 MgO (15 CaMgSiO$_4$)	25	25	15	15
2	80 MgO (15 CaMgSiO$_4$·5Cr$_2$O$_3$)	40	30	16	15
3	80 MgO (15 CaMgSiO$_4$·5Fe$_2$O$_3$)	20	20	16	16
4	85 MgO (15 Ca$_2$Fe$_2$O$_5$)	15	15	11	12

*See Figure 4.36 to Figure 4.38.

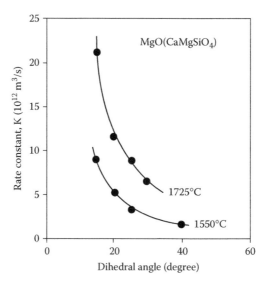

FIGURE 4.38 Effect of dihedral angle on the grain growth rate constant.

in the grain growth rate. Figure 4.38 shows that the measured grain growth rate constant for liquid-phase-sintered MgO compositions (Table 4.4) decreases with increasing dihedral angle. The microstructures of the compositions 2 and 3 should also be compared. Under nearly identical conditions, composition 3, with the lower dihedral angle, shows a larger grain size and less solid–solid contact (Figure 4.36b) than composition 2, which has the higher dihedral angle (Figure 4.36c).

Decreasing the volume fraction of liquid has the effect of raising the rate of grain growth [62], which is consistent with diffusion through the liquid being the rate-controlling mechanism. With decreasing liquid volume, the diffusion distance decreases, so matter transport is faster, leading to an increase in the growth rate. Several theories have been proposed to explain the dependence of grain growth on the volume fraction of liquid [16,53,63]. These analyses result in the same functional dependence of the grain size on time (Equation 4.41), but with a modified rate constant to account for the shorter diffusion distance with decreasing liquid volume. The most successful modification gives [31]

$$K = K_I + \frac{K_L}{V_L^{2/3}} \qquad (4.42)$$

where $K_I + K_L$ represents the rate constant at infinite dilution, K_L is a microstructure-dependent parameter, and V_L is the volume fraction of liquid.

4.7 NUMERICAL MODELING OF LIQUID-PHASE SINTERING

Because of the additional complexity resulting from the presence of the liquid phase, as well as the amount of computing power that would be required, numerical modeling of the microstructural evolution in liquid-phase sintering is less well developed than for solid-state sintering. The numerical models involve the assumed densification mechanisms, the kinetic laws of sintering and grain growth, and some characteristics of the microstructure. An example is the constitutive model developed by Svoboda, Riedel, and Gaebel [64], which incorporates rearrangement of the solid particles when the liquid is formed, grain shape accommodation by contact flattening, and filling of large pores and grain coarsening in the later stages of sintering. By fitting the predictions of the model to experimental data for liquid-phase sintering of Si_3N_4, it was found that after only a few percent of densification by rearrangement, the contact-flattening mechanism started to dominate and contributed by far the greatest part to the densification of this system. As indicated earlier, a recent model predicts that pore filling driven by grain growth is the dominant densification mechanism [50]. Which of the two mechanisms dominates appears to depend on the parameters of the system, such as the volume fraction of liquid phase and the dihedral angle.

4.8 HOT PRESSING WITH A LIQUID PHASE

Densification in the presence of a liquid phase can also be used with hot pressing. The chemical potential of the atoms under the contact surfaces increases with the applied stress, so matter transport from the contact regions to the pores is enhanced, leading to an increase in the densification rate. Hot pressing with a liquid phase was used by Bowen et al. [65] to densify Si_3N_4 powder containing MgO as the liquid-producing additive, which reacts with the SiO_2-rich surface layer on the Si_3N_4 powder to produce a eutectic liquid at ~1550°C. The measured densification kinetics are consistent with a solution-precipitation mechanism in which diffusion through the liquid is rate controlling. They can be described by an equation similar to Coble's equation for grain boundary diffusion–controlled solid-state sintering (see Chapter 2), provided that the grain boundary thickness, δ_{gb}, is replaced by the thickness of the liquid layer, δ_L; the grain boundary diffusion coefficient, D_{gb}, is replaced by the diffusion coefficient for the solute in the liquid, D_L; and the solid–vapor interfacial energy, γ_{sv}, is replaced by the liquid–vapor interfacial energy, γ_{lv}. When the applied pressure, p_a, is much greater than the capillary pressure due to the liquid meniscus, the densification rate can be written

$$\frac{1}{\rho}\frac{d\rho}{dt} = \frac{AD_L\delta_L\Omega}{G^3kT}p_a\phi \tag{4.43}$$

where A is a constant that depends on the geometry, Ω is the atomic volume of the rate-controlling species, G is the grain size, k is the Boltzmann constant, T is the absolute temperature, and ϕ is the stress intensification factor.

Figure 4.39 shows an example of the data used to test the validity of Equation 4.43 for the hot pressing of Si_3N_4(MgO) at 1650°C. After some threshold MgO content of ~1 wt%, the densification rate at a fixed relative density (0.70) increases linearly with the amount of MgO additive. If the MgO reacts with the available SiO_2 on the Si_3N_4 grains to form a composition close to the $MgSiO_3$–SiO_2 eutectic, then the amount of liquid phase is approximately proportional to the MgO content. The thickness of the liquid layer is proportional to the amount of liquid, and, thus, to the amount of MgO, so Figure 4.39 indicates that the densification rate is proportional to δ_L.

4.9 USE OF PHASE DIAGRAMS IN LIQUID-PHASE SINTERING

Phase equilibrium diagrams (or simply phase diagrams) play an important role in the selection of powder compositions and sintering conditions in liquid-phase sintering. Although the diagrams give the phases present under equilibrium conditions, the reaction kinetics during liquid-phase sintering are often too fast for equilibrium to be achieved, so the phase diagrams should serve only as a guide.

For a binary system consisting of a major component (referred to as the base, B) and a liquid-producing additive, A, the first task is to determine which additives will form a liquid phase

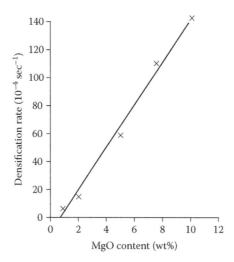

FIGURE 4.39 Densification rate at a relative density of 0.70 versus MgO additive content for hot pressing of the $Si_3N_4(MgO)$ system at 1650°C.

with the major component and under what conditions. For most systems, phase diagrams will be available in reference books [66]; if the desired diagram or region of the diagram is not available, considerable time is required to construct it. Figure 4.40 shows an idealized binary phase diagram that indicates the desirable composition and temperature characteristics for liquid-phase sintering. In addition to the solubility requirements discussed earlier, a desirable feature is a large difference in melting temperature between the eutectic and the base. The composition of the system should also be chosen away from the eutectic composition so that the liquid volume increases slowly with temperature, preventing the sudden formation of all of the liquid at or near the eutectic temperature. In practice, a typical sintering temperature would be chosen somewhat above the eutectic temperature, with a composition in the $L + S_2$ region.

4.9.1 ZINC OXIDE

Zinc oxide is liquid-phase sintered with ~0.5 mol% Bi_2O_3 (and small additions of other oxides) for the production of varistors. The phase diagram for the $ZnO(Bi_2O_3)$ system (Figure 4.41) shows the

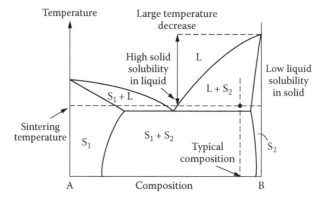

FIGURE 4.40 Model binary phase diagram showing the composition and sintering temperature associated with liquid-phase sintering in the $L + S_2$ phase field. The favorable characteristics for liquid-phase sintering include a suppression of the melting temperature, high solid solubility in the liquid, and low liquid solubility in the solid.

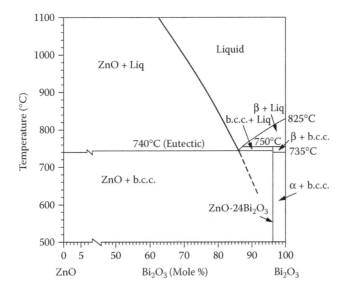

FIGURE 4.41 ZnO-Bi$_2$O$_3$ phase diagram showing limited solid solubility of Bi$_2$O$_3$ in ZnO and the formation of a eutectic containing 86 mol% Bi$_2$O$_3$ at 740°C. (From Safronov, G.M., Batog, V.N., Steoanyuk, T.V., and Fedorov, P.M., *Russian J. Inorg. Chem.*, 16, 460, 1971.)

formation of a Bi-rich liquid phase above the eutectic temperature of 740°C [67]. Samples that have been quenched from the sintering temperature show a liquid film that completely penetrates the grain boundaries (Figure 4.14c), indicating a zero dihedral angle. On the other hand, during slow cooling, the liquid and solid compositions change, with precipitation of the principal phase, leading to a substantial increase in the dihedral angle and to a nonwetting configuration (Figure 4.42).

4.9.2 SILICON NITRIDE

Silicon nitride is one of the ceramics best suited for mechanical engineering applications at high temperature. The crystal structure of Si$_3$N$_4$ has a high degree of covalent bonding, so diffusional mass transport by solid-state diffusion is very slow. Si$_3$N$_4$ powders are, therefore, normally converted

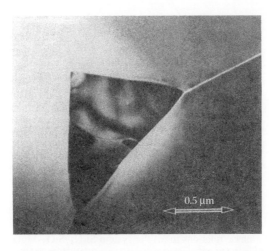

FIGURE 4.42 ZnO-Bi$_2$O$_3$ varistor composition, annealed for 43.5 h at 610°C, resulting in a nonwetting secondary phase. (Courtesy Y.-M. Chiang.)

into a dense body by liquid-phase sintering or by pressure sintering with a liquid phase. Much of the progress made in the fabrication and application of Si_3N_4 can be traced to careful work in the construction and use of the phase diagrams.

Si_3N_4 powders commonly contain 1–5 wt% SiO_2 as a surface oxidation layer. At a high enough temperature, an oxide additive reacts with the SiO_2 to form a silicate liquid that aids densification. On cooling, the silicate liquid often forms an amorphous intergranular glass. MgO (5–10 wt%) was one of the first successful additives used in the fabrication of Si_3N_4. It forms a eutectic liquid with SiO_2 at ~1550°C, but the sintering temperature can be anywhere between 100°C–300°C higher than the eutectic temperature. Figure 4.43a shows the identified tie lines in the subsolidus system Si_3N_4–SiO_2–MgO and the compositions used in their identification [68]. If equilibrium could be achieved during cooling, the fabricated materials would contain two phases if the composition lies

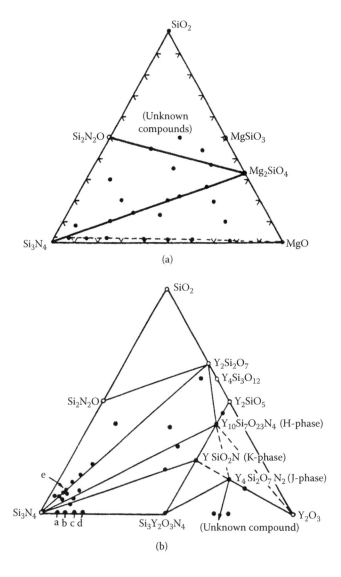

FIGURE 4.43 Experimental phase relations in (a) the system Si_3N_4–SiO_2–MgO determined from specimens hot pressed between 1500°C and 1750°C and (b) the system Si_3N_4–SiO_2–Y_2O_3 determined from specimens hot pressed between 1600°C and 1750°C. The filled circles represent the compositions examined. (Courtesy of F.F. Lange.)

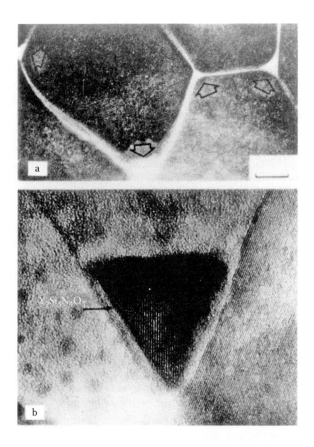

FIGURE 4.44 Transmission electron micrographs of (a) commercial sample of hot pressed $Si_3N_4(MgO)$, showing continuous glassy phase between β-Si_3N_4 grains (dark-field, bar = 50 nm) and (b) hot pressed $Si_3N_4(Y_2O_3)$ showing $Y_2Si_3O_3N_4$ grain surrounded by β-Si_3N_4 grains, with a glass phase between the crystalline phases (lattice fringe spacing for the β-Si_3N_4 grain at lower left is 0.663 nm). (Courtesy of D.R. Clarke.)

on the tie lines, or three phases if the composition lies within one of the compatibility triangles: $Si_3N_4 + MgO + Mg_2SiO_4$ or $Si_3N_4 + Si_2N_2O + Mg_2SiO_4$. Commonly, all of the magnesium is incorporated into the continuous glassy phase shown in Figure 4.44a, so only two phases are observed. The glassy phase has a relatively low softening temperature, and this causes a severe reduction in the high-temperature creep resistance of the material.

The use of Y_2O_3 as the additive instead of MgO leads to the formation of a eutectic liquid at a higher temperature (~1660°C). Although higher sintering temperatures are required, an improvement in the high-temperature creep resistance of the fabricated material is also observed. The phase relations for the Si_3N_4–SiO_2–Y_2O_3 system are shown in Figure 4.43b, along with the compositions used in their identification [69]. The system contains three quaternary Y–Si–O–N compounds compatible with Si_3N_4 that crystallize on cooling. Figure 4.44b shows the $Y_2Si_3O_3N_4$ phase that crystallized on cooling at a three- (or four-) grain junction for a material with a composition on the Si_3N_4–$Y_2Si_3O_3N_4$ tie line. It is separated from the surrounding Si_3N_4 grains by a thin residual glass film. The conversion of the glassy phase to a crystalline phase leads to a significant improvement in the creep resistance. Many Si_3N_4 ceramics are now sintered with an additive consisting of a combination of Y_2O_3 and Al_2O_3 [12,13]. On cooling, the liquid phase crystallizes to an yttrium–aluminum garnet phase. The fabricated Si_3N_4 shows improved high-temperature creep resistance as well as better oxidation resistance compared to the material fabricated with Y_2O_3 as the additive.

4.10 ACTIVATED SINTERING

When powder compacts containing minor amounts of a eutectic-forming additive are sintered, enhanced densification rates, compared to the pure powder, may be observed well before the eutectic temperature is reached. This effect is commonly referred to as *activated sintering*. Examples are found in the sintering of MgF_2 with CaF_2 additive [70], Al_2O_3 with CaO [71], ZnO with Bi_2O_3 [72], and several metallic systems, such as W with Pd, Ni, Pt, Co, or Fe [73]. Figure 4.45 shows the shrinkage and grain size for pure MgF_2 with up to 5 wt% CaF_2 during constant heating rate sintering. In this system, the eutectic temperature is 980°C, and the melting points of CaF_2 and MgF_2 are 1410°C and 1252°C, respectively. Enhanced densification rates are observed as much as 200°C below the eutectic temperature. The enhancing effect appears noticeable when as little as 0.1 wt% CaF_2 is added, and it saturates above 1 wt%. Grain growth rates are increased as a result of the CaF_2 addition, but at 900°C they are still relatively modest.

Generally, there is no clear difference in the principles involved in activated sintering and liquid-phase sintering other than that the activated system has a small concentration of additive (typically less than 1 wt%) and a sintering temperature below the eutectic temperature [74]. In some systems showing enhanced subeutectic densification, the presumption that no liquid is present may be in error due to impurities that lower the solidus temperature below the apparent eutectic temperature of the system, along with difficulties in detecting the liquid film.

Although the process of activated sintering is not clear, the enhanced subeutectic densification rates are often observed in systems where the additive segregates strongly to the grain boundary, presumably enhancing grain boundary transport rates significantly. The additive must also form a low-melting phase or eutectic with the major component and must have a large solubility for the major component. This phenomenon may be understood if one considers that diffusion rates scale approximately as the absolute temperature of the melting point of a solid. If the grain boundaries are rich in the eutectic-forming additive, then their relative melting points should be considerably below that of the major component, and their transport rates should be accordingly enhanced.

In ZnO with Bi_2O_3 additive, Bi-enriched intergranular amorphous films (~1 nm thick) are observed to form concurrently with activated sintering (Figure 4.46), and accelerated mass transport through these amorphous films has been suggested to be responsible for the activated sintering in

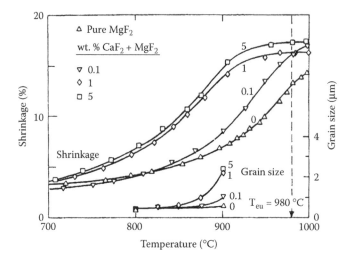

FIGURE 4.45 Shrinkage and grain size data for pure MgF_2 powder compacts with up to 5 wt% CaF_2 during constant heating rate sintering at 4.5°C/min. The eutectic temperature is 980°C and theoretical density corresponds to 20.1% shrinkage. (Courtesy of L.C. De Jonghe.)

FIGURE 4.46 High-resolution electron micrograph of Bi-enriched intergranular amorphous film in ZnO + 0.58 mol% Bi_2O_3 sintered at 4°C/min and air-quenched from 700°C. (Courtesy of Y.-M. Chiang.)

this system [72]. It is possible that intergranular films of a similar nature may also be responsible for the enhanced subeutectic densification in other systems, but experimental observations are currently lacking. Studies on the activated sintering of W indicate that diffusion through the intergranular layer is the rate-limiting step. Figure 4.47 shows that the degree of activated sintering depends on the additive; it results from the different activation energies for diffusion of W through the activator-rich intergranular phase [73].

4.11 VITRIFICATION

Vitrification is the term used to describe liquid-phase sintering in which densification is achieved by the viscous flow of a sufficient amount of liquid to fill up the pore spaces between the solid grains [75]. The driving force for vitrification is the reduction of solid–vapor interfacial energy due to the flow of the liquid to cover the solid surfaces. Vitrification is the common firing method for traditional clay-based ceramics, sometimes referred to as silicate systems. The process involves physical and chemical changes (e.g., liquid formation, dissolution, crystallization) as well as shape

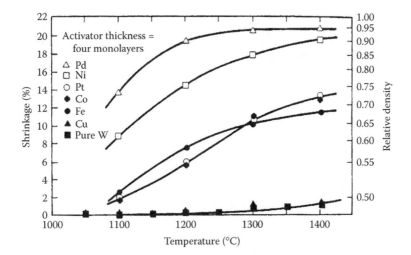

FIGURE 4.47 Activated sintering data for W particles treated with various transition metal additives. In each case the W was sintered for 1 h in hydrogen at the indicated temperature after being combined with one of the additives. (Courtesy of R.M. German.)

changes (e.g., shrinkage and deformation). A viscous silicate glass forms at the firing temperature and flows into the pores under the action of capillary forces, but it also provides some cohesiveness to the system to prevent significant distortion under the force of gravity. On cooling, a dense solid product results, with the glass gluing the solid particles together.

4.11.1 THE CONTROLLING PARAMETERS

The amount of liquid formed at the firing temperature and the viscosity of the liquid must be such that the required density (commonly full density) is achieved within a reasonable time without the sample's deforming under the force of gravity. As discussed earlier, the amount of liquid required to produce full densification by vitrification depends on the packing density achieved by the solid grains after rearrangement. In real powder systems, the use of a particle size distribution to improve the packing density, coupled with the occurrence of a limited amount of solution-precipitation, means that the amount of liquid required for vitrification is commonly 25–30 vol%. As in the case of liquid-phase sintering, the formation of the liquid must be controlled to prevent sudden formation of a large volume of liquid that would lead to distortion of the body under the force of gravity. The composition of the system should be chosen away from any eutectic composition.

We require a high enough densification rate of the system so that vitrification is completed within a reasonable time (less than a few hours), as well as a high ratio of the densification rate to the deformation rate so that densification is achieved without significant deformation of the article. These requirements determine, to a large extent, the firing temperature and the composition of the powder mixture, which control the viscosity of the liquid. The models for viscous sintering of a glass (see Chapter 2) predict that the densification rate depends on three major variables: the surface tension of the glass γ_{sv}, the viscosity of the glass η, and the pore radius r. Assuming that r is proportional to the particle radius, a, then the dependence of the densification rate on these parameters can be written

$$\dot{\rho} \sim \frac{\gamma_{sv}}{\eta a} \tag{4.44}$$

In many silicate systems, the surface tension of the glassy phase does not change significantly with composition, and the change in surface tension within the limited range of firing temperatures is also small. On the other hand, the particle size has a significant effect, with the densification rate increasing inversely with the particle size. However, by far the most important variable is the viscosity. The dependence of the viscosity of a glass on temperature is well described by the *Vogel-Fulcher equation*:

$$\eta = \eta_o \exp\left(\frac{C}{T - T_o} \right) \tag{4.45}$$

where T is the absolute temperature and η_o, C, and T_o are constants. This dependence is very strong; for example, the viscosity of soda-lime glass can decrease typically by a factor of ~1000 for an increase in temperature of only ~100°C. The glass viscosity also changes significantly with composition. The rate of densification can, therefore, be increased significantly by changing the composition, increasing the temperature, or some combination of the two, to reduce the viscosity. However, the presence of a large volume of liquid during vitrification means that if the viscosity is too low, the sample will deform easily under the force of gravity. Thus, the rate of densification relative to the rate of deformation must also be considered. If the ratio of the densification rate to the deformation rate is large, densification without significant deformation will be achieved.

The deformation rate is related to the applied stress, σ, and the viscosity by the expression

$$\dot{\varepsilon} = \sigma/\eta \qquad (4.46)$$

The force due to gravity exerted on a particle of mass m is given by $W = mg$, where g is the acceleration due to gravity. The mass of the particle varies as a^3, and the area over which the force acts is proportional to the area a^2 of the particle. The stress, therefore, varies as a. The ratio of the densification rate to the deformation rate varies according to

$$\frac{\dot{\rho}}{\dot{\varepsilon}} \sim \frac{1}{\eta a}\frac{\eta}{a} \sim \frac{1}{a^2} \qquad (4.47)$$

According to Equation 4.47, the ratio of the densification rate to the deformation rate of the system is enhanced as the particle size decreases. Therefore, the best way to achieve high densification without significant deformation is to use a fine particle size. Many successful silicate systems satisfy this requirement in that the compositions contain a substantial amount of clays that are naturally fine grained.

4.11.2 Vitrification of Silicate Systems

Triaxial whiteware compositions, such as porcelains and sanitary ware, are important silicate systems produced by vitrification. They are commonly prepared from powder mixtures containing three components. A typical porcelain composition is as follows:

50 wt% clay. For a clay consisting of predominantly kaolinite, the most common clay mineral, with a formula $Al_2(Si_2O_5)(OH)_4$, the composition of the clay corresponds to ~45 wt% Al_2O_3 and ~55 wt% SiO_2.

25 wt% feldspar. Feldspar is an alkali-containing mineral that acts as a flux. The alkali (often K^+) serves to lower the temperature at which the viscous liquid forms. A common feldspar has the formula $KAlSi_3O_8$.

25 wt% silica, present as quartz. The quartz acts as a filler to reduce the amount of shrinkage during firing and also serves to reduce the thermal expansion coefficient of the porcelain.

This and similar compositions lie in the primary mullite phase field in the ternary $K_2O–SiO_2–Al_2O_3$ phase diagram (Figure 4.48). Typical firing conditions are in the range 1200°C–1400°C. Between 1200°C and 1600°C, the equilibrium phases are mullite and liquid [76]. An isothermal section of the phase diagram (Figure 4.49) shows the equilibrium phases at 1200°C. At this temperature, the liquid has the composition of ~75 wt% SiO_2, ~12.5 wt% K_2O, and ~12.5 wt% Al_2O_3. In practice, the approach to equilibrium is incomplete, and only a small part of the SiO_2 present as quartz enters the liquid phase. The amount of SiO_2 that dissolves does not have a large effect on the amount or composition of the liquid. The cooled material contains mullite grains, a glass, and residual quartz particles. Triaxial porcelains are white, dense, and translucent and are used for the production of pottery, tiles, and insulators. Figure 4.50 shows the microstructure of a sanitary whiteware body that is commonly fired at ~1225°C [77]. The reflected light microscopy reveals the general distribution of the phases.

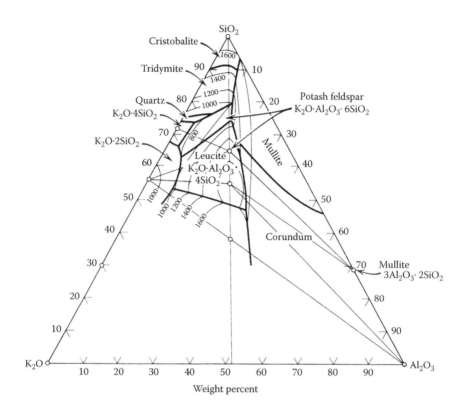

FIGURE 4.48 Phase relations for the ternary system $K_2O–Al_2O_3–SiO_2$. (From Schairer, J.F., and Bowen, N.L., Melting relations in the systems $Na_2O–Al_2O_3–SiO_2$ and $K_2O–Al_2O_3–SiO_2$, *Am. J. Sci.*, 245, 193, 1947. With permission.)

4.12 CONCLUDING REMARKS

Liquid-phase sintering is a common sintering process used in the production of a wide variety of ceramic materials. The liquid enhances densification through easier rearrangement of the grains and faster matter transport through the liquid. Liquid-phase sintering is divided conceptually into three overlapping stages defined in terms of the dominant mechanism operating in each stage: rearrangement, solution-precipitation, and Ostwald ripening. Rearrangement is typically fast, and models have been proposed to predict the shrinkage of the system as a function of time following its completion. The kinetics of densification are of interest because, together with grain growth, they form an integral part of microstructural evolution. Successful exploitation of liquid-phase sintering depends on controlling several material and processing parameters and requires a low contact angle, low dihedral angle, high solid solubility of the solid in the liquid, homogeneous packing of the particulate solid, homogeneous distribution of the liquid-producing additive, and fine particle size. Phase diagrams play an important role in the selection of compositions and sintering conditions for liquid-phase sintering. Liquid films that appear to have an equilibrium thickness have been observed at the grain boundaries of several ceramics, and this equilibrium film thickness has been explained in terms of charge interactions between surfaces, as well as structural and chemical forces. A process that is related to liquid-phase sintering is activated sintering, in which minor amounts of additives segregate to the grain boundaries and can enhance densification rates at temperatures well below any eutectic temperature in the system. If sufficient liquid is present, rearrangement by itself can lead to a fully dense material. This type of liquid-phase sintering, referred to as vitrification, is often used in the production of traditional clay-based ceramics.

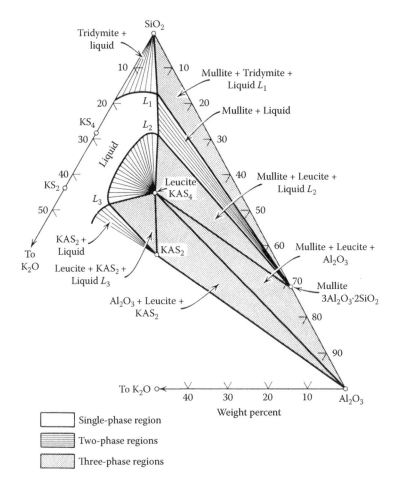

FIGURE 4.49 Isothermal cut in the K_2O–Al_2O_3–SiO_2 phase diagram at 1200°C. (From Kingery, W.D., Bowen, H.K., and Uhlmann, D.R., *Introduction to Ceramics,* 2nd ed., Wiley, New York, 1976. With permission.)

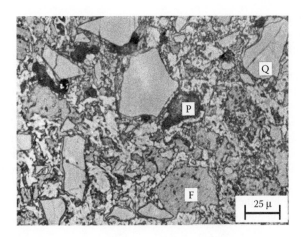

FIGURE 4.50 Reflected light micrograph of commercial whiteware body composed of clay, feldspar, and quartz, showing a microstructure containing some residual feldspar (F), porosity (P), and quartz (Q). (From Chu, G.P.K., Microstructure of complex ceramics, in *Ceramic Microstructures*, Fulrath, R.M., and Pask, J.A., Eds., Wiley, New York, 1968, p. 828.)

PROBLEMS

4.1   For each of the stages associated with classic liquid-phase sintering, give the name of
      the stage, the dominant processes, and the main microstructural changes.

4.2   Determine the range of contact angles for the following conditions:

      (a) $\gamma_{lv} = \gamma_{sv} < \gamma_{sl}$
      (b) $\gamma_{lv} > \gamma_{sv} > \gamma_{sl}$
      (c) $\gamma_{lv} > \gamma_{sv} = \gamma_{sl}$

4.3   Given the following interfacial energies (in units of J/m^2) for liquid-phase sintering: $\gamma_{sv} =$
      1.0, $\gamma_{lv} = 0.8$, $\gamma_{ss} = 0.75$, and $\gamma_{sl} = 0.35$, determine:

      (a) the solid–liquid contact angle
      (b) whether the liquid will penetrate the grain boundaries

4.4   State the relationship showing how the dihedral angle ψ is related to the values of the
      grain boundary energy γ_{ss} and the solid–liquid interfacial energy γ_{sl} for a polycrystalline
      solid in contact with a liquid.

      Consider two spheres with the same radius a in contact during liquid-phase sintering.
      Derive an expression for the equilibrium radius of the neck X formed between the spheres
      when the dihedral angle is ψ. Using the relationship between the dihedral angle and the
      interfacial energies, plot a graph of the ratio X/a versus γ_{ss}/γ_{sl}.

4.5   Wetting of the solid by the liquid phase is a necessary criterion for efficient liquid-phase
      sintering. Discuss the wetting phenomena that are generally relevant to the liquid-phase
      sintering of ceramic systems.

4.6   Amorphous films with an equilibrium thickness of 0.5–2 nm are often observed between
      the grains in ceramics that have been fabricated by liquid-phase sintering. Discuss the
      explanations that have been put forward for the stabilization of the film thickness.

4.7   Briefly discuss whether each of the following factors is likely to promote shrinkage or
      swelling during the initial stage of liquid-phase sintering:

      (a) high solid solubility in the liquid
      (b) low liquid solubility in the solid
      (c) large contact angle
      (d) small dihedral angle
      (e) high green density of the compact
      (f) large particle size of the liquid-producing additive

4.8   Derive Equation 4.38 for the shrinkage during the solution-precipitation stage of liquid-
      phase sintering when the phase boundary reaction is the rate-controlling mechanism.

4.9   Estimate the factor by which the densification rate during hot pressing is increased by the
      presence of a liquid phase relative to a similar system without a liquid phase, in which
      densification occurs by a grain boundary diffusion mechanism. Assume that a continuous
      wetting grain boundary film with a thickness $\delta_L = 50$ nm forms during liquid-phase sintering
      and that the diffusion coefficient through the liquid $D_L = 100\, D_{gb}$, where D_{gb} is the diffusion
      coefficient for grain boundary diffusion in the absence of the liquid.

4.10 Discuss how each of the following variables is expected to influence the densification
      and microstructural evolution during liquid-phase sintering of Si$_3$N$_4$:

      (a) the composition of the liquid-producing additive
      (b) the volume fraction of the liquid phase
      (c) the sintering temperature
      (d) the particle size of the Si$_3$N$_4$ powder
      (e) the applied pressure

REFERENCES

1. Kim, H.Y., Lee, J.A., and Kim, J.J., Densification behavior of fine alumina and coarse alumina compacts during liquid-phase sintering with the addition of talc, *J. Am. Ceram. Soc.*, 83, 3128, 2000.
2. Buist, D.S., Jackson, C., Stephenson, I.M., Ford, W.F., and White, J., The kinetics of grain growth in two-phase (solid–liquid) systems, *Trans. Br. Ceram. Soc.*, 64, 173, 1965.
3. Hart, P.E., Atkin, R.B., and Pask, J.A., Densification mechanisms in hot-pressing of magnesia with a fugitive liquid, *J. Am. Ceram. Soc.*, 53, 83, 1970.
4. Kim, J., Kimura, T., and Yamaguchi, T., Effect of bismuth oxide content on the sintering of zinc oxide, *J. Am. Ceram. Soc.*, 72, 1541, 1989.
5. Dey, D., and Bradt, R.C., Grain growth of ZnO during Bi_2O_3 liquid-phase sintering, *J. Am. Ceram. Soc.*, 75, 2529, 1992.
6. Hennings, D., Janssen, R., and Reynen, P.J.L., Control of liquid-phase-enhanced discontinuous grain growth in barium titanate, *J. Am. Ceram. Soc.*, 70, 23, 1987.
7. Guha, J.P., and Anderson, H.U., Reactions during sintering of barium titanate with lithium fluoride, *J. Am. Ceram. Soc.*, 69, C193, 1986.
8. Hausonne, J.M., Desgardin, G., Bajolet, P., and Raveau, B., Barium titanate perovskite sintered with lithium fluoride, *J. Am. Ceram. Soc.*, 77, 801, 1983.
9. Kingery, W.D., Bowen, H.K., and Uhlmann, D.R., *Introduction to Ceramics*, 2nd ed., Wiley, New York, 1976, Chap. 10.
10. Brook, R.J., Preparation and electrical behavior of zirconia ceramics, in *Advances in Ceramics, Vol. 3, Science and Technology of Zirconia*, Heuer, A.H., and Hobbs, L.W., Eds., The American Ceramic Society, Westerville, OH, 1981, p. 272.
11. Terwilliger, G.R., and Lange, F.F., Pressureless sintering of silicon nitride, *J. Mater. Sci.*, 10, 1169, 1975.
12. Riley, F.L., Silicon nitride and related materials, *J. Am. Ceram. Soc.*, 83, 245, 2000.
13. Suttor, D., and Fischman, G.S., Densification and sintering kinetics in sintered silicon nitride, *J. Am. Ceram. Soc.*, 75, 1063, 1992.
14. Kim, D.H., and Kim, C.H., Toughening behavior of silicon carbide with additions of yttria and alumina, *J. Am. Ceram. Soc.*, 73, 1431, 1990.
15. Kingery, W.D., Niki, E., and Narasimhan, M.D., Sintering of oxide and carbide-metal compositions in presence of a liquid phase, *J. Am. Ceram. Soc.*, 44, 29, 1961.
16. German, R.M., *Liquid Phase Sintering*, Plenum Press, New York, 1985.
17. German, R.M., *Sintering Theory and Practice*, Wiley, New York, 1996.
18. Weirauch, D.A., Jr. and Ziegler, D.P., An improved method for the determination of the surface tension of silicate melts, Proceedings of the 12th University Conference on Glass Science, Alfred, New York, 1993.
19. Cannon, R.M., Saiz, E., Tomsia, A.P., and Carter, W.C., Reactive wetting taxonomy, *Mater. Res. Soc. Symp. Proc.*, 357, 279, 1995.
20. Beeré, W., A unifying theory of the stability of penetrating liquid phases and sintering pores, *Acta Metall.*, 23, 131, 1975.
21. Park, H.H., and Yoon, D.N., Effect of dihedral angle on the morphology of grains in a matrix phase, *Met. Trans. A*, 16, 923, 1985.
22. Smith, C.S., Grains, phases, and interfaces: an interpretation of microstructure, *Trans. AIME*, 175, 15, 1948.
23. Heady, R.B., and Cahn, J.W., An analysis of the capillary forces in liquid-phase sintering of spherical particles, *Metall. Trans.*, 1, 185, 1970.
24. Hwang, K.S., German, R.M., and Lenel, F.V., Capillary forces between spheres during agglomeration and liquid-phase sintering, *Metall. Trans. A*, 18, 11, 1987.
25. Huppmann, W.J., and Riegger, H., Modeling rearrangement processes in liquid-phase sintering, *Acta Metall.*, 23, 965, 1975.
26. Kipphut, C.M., Bose, A., Farooq, S., and German, R.M., Gravity and configurational energy induced microstructural changes in liquid-phase sintering, *Metall. Trans. A*, 19, 1905, 1988.
27. German, R.M., Limitations in net shaping by liquid-phase sintering, *Advances in Powder Metallurgy*, Vol. 4, Metal Powder Industries Federation, Princeton, NJ, 1991, p. 183.
28. Raman, R., and German, R.M., A mathematical model for gravity-induced distortion during liquid-phase sintering, *Metall. Mater. Trans. A*, 26, 653, 1995.

29. Upadhyaya, A., and German, R.M., Shape distortion in liquid-phase sintered tungsten heavy alloys, *Metall. Mater. Trans. A*, 29, 2631, 1998.

30. Norton, F.H., *Elements of Ceramics*, 2nd ed., Addison-Wesley, Reading, MA, 1974.

31. German, R.M., Microstructure of the gravitationally settled region in a liquid-phase sintered dilute tungsten heavy alloy, *Metall. Mater. Trans. A*, 26, 279, 1995.

32. German, R.M., and Liu, Y., Grain agglomeration in liquid-phase sintering, *J. Mater. Synthesis & Processing*, 4, 23, 1996.

33. Lange, F.F., Liquid-phase sintering: are liquids squeezed out from between compressed particles? *J. Am. Ceram. Soc.*, 65, C23, 1982.

34. Eley, D.D., *Adhesion*, Oxford University Press, Oxford, UK, 1961, p. 118.

35. Clarke, D.R., On the equilibrium thickness of intergranular glass phases in ceramic materials, *J. Am. Ceram. Soc.*, 70, 15, 1987.

36. Derjaguin, B.V., and Churaev, N.V., Structural component of disjoining pressure, *J. Colloid Interface Sci.*, 49, 249, 1974.

37. Clarke, D.R., Shaw, T.M., Philipse, A.P., and Horn, R.G., Possible electrical double-layer contribution to the equilibrium thickness of intergranular glass films in polycrystalline ceramics, *J. Am. Ceram. Soc.*, 76, 1201, 1993.

38. Rahaman, M.N., *Ceramic Processing*, CRC Press, Boca Raton, FL, 2006.

39. Tanaka, I., Kleebe, H.J., Cinibulk, M.K., Bruley, J., Clarke, D.R, and Ruhle, M., Calcium concentration dependence of the intergranular film thickness in silicon nitride, *J. Am. Ceram. Soc.*, 77, 911, 1994.

40. Gu, H., Pan, X., Cannon, R.M., and Ruhle, M., Dopant distribution in grain-boundary films in calcia-doped silicon nitride ceramics, *J. Am. Ceram. Soc.*, 81, 3125, 1998.

41. Brydson, R., Chen, S.C., Riley, F.L., Milne, S.J., Pan, X., and Ruhle, M., Microstructure and chemistry of intergranular glassy films in liquid-phase sintered alumina, *J. Am. Ceram. Soc.*, 81, 369, 1998.

42. Wang, H., and Chiang, Y.M., Thermodynamic stability of intergranular amorphous films in bismuth-doped zinc oxide, *J. Am. Ceram. Soc.*, 81, 89, 1998.

43. Kwon, O.-J., and Yoon, D.N., Closure of isolated pores in liquid-phase sintering of W-Ni, *Internat. J. Powder Metall. & Powder Technol.*, 17, 127, 1981.

44. Shaw, T.M., Liquid redistribution during liquid-phase sintering, *J. Am. Ceram. Soc.*, 69, 27, 1986.

45. Kingery, W.D., Densification during sintering in the presence of a liquid phase, I. Theory, *J. Appl. Phys.*, 30, 301, 1959.

46. Kang, S.J.L., Kaysser, W.A., Petzow, G., and Yoon, D.N., Elimination of pores during liquid-phase sintering of Mo-Ni, *Powder Metall.*, 27, 97, 1984.

47. Huppmann, W.J., Riegger, H., Kaysser, W.A., Smolej, V., and Pejovnik, S., The elementary mechanisms of liquid-phase sintering, I. Rearrangement, *Z. Metallkunde*, 70, 707, 1979.

48. Fortes, M.A., The kinetics of powder densification due to capillary forces, *Powder Metall. Internat.*, 14, 96, 1982.

49. Lee, S.M., Chaix, J.M., Martin, C.L., Allibert, C.H., and Kang, S.J.L., Computer simulation of particle rearrangement in the presence of liquid, *Met. Mater.*, 5, 197, 1999.

50. Lee, S.M., and Kang, S.J.L., Theoretical analysis of liquid-phase sintering: pore filling theory, *Acta Mater.*, 46, 3691, 1998.

51. Yoon, D.N., and Huppmann, W.J., Grain growth and densification during liquid-phase sintering of W-Ni, *Acta Metall.*, 27, 693, 1979.

52. Takajo, S., Kaysser, W.A., and Petzow, G., Analysis of particle growth by coalescence during liquid-phase sintering, *Acta Metall.*, 32, 107, 1984.

53. Vorhees, P.W., Ostwald ripening of two-phase mixtures, *Ann. Rev. Mater. Sci.*, 22, 197, 1992.

54. Kingery, W.D., and Narasimhan, M.D., Densification during sintering in the presence of a liquid phase, II. Experimental, *J. Appl. Phys.*, 30, 307, 1959.

55. Huppmann, W.J., and Riegger, H., Liquid-phase sintering of the model system W–Ni, *Internat. J. Powder Met. Powder Technol.*, 13, 243, 1977.

56. Kaysser, W.A, Takajo, S., and Petzow, G., Particle growth by coalescence during liquid-phase sintering of Fe–Cu, *Acta Metall.*, 32, 115, 1984.

57. Huppmann, W.J., and Petzow, G., The elementary mechanisms of liquid-phase sintering, in *Sintering Processes, Mater. Sci. Res.* Vol. 13, Kuczynski, G.C., Ed., Plenum Press, New York, 1980, p. 189.

58. Yoon, D.N., and Huppmann, W.J., Chemically driven growth of tungsten grains during sintering in liquid nickel, *Acta Metall.*, 27, 973, 1979.

59. Kang, S-J.L., Kim, K.-H., and Yoon, D.N., Densification and shrinkage during liquid-phase sintering, *J. Am. Ceram. Soc.*, 74, 425, 1991.

60. Park, H.H., Cho, S.-J., and Yoon, D.N., Pore filling process during liquid-phase sintering, *Metall. Trans. A*, 15A, 1075, 1984.

61. Park, H.-H., Kwon, O.-J., and Yoon, D.N., The critical grain size for liquid flow into pores during liquid-phase sintering, *Metall. Trans. A*, 17, 1915, 1986.

62. Kang, T.-K., and Yoon, D.N., Coarsening of tungsten grains in liquid nickel-tungsten matrix, *Metall. Trans. A*, 9, 433, 1978.

63. Yang, S.C., Manni, S.S., and German, R.M., The effect of contiguity on growth kinetics in liquid-phase sintering, *J. Metals*, 42, 16, 1990.

64. Svoboda, J., Riedel, H., and Gaebel, R., A model for liquid-phase sintering, *Acta Mater.*, 44, 3215, 1996.

65. Bowen, L.J., Weston, R.J., Carruthers, T.G., and Brook, R.J., Hot-pressing and the α–β phase transformation in silicon nitride, *J. Mater. Sci.*, 13, 341, 1978.

66. Levin, E.M., Robbins, C.R., and McMurdie, H.W., *Phase Diagrams for Ceramists*, The American Ceramic Society, Columbus, OH, 1964.

67. Safronov, G.M., Batog, V.N., Steoanyuk, T.V., and Fedorov, P.M., *Russian J. Inorg. Chem.*, 16, 460, 1971.

68. Lange, F.F., Phase relations in the system Si_3N_4–SiO_2–MgO and their interrelation with strength and oxidation, *J. Am. Ceram. Soc.*, 61, 53, 1978.

69. Lange, F.F., Singhal, S.C., and Kuznicki, R.C., Phase relations and stability studies in the Si_3N_4–SiO_2–Y_2O_3 pseudoternary, *J. Am. Ceram. Soc.*, 60, 249, 1977.

70. Hu, S.C., and De Jonghe, L.C., Pre-eutectic densification in MgF_2–CaF_2, *Ceramics Internat.*, 9, 123, 1983.

71. Wu, S.J., De Jonghe, L.C., and Rahaman, M.N., Subeutectic densification and second-phase formation in Al_2O_3–CaO, *J. Am. Ceram. Soc.*, 68, 385, 1985.

72. Luo, J., Wang, H., and Chiang, Y.M., Origin of solid-state activated sintering in Bi_2O_3-doped ZnO, *J. Am. Ceram. Soc.*, 82, 916, 1999.

73. German, R.M., and Munir, Z.A., Enhanced low-temperature sintering of tungsten, *Metall. Trans. A*, 7, 1873, 1976.

74. German, R.M., A quantitative theory of diffusional activated sintering, *Sci. Sintering*, 15, 27, 1983.

75. Cambier, F., and Leriche, A., Vitrification, in *Materials Science and Technology*, Vol. 17B, Cahn, R.W., Haasen, P., and Kramer, E.J., Eds., VCH, New York, 1996, p. 123.

76. Schairer, J.F., and Bowen, N.L., Melting relations in the systems Na_2O–Al_2O_3–SiO_2 and K_2O–Al_2O_3–SiO_2, *Am. J. Sci.*, 245, 193, 1947.

77. Chu, G.P.K., Microstructure of complex ceramics, in *Ceramic Microstructures*, Fulrath, R.M., and Pask, J.A., Eds., Wiley, New York, 1968, p. 828.

5 Special Topics in Sintering

5.1 INTRODUCTION

The effects of variables such as grain size, temperature, and applied pressure on sintering are well understood. For powder systems that approach the assumptions of the theoretical models (i.e., homogeneously packed spherical particles with a narrow size distribution), the sintering kinetics can be well described by the models. However, some important issues remain, and these need to be examined in terms of their ability to limit the attainment of high density with controlled microstructure.

The structure of real powder compacts is never completely homogeneous. Various types of inhomogeneities, such as variations in packing density, a distribution of pore sizes, and a distribution of particle sizes, lead to *differential densification* during sintering, where different regions of the powder compact sinter at different rates. Differential densification leads both to the development of *transient stresses* that reduce the densification rate and to the growth of microstructural flaws that limit the engineering properties of the ceramic. The quantification and control of inhomogeneities in the powder compact, as well as techniques for limiting the magnification of inhomogeneities during sintering, form key issues for microstructural control.

In composites, inhomogeneities are a necessary feature of the microstructural design of the material. *Rigid inclusions* such as particles, platelets, or fibers are incorporated into the material to improve its properties, but they invariably create severe problems for sintering by retarding the densification of the powder matrix. The retardation of densification is often observed to be much more severe in polycrystalline matrices than in glass matrices. Key issues in the sintering of powder compacts with rigid inclusions include the mechanisms responsible for the reduced densification and how the sintering difficulties can be alleviated.

The sintering of *adherent films* on a rigid substrate is required in several important applications. The substrate provides an external constraint, preventing shrinkage in the plane of the substrate, so all of the shrinkage occurs in the direction perpendicular to the plane of the film. Tensile stresses develop in the plane of the film that lower the densification rate of the constrained film relative to that of a free, unconstrained film. If high enough, they can lead to delamination of the film from the substrate or to cracking of the film.

A continuous second phase in polycrystalline ceramics, such as a continuous pore phase or fibers in a composite, may undergo capillary-induced shape changes, leading to breakup and the formation of discrete pores or particles. The physical distribution of secondary phases has important consequences for the properties of ceramics, so an understanding of the *morphological stability* of continuous phases in ceramics is important. The stability of thin polycrystalline films must also be considered, because under certain conditions they may become unstable and break up into islands, uncovering the substrate.

Solid solution additives or dopants provide an effective approach for the fabrication, by sintering, of ceramics with high density and controlled grain size, but the role of the dopant is not well understood. This gap in understanding limits the applicability of the approach. The major roles of the dopant and the parameters that control the effectiveness of a dopant in a given role form important issues in the solid solution additive approach.

In the production of multicomponent ceramics from a mixture of powders, a chemical reaction occurring between the starting materials during sintering can introduce further problems for microstructural control. Crystallization of an amorphous material during sintering can also lead to sintering difficulties. The parameters that control the separate processes and the interaction of densification with a chemical reaction or with crystallization must be understood to provide a basis for selecting the appropriate processing conditions for sintering.

5.2 INHOMOGENEITIES AND THEIR EFFECTS ON SINTERING

Microstructural inhomogeneities present in the powder compact generally have an adverse effect on sintering. The characteristics of the powder and the forming method control the extent of the structural inhomogeneities in the green compact [1]. In general, we can expect various types of structural inhomogeneity, such as variations in packing density, a distribution in pore sizes, and a distribution in particle sizes. If the powder contains impurities, chemical inhomogeneity in the form of variations in composition will also be present. A common feature of such inhomogeneities is that they become exaggerated during sintering, often leading to a reduction of the densification rate and to the development of large pores or crack-like voids during sintering. It is, therefore, important to understand how inhomogeneities influence sintering and how their effects on sintering can be controlled.

5.2.1 DIFFERENTIAL DENSIFICATION

Structural inhomogeneities lead to some regions of the powder compact densifying at different rates from other regions, a process referred to as *differential densification*. The regions densifying at different rates interact with each other, and this interaction leads to the development of transient stresses during sintering. The stresses are said to be transient because they diminish rapidly when the densification is complete or when the sample is cooled to temperatures where no densification takes place, but in many cases some residual stress may remain after densification. The development of transient stresses during sintering is analogous to the development of thermal stresses in materials that differ in thermal expansion coefficients. In sintering, the volumetric strain rate takes the place of the thermal strain.

Stresses due to differential densification have been analyzed for a model consisting of a spherical inhomogeneity surrounded by a uniform powder matrix [2,3]. The stress system is analogous to that of a thermal stress problem in which a spherical core is surrounded by a cladding with a different thermal expansion coefficient. If the inhomogeneity (e.g., a hard, dense agglomerate) shrinks more slowly than the surrounding matrix, a hydrostatic backstress is generated in the matrix. This backstress opposes the sintering stress and leads to a reduction in the densification rate of the powder matrix (Figure 5.1a). Differential densification may also lead to the growth of preexisting flaws in the body. For the case of the spherical inhomogeneity that shrinks more slowly than the powder matrix, a circumferential (or hoop) stress is also set up in the matrix, which, if high enough, can lead to the growth of crack-like voids in the matrix (Figure 5.1b). Microstructural flaws can also result if the inhomogeneity shrinks faster than the surrounding matrix. In this case, the inhomogeneity can shrink away from the matrix, creating a circumferential void (Figure 5.1c). Flaws such as those sketched in Figure 5.1b and Figure 5.1c have been observed in some sintered articles produced from heterogeneous powder compacts [4].

Another type of microstructural flaw produced by differential densification has been observed in irregular two-dimensional arrays of monosize copper spheres [5]. An example of the microstructural evolution in such arrays is shown in Figure 5.2. A detailed statistical analysis of the evolution of a similar type of array revealed that rearrangement of the particles occurred as a result of differential densification [6]. The regions that densify faster exert tensile stresses on the neighboring regions that are densifying more slowly. If the stresses are larger than and opposed to the sintering stress, pores at that location will grow rather than shrink, despite the fact that the compact undergoes an overall shrinkage. In three dimensions, movement of the particles is more constrained because

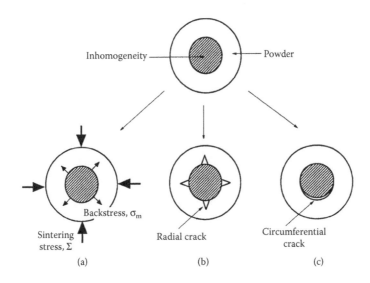

FIGURE 5.1 Schematic diagram illustrating the effects of inhomogeneities in a powder compact. Reduced densification (a) and the growth of radial flaws (b) due to an inhomogeneity that sinters more slowly than the surrounding region, and (c) a circumferential flaw due to an inhomogeneity that sinters at a faster rate than the surrounding region.

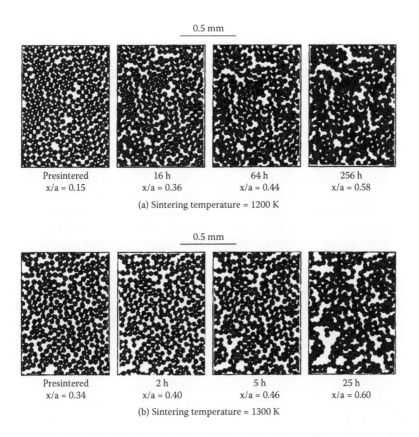

FIGURE 5.2 Evolution of the structure of two planar arrays of copper spheres during sintering. Two different starting densities, expressed as x/a, are shown, where x is the neck radius between particles of radius a.

of the higher coordination number, so particle rearrangement due to differential densification is expected to be lower than that observed in the two-dimensional arrays.

5.2.2 CONTROL OF INHOMOGENEITIES

It is clear from Figure 5.1 and Figure 5.2 that differential densification can have a significant effect on densification and microstructural evolution. To reduce the undesirable effects of differential densification, we must reduce the extent of inhomogeneity present in the green compact through control of the powder quality and the forming method [1]. Important issues in the control of inhomogeneities in the green compact are the quantification of the extent of inhomogeneity in a powder compact, methods for reducing or averting inhomogeneities, and mechanisms for reversing structural inhomogeneities.

The quantification of inhomogeneities may allow us to determine what level of inhomogeneity can be tolerated in a given process. However, such quantification is difficult and has been discussed only at a very qualitative level. The inhomogeneity depends on the scale of observation, in that the structure appears more uniform as the scale of observation becomes coarser.

One approach to reducing structural inhomogeneities in a powder compact involves the application of pressure during the compaction stage (e.g., by cold isostatic pressing) or during sintering (e.g., by hot pressing). Another approach is to attempt to produce a uniform pore size that is smaller than the particle size through colloidal consolidation or other methods. This approach has shown clear benefits, as exemplified by the work of Rhodes [7] and Yeh and Sacks [8] in which the effect of particle packing on sintering was investigated. Using the same Al_2O_3 powder, which had a narrow size distribution, Yeh and Sacks prepared powder compacts by slip casting a well-stabilized aqueous suspension (pH = 4) and a flocculated suspension (pH = 9). The structural variation in the two types of powder compacts is shown in Figure 5.3. During sintering under the same conditions (1340°C), the more homogeneous compact with the higher packing density (pH = 4) had a higher densification rate and reached a higher final density than the less homogeneous compact (Figure 5.4).

5.2.3 CORRECTION OF INHOMOGENEITIES

Faced with the problem of inhomogeneities normally present in real powder compacts, it is worth examining whether sintering procedures are available for reversing such inhomogeneities and the extent to which the inhomogeneities can be reversed by each procedure.

An example where some homogenization of the microstructure has been observed comes from the sintering of compacts formed from mixtures of powders with particles of two different sizes, i.e., powders with a bimodal particle size distribution [9]. If each fraction in the mixture is considered to densify in the same way as it would independently of the other, we may expect the densification of the mixture to obey a simple rule of mixtures, in which case the densification of the overall system is some weighted average of the densification of the separate powder fractions. For Al_2O_3 powder in which the particle size of the coarse fraction was approximately 10 times that of the fine fraction, the densification behavior was indeed found to approximately obey a rule of mixtures. An interesting observation was that grain growth occurred only in the fine powder fraction, allowing the microstructure to become more homogeneous than that of the initial powder compact.

A common type of inhomogeneity present in ceramic powder compacts consists of agglomerates that lead to density variations and to a distribution of the pore sizes. At one extreme of the pore size range, we have pores that are large compared to the grain size of the powder system, which would normally be characteristic of the pores between the agglomerates. These pores are surrounded by a large number of grains, so that the pore coordination number N, defined as the number of grains surrounding the pore, is large. At the other extreme, small pores with low values of N will exist, which would be characteristic of the pores in the well-packed regions of the agglomerates. As discussed in Chapter 3, according to the thermodynamic analysis of Kingery and Francois [10], pores with N greater than some critical value N_C will grow, whereas those pores with $N < N_C$ will shrink. An important

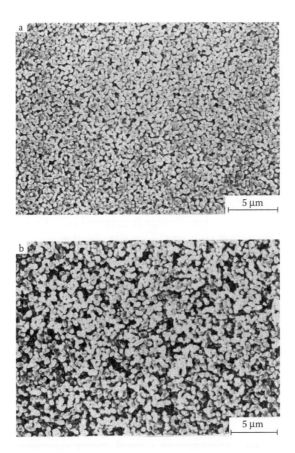

FIGURE 5.3 Scanning electron micrographs of the top surfaces of slip-cast samples of Al$_2$O$_3$ formed from (a) well-dispersed (pH 4) and (b) flocculated (pH 9) suspensions. Note the difference in packing homogeneity between the samples. (From Yeh, T.-S., and Sacks, M.D., Effect of green microstructure on sintering of alumina, *Ceramic Trans.*, 7, 309, 1990. With permission.)

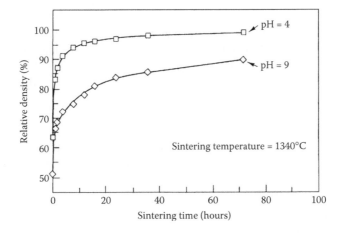

FIGURE 5.4 Relative density versus sintering time at 1340°C for the two samples described in Figure 5.3. The sample with the more homogeneous packing (prepared from the suspension at pH 4) has a higher density at any given time. (From Yeh, T.-S., and Sacks, M.D., Effect of green microstructure on sintering of alumina, *Ceramic Trans.*, 7, 309, 1990. With permission.)

question relating to the correction of inhomogeneities during sintering is whether the growth of the large pores can be reversed, thereby reversing the progress toward a more heterogeneous structure.

5.2.4 Shrinkage of Large Pores

As sketched in Figure 5.5, assuming that the grains grow faster than the pore, grain growth can eventually lead to a situation where N for the large pore becomes less than N_C. When this situation occurs, shrinkage of the large pore will occur, and the progress toward a wide distribution in pore sizes (i.e., increased inhomogeneity) will be reversed. This suggests that normal grain growth might be beneficial for the sintering of heterogeneous powder compacts, since it leads to a reduction in N [11]. Abnormal grain growth should still be avoided in this concept. However, we must also consider the kinetics of achieving the required density. Although grain growth can indeed lead to a reduction in N, the kinetics of densification can often be decreased so dramatically by the larger grains that no long-term benefit for densification is achieved [12]. Grain growth during sintering can, therefore, provide only a short-term benefit for densification.

The ability of a limited amount of grain growth to homogenize a microstructure has been clearly demonstrated [13]. Figure 5.6 shows a sequence of microstructures of an MgO powder compact sintered at 1250°C to various densities. The initial compact contains clearly identifiable large pores in a fine-grained matrix. As sintering proceeds, grain growth and pore coalescence cause the microstructure to become more homogeneous. At a relative density of ~0.65, grain growth has proceeded to such an extent that it becomes thermodynamically feasible for the large pores in the original compact to shrink. The onset of shrinkage of the large pores coincides with an increase in the measured densification rate of the compact (Figure 5.7). Densification is relatively fast in this system, so that the sacrifice in the densification rate produced by the limited grain growth to improve the homogeneity still leaves reasonable sintering rates to achieve a high final density.

A limited amount of precoarsening, that is, coarsening at temperatures lower than the onset of densification, can also improve the microstructural homogeneity of powder compacts, providing benefits for densification and microstructural control during subsequent sintering. The overall sintering procedure, consisting of a precoarsening step followed by the sintering step, is referred to as *two-step sintering*. It has been successfully applied to powder compacts of MgO, ZnO, and Al$_2$O$_3$ [14–16]. As an example, Al$_2$O$_3$ powder compacts precoarsened for 50 h at 800°C have a more uniform pore size distribution than the green compact (Figure 5.8). In subsequent sintering, the microstructure develops in a more homogeneous manner, leading to a higher final density, a smaller average grain size, and a narrower distribution of grain sizes (Figure 5.9). During the precoarsening step, coarsening by surface diffusion, vapor transport, or a combination of these two mechanisms leads to a more uniform microstructure by an Ostwald ripening process. This improvement in the microstructural homogeneity more effectively offsets the sacrifice in the initial densification rate resulting from the precoarsening. Enhanced neck formation during precoarsening also

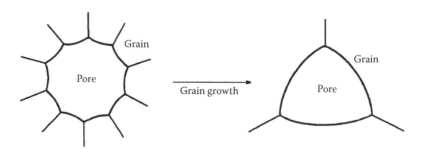

FIGURE 5.5 Schematic diagram illustrating the reduction in the pore coordination number as a result of normal grain growth.

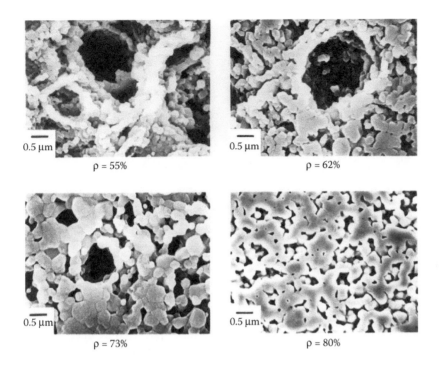

FIGURE 5.6 Microstructural evolution of an inhomogeneous MgO powder compact sintered to various densities at 1250°C. Note the trend towards homogenization of the microstructure.

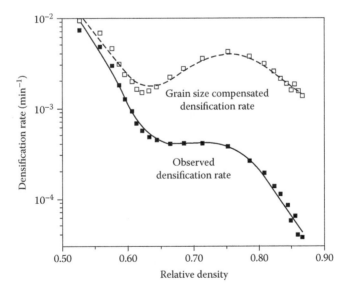

FIGURE 5.7 Observed densification rate and the grain size compensated densification rate versus relative density for the MgO powder compact described in Figure 5.6. The grain size compensated densification rate is a measure of the densification rate at approximately the same grain size (equal to the initial grain size). After sufficient grain growth and microstructural homogenization ($\rho \approx 0.65$), the large pores originally present in the powder compact can shrink, leading to an increase in the densification rate.

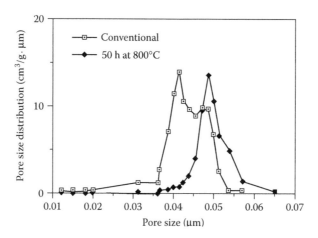

FIGURE 5.8 Pore-size distribution as a function of pore size obtained from mercury porosimetry for an Al$_2$O$_3$ powder compact heated at 10°C/min to 800°C (conventional) and a similar compact heated for 50 h at 800°C.

produces a stronger compact that is better able to resist differential densification. The combination of a more uniform microstructure and a stronger compact produced by the precoarsening step is responsible for the microstructural refinement in the subsequent sintering step. These examples show that a limited amount of coarsening can have a beneficial effect on sintering, but the approach is not used practically because it requires careful process optimization and a longer heating cycle.

Shrinkage of large pores in a dense polycrystalline matrix has also been analyzed using kinetic models. Evans and Hsueh [17] assumed that the polycrystalline material can be regarded as a continuum subject to viscoelastic deformation as a result of the sintering stress. Large pores are predicted to shrink slowly because of the small sintering stress, but a rapid increase in the densification rate is expected to occur when the pore size becomes on the order of the grain size, as may be observed in Figure 5.7.

Pan, Ch'ng, and Cocks [18] used computer simulations to study the sintering kinetics of a large pore in a dense polycrystalline matrix. According to the simulations, the thermodynamic barrier to the closure of a large pore predicted by the analysis of Kingery and Francois [10] results from the assumption that the large pore is surrounded by identical grains and from the imposed kinetic route

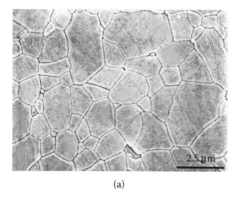

(a)

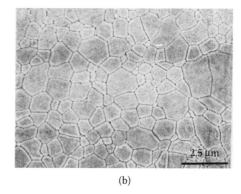

(b)

FIGURE 5.9 Scanning electron micrographs of Al$_2$O$_3$ compacts after sintering at 4°C/min to 1450°C. (a) Conventional sintering (relative density = 0.98; grain size = 1.5 ± 0.2 μm) and (b) two-step sintering with a precoarsening step (relative density = 0.99; grain size = 1.2 ± 0.1 μm).

that the grains move simultaneously into the pore as it shrinks. The simulations indicate that, in general, the elimination of a large pore from a dense polycrystalline solid does not necessarily have a free energy barrier, and a large pore will shrink even though N is greater than N_C. Figure 5.10a shows the simulated microstructure at three different times for a large pore surrounded by 24 identical grains. In the model, matter transport was assumed to occur by the coupled mechanisms of grain boundary diffusion, surface diffusion, and grain boundary migration. Matter transported into the pores occurred by grain boundary diffusion, whereas surface diffusion acted to transport the deposited mass over the surface of the pores. The following material parameters were assumed: the ratio of the surface and grain boundary diffusion coefficients $D_s/D_{gb} = 1$, the ratio of the surface and grain boundary energies $\gamma_{sv}/\gamma_{gb} = 3$, and the ratio of the grain boundary length to the pore radius

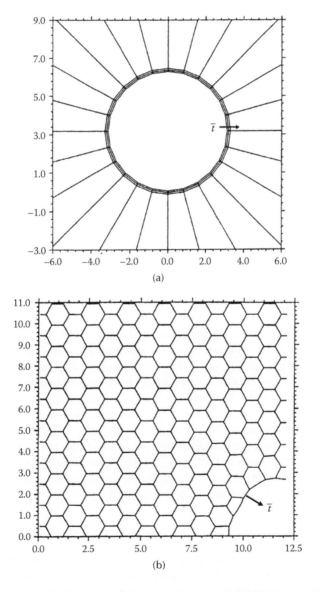

(a)

(b)

FIGURE 5.10 Microstructural evolution of a large pore with a pore coordination number of 24: (a) the pore is surrounded by identical grains, and (b) the pore is surrounded by hexagonal grain matrix in which the grains at the pore surface are not identical.

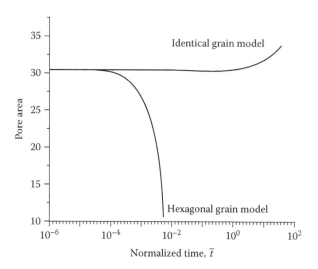

FIGURE 5.11 Comparison of the numerical calculated pore area as a function of time for a pore surrounded by identical grains and by hexagonal grains. The pore coordination number is 24 in both cases, and the material parameters are given in the text. (Courtesy of J. Pan.)

$X/r = 4$. The simulations show an increase in the pore size with time, in agreement with the analysis of Kingery and Francois. In comparison, the simulated microstructures for a large pore in a matrix of hexagonal grains, in which the grains at the pore surface are not identical, show a decrease in the pore size at longer times (Figure 5.10b). In this model, it was assumed that $D_s/D_{gb} = 1$, $\gamma_{sv}/\gamma_{gb} = 3$, and $M_b G_o^2 kT/D_{gb} = 10$, where G_o is the initial grain size, k is the Boltzmann constant, and T is the absolute temperature. In both models, the pore coordination number and material parameters were the same, except that the grain boundary mobility M_b is not needed in the identical grain model. Figure 5.11 compares the pore area as a function of time for the two different models. After a period of almost no change, the pore size for the hexagonal grain model decreases rather rapidly.

Flinn et al. [19] observed the evolution of large pores in a fine-grained Al_2O_3 powder matrix in experiments designed to study mechanical failure due to the presence of pores. Spherical pores were introduced into the matrix by dispersing a low volume fraction of uniform polystyrene spheres (diameters of 25 μm, 50 μm, 80 μm, and 120 μm) into a colloidal suspension of fine Al_2O_3 powder (particle size of 0.25 μm), followed by slip casting to form the green compact and subsequent heat treatment to burn out the spheres. The large pores shrank continuously but slowly during sintering of the compact, such that the radius of the large pores decreased linearly with the macroscopic shrinkage of the powder compact. The shrinkage of the large pores is in contradiction to the thermodynamic analysis of Kingery and Francois [10], but this does not mean that the observations support the numerical simulations of Pan et al. [18]. The small shrinkage of the large pores can be well explained by the requirement that the microscopic strain in the region surrounding the large pore must be compatible with the macroscopic strain of the powder compact.

5.3 CONSTRAINED SINTERING I: RIGID INCLUSIONS

Almost all sintering systems are constrained to some extent. For example, the sintering of a pure single-phase powder is constrained by inhomogeneities (such as agglomerates) present in the powder compact. However, the term *constrained sintering* is commonly taken to describe sintering in which the constraint is deliberately imposed and forms a necessary feature of the system. An identical system in which the constraint is absent is said to be the free or unconstrained system.

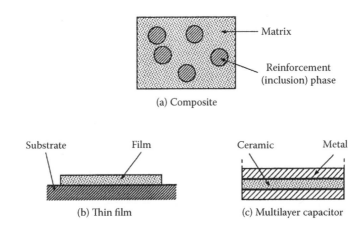

(a) Composite

(b) Thin film   (c) Multilayer capacitor

FIGURE 5.12 Examples of constrained sintering: (a) a composite in which inclusions prevent the powder matrix from sintering freely; (b) an adherent thin film for which sintering in the plane of the film is inhibited by the substrate; (c) a multilayer in which the metal and ceramic layers sinter at different rates.

The constraint may be internal to the sintering system, as exemplified by ceramic matrix composites in which a dense, rigid second phase in the form of particles, whiskers, platelets, or continuous fibers is incorporated into the powder matrix (Figure 5.12a), or external to the sintering system, as found in the sintering of a thin adherent film on a rigid substrate (Figure 5.12b), or the sintering of laminated substrates for electronic applications (Figure 5.12c).

In the fabrication of composites, the reinforcing phase is commonly incorporated into the powder matrix, and the mixture is formed into a shaped article that is densified to produce the final article. A key issue for achieving the required density and microstructure of the composite is the influence of the rigid reinforcing phase on the sintering of the porous powder matrix. To simplify the treatment, we adopt a model system in which the reinforcing phase consists of coarse rigid inclusions that are spherical in shape and randomly distributed in a fine-grained, porous matrix.

5.3.1 Volume Fraction of Inclusions

The volume fraction of the inclusions is a key parameter that influences the sintering (and properties) of composites. It is defined as the volume of the inclusions divided by the total volume of the composite (solid plus pores). Suppose that the composite consists of a porous powder matrix with a relative density ρ_m and rigid, fully dense inclusions (relative density $\rho_i = 1$). As the system densifies (Figure 5.13), the total volume decreases, so the volume fraction of the inclusions increases.

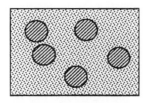

Matrix relative density = ρ_m
Inclusion volume fraction = v_i

(a) Porous composite

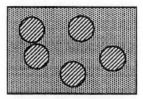

Matrix relative density = 1
Inclusion volume fraction = v_f

(b) Dense composite

FIGURE 5.13 Matrix and inclusion parameters in the sintering of composites.

Commonly, the volume fraction of the inclusions v_f is determined on the basis of the fully dense composite, such that

$$v_f + v_{mf} = 1 \qquad (5.1)$$

where v_{mf} is the volume fraction of the matrix in the fully dense composite. If v_i is the volume fraction of the inclusions when the relative density of the matrix is ρ_m (< 1), then v_i is related to v_f by

$$v_i = \frac{\rho_m}{\rho_m + (1 - v_f)/v_f} \qquad (5.2)$$

Therefore, knowing v_f, we can determine v_i at any value of ρ_m.

5.3.2 DENSIFICATION RATE OF THE COMPOSITE AND THE MATRIX

Normally we measure the density of the composite, but in many cases we may need to determine the density of the matrix phase of the composite in order to compare its densification kinetics with those of the free matrix. If D_c and D_m are the densities (i.e., mass/volume) of the composite and the matrix, respectively, then by simple geometry

$$D_m = D_c \frac{(D_{co} - v_{io} D_i)}{(D_{co} - v_{io} D_c)} \qquad (5.3)$$

where D_i is the density of the inclusions, and D_{co} and v_{io} are the initial values of D_c and v_i, respectively. Differentiating Equation 5.3, followed by some manipulation, gives

$$\frac{1}{D_m} \frac{dD_m}{dt} = \frac{1}{D_c} \frac{dD_c}{dt} \left(\frac{D_{co}}{D_{co} - v_{io} D_c} \right) \qquad (5.4)$$

If ρ_c and ρ_m are the relative densities of the composite and the matrix respectively, Equation 5.4 can be written as

$$\frac{1}{\rho_m} \frac{d\rho_m}{dt} = \frac{1}{\rho_c} \frac{d\rho_c}{dt} \left(\frac{\rho_{co}}{\rho_{co} - v_{io} \rho_c} \right) \qquad (5.5)$$

Equation 5.4 and Equation 5.5 allow for the determination of the matrix densification rate from the measured densification rate of the composite.

5.3.3 THE RULE OF MIXTURES

The rule of mixtures assumes the densification of the composite to be a weighted average of the independent densification rates of the matrix and of the inclusions; that is, it assumes that in the composite, each phase densifies in the same way as it would independently, by itself. If, for example, we consider the linear densification rate $\dot{\varepsilon}$, defined as one third the volumetric densification rate $(1/\rho)d\rho/dt$, then according to the rule of mixtures

$$\dot{\varepsilon}_c^{rm} = \dot{\varepsilon}_{fm}(1 - v_i) + \dot{\varepsilon}_i v_i = \dot{\varepsilon}_{fm}(1 - v_i) \qquad (5.6)$$

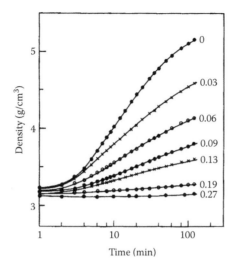

FIGURE 5.14 Density versus sintering time at 700°C for a polycrystalline ZnO powder containing different volume fractions of rigid SiC inclusions.

where $\dot{\varepsilon}_c^{rm}$ is the composite densification rate predicted by the rule of mixtures, $\dot{\varepsilon}_{fm}$ is the densification rate of the free matrix, and $\dot{\varepsilon}_i$ is the densification rate of the inclusions, taken to be zero. According to Equation 5.6, the ratio $\dot{\varepsilon}_c^{rm}/\dot{\varepsilon}_{fm}$ varies linearly as $(1 - v_i)$.

When the sintering data for ceramic matrices with controlled amounts of rigid inclusions are compared with the predictions of the rule of mixtures, dramatic deviations are found, particularly for polycrystalline powder matrices. Figure 5.14 shows the sintering data for polycrystalline ZnO powder matrix (particle size of approximately 0.4 μm) with different amounts of coarse, inert SiC inclusions (size approximately 14 μm) [20]. The inclusions severely reduce the densification of the composite relative to that for the free ZnO matrix; for inclusion volume fractions v_i greater than ~20 vol%, densification is almost completely inhibited. Data for a soda-lime glass powder (particle size of approximately 4 μm) containing different volume fractions of coarse, inert SiC inclusions (particle size of approximately 35 μm) are shown in Figure 5.15 [21]. The effect of the inclusions

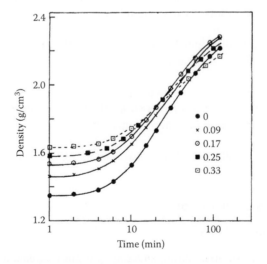

FIGURE 5.15 Density versus sintering time at 600°C for a soda-lime glass powder containing different volume factions of rigid SiC inclusions.

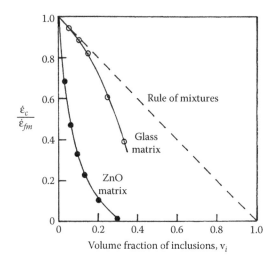

FIGURE 5.16 Comparison of the sintering data for the composites described in Figure 5.14 and Figure 5.15 with the predictions of the rule of mixtures. The densification rate of the composite relative to that for the free matrix is plotted versus the volume fraction of inclusions. Note the strong deviation from the predictions of the rule of mixtures for the ZnO matrix even at low inclusion content.

on the densification of the glass matrix is considerably smaller than that for the polycrystalline ZnO matrix.

The data of Figure 5.14 and Figure 5.15 can be manipulated to produce results suitable for comparison with predictions of the rule of mixtures. By fitting smooth curves to the data and differentiating, we can find the densification rate at any time. Figure 5.16 shows the results of the comparison. Dramatic deviations are observed for the ZnO matrix composite for v_i as low as 3–6 vol%. The glass matrix composite shows good agreement with the predictions of the rule of mixtures for $v_i < {\sim}20$ vol%, but the deviations become increasingly severe at higher v_i values.

Faced with the data showing dramatic deviations from the predictions, it may be argued that the rule of mixtures is inadequate to describe the data because the assumptions are too simplistic. For example, it ignores key factors, such as the following:

Transient stresses due to differential sintering between the inclusions and the matrix may arise, and, if large enough, they can reduce the densification rate and cause microstructural damage.

The inclusions will interfere and eventually form a touching, *percolating network* at some critical value of v_i. The formation of a network will inhibit densification, and, if the network is rigid enough, densification may stop completely.

We will briefly analyze the influence of these two factors for explaining the dramatic deviations from the rule of mixtures.

5.3.4 Transient Stresses During Sintering

5.3.4.1 Composite Sphere Model

To determine the stresses that arise during sintering, we must assume a geometrical model. For well-dispersed inclusions, a model that has been commonly used is the *composite sphere*, in which the core represents the inclusion and the cladding represents the porous matrix. As sketched in

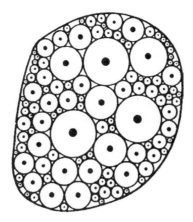

FIGURE 5.17 The composite sphere model. A composite containing spherical inclusions is conceptually divided into composite spheres (cross-sectional view). The core of each sphere is an inclusion, and the outer radius of the cladding of the sphere is chosen such that each sphere has the same volume fraction of inclusion as the whole body.

Figure 5.17, the composite containing a volume fraction v_i of inclusions is conceptually divided into composite spheres, each having a volume fraction v_i of core. We can consider a single composite sphere, with the assumption that the properties of the composite sphere are representative of the whole composite. When the inclusion is much larger than the particle size of the matrix, as in most practical composites, the matrix can be regarded, to a good approximation, as a continuum. With this assumption, phenomenological constitutive equations can be used for the matrix, and microscopic considerations can be neglected.

5.3.4.2 Stress Components

A shrinking cladding (matrix) around a core (inclusion) gives rise both to compressive stresses within the core and to a compressive radial stress and tangential tensile stresses within the cladding [22]. If σ_r, σ_θ, and σ_ϕ are the components of the stress in spherical coordinates (Figure 5.18), all three stress components in the inclusion are equal:

$$\sigma_{ri} = \sigma_{\theta i} = \sigma_{\phi i} = \sigma_i \tag{5.7}$$

where σ_i is the mean hydrostatic stress in the inclusion. Therefore, the inclusion is under purely hydrostatic stresses. In the convention that compressive stresses and strains are negative quantities, σ_i is negative. The stresses in the matrix are given by

$$\sigma_{rm} = -\frac{v_i \sigma_i}{(1-v_i)}\left(1 - \frac{b^3}{r^3}\right) \tag{5.8}$$

$$\sigma_{\theta m} = \sigma_{\phi m} = -\frac{v_i \sigma_i}{(1-v_i)}\left(1 + \frac{b^3}{r^3}\right) \tag{5.9}$$

where a and b are the inner and outer radii, respectively, of the composite sphere, and v_i is the volume fraction of inclusions, given by

$$v_i = a^3/b^3 \tag{5.10}$$

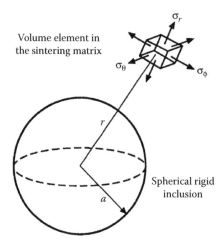

FIGURE 5.18 The stress components in a volume element at a distance r in the sintering matrix of a composite sphere. The radius of the inclusion is a, and the outer radius of the cladding (the matrix) is b.

The stresses in the matrix have their maximum values at the inclusion/matrix interface ($r = a$) and decrease as $1/r^3$. If $\sigma_{rm}(a)$ is the radial stress and $\sigma_{\theta m}(a)$ is the tangential (hoop) stress in the matrix at the interface, then from Equation 5.8, Equation 5.9, and Equation 5.10, we obtain

$$\sigma_{rm}(a) = \sigma_i \tag{5.11}$$

$$\sigma_{\theta m}(a) = -\sigma_i \frac{(1 + 2v_i)}{2(1 - v_i)} \tag{5.12}$$

The mean hydrostatic stress in the matrix is

$$\sigma_m = \frac{1}{3}(\sigma_{rm} + \sigma_{\theta m} + \sigma_{\phi m}) = -\sigma_i \left(\frac{v_i}{1 - v_i} \right) \tag{5.13}$$

According to Equation 5.13, the mean hydrostatic stress in the matrix is uniform, i.e., it is independent of r. Furthermore, the matrix is under both hydrostatic and tangential stresses. Since σ_i is compressive (negative), σ_m is tensile (positive), so it opposes the compressive sintering stress. Because of this opposition to the sintering stress, σ_m is often called a *backstress*.

5.3.4.3 Effect of Transient Stresses on Sintering

The densification rate of the matrix is affected by the hydrostatic component of the stress in the matrix. According to Equation 2.124, the linear densification rate (or strain rate) of the free matrix can be written as:

$$\dot{\varepsilon}_{fm} = \frac{1}{3} \left(\frac{\Sigma}{K_m} \right) \tag{5.14}$$

where Σ is the sintering stress and K_m is the bulk viscosity. Both $\dot{\varepsilon}_{fm}$ and Σ are negative quantities. For the matrix phase of the composite, the linear densification rate is

$$\dot{\varepsilon}_m = \frac{(\Sigma + \sigma_m)}{3K_m} = \frac{1}{3K_m}\left[\Sigma - \sigma_i\left(\frac{v_i}{1 - v_i}\right)\right] \tag{5.15}$$

The backstress σ_m opposes Σ, so it reduces the densification rate of the matrix. The tensile hoop stress $\sigma_{\theta m}$ may influence the growth of radial cracks in the matrix (Figure 5.1b). According to Equation 5.15, if the transient stresses were the only significant factor, the important effects on sintering would be determined by the value of σ_i. The calculation of σ_i therefore forms an important part in many analyses.

5.3.4.4 Calculation of Transient Stresses and Strain Rates

When one region of a sample (e.g., the matrix) shrinks at a different rate from a neighboring region (e.g., the inclusion), transient stresses are generated. Viscous flow (or creep) will always seek to relieve the stresses, so it is reasonable to assume that the calculation of the stresses would require a time-dependent, viscoelastic solution, where the response of the material to a stress consists of an elastic strain and a time-dependent strain due to viscous flow. However, the elastic strain is much smaller than the strains observed during sintering, so the observed deformation results almost entirely from viscous flow. A viscoelastic solution is not necessary. Instead, we can consider only the purely viscous response of the porous sintering material.

Viscoelastic models [3,23] that have been developed for sintering with rigid inclusions predict large values for the interfacial stresses and backstresses, with σ_i as high as 100 times Σ. The models assume that the materials show linear viscoelastic behavior and utilize the viscoelastic analogy [24,25] for calculating the time-dependent stresses and strain rates from the elastic solutions. However, Bordia and Scherer [26–28] have pointed out the inadequacy of applying the viscoelastic analogy to sintering materials that often do not show linear viscoelastic behavior.

Scherer [29] developed a model based on the purely viscous response of the sintering material. In this case, the calculation of the time-dependent stresses and strain rates is greatly facilitated by assuming the elastic–viscous analogy, where the equations for the time-dependent viscous response are found from those for the elastic solution [24]. Scherer's model predicts relatively small backstresses. The first stage in the calculation is to choose a constitutive equation that relates the applied stresses to the resulting strains. For an elastic material, the behavior is described by two independent elastic constants, such as the shear modulus G and the bulk modulus K. The constitutive equation for an isotropic linear elastic solid has the form [22]

$$\varepsilon_x = \varepsilon_f + \frac{1}{E}[\sigma_x - \nu(\sigma_y + \sigma_z)] \tag{5.16}$$

$$\varepsilon_y = \varepsilon_f + \frac{1}{E}[\sigma_y - \nu(\sigma_x + \sigma_z)] \tag{5.17}$$

$$\varepsilon_z = \varepsilon_f + \frac{1}{E}[\sigma_z - \nu(\sigma_x + \sigma_y)] \tag{5.18}$$

where ε_x, ε_y, and ε_z are the strains, and σ_x, σ_y, and σ_z are the stresses in the x, y, and z directions, respectively; E is the Young's modulus; ν is the Poisson's ratio; and ε_f is the free strain, i.e., the strain that would be produced in the absence of local stresses. E and v are an alternative pair of elastic constants that are related to G and K by

$$G = \frac{E}{2(1+v)} \tag{5.19}$$

$$K = \frac{E}{3(1-2v)} \tag{5.20}$$

For an isotropic, linearly viscous, incompressible material, the constitutive equation is easily obtained from Equation 5.16, Equation 5.17, and Equation 5.18 by invoking the elastic–viscous analogy: the strain is replaced by the strain rate, E is replaced by the shear viscosity η, and v becomes 1/2 (for an incompressible material), giving (for the x-direction):

$$\dot{\varepsilon}_x = \dot{\varepsilon}_f + \frac{1}{3\eta}\left[\sigma_x - \frac{1}{2}(\sigma_y + \sigma_z)\right] \tag{5.21}$$

where the dot denotes the derivative with respect to time. For a porous material, Equation 5.21 must be modified to allow for the compressibility of the pores; that is

$$\dot{\varepsilon}_x = \dot{\varepsilon}_f + \frac{1}{E_m}[\sigma_x - v_m(\sigma_y + \sigma_z)] \tag{5.22}$$

where E_m is the uniaxial viscosity of the matrix, which varies from 0 to 3η as the relative density of the matrix ρ_m goes from 0 to 1, and v_m is the Poisson's ratio of the matrix, which varies from 0 to 1/2 as ρ_m goes from 0 to 1. As ρ_m approaches 1, Equation 5.22 becomes identical to Equation 5.21. By analogy with Equation 5.19 and Equation 5.20, the shear viscosity, G_m, and the bulk viscosity, K_m, are related by

$$G_m = \frac{E_m}{2(1+v_m)} \tag{5.23}$$

$$K_m = \frac{E_m}{3(1-2v_m)} \tag{5.24}$$

Examination of Equation 5.22, Equation 5.23, and Equation 5.24 indicates that the solutions for the sintering material are obtained from those for the elastic material, Equation 5.16 to Equation 5.20, by replacing G and K by G_m and K_m, respectively, and replacing the strains by the strain rates.

For the composite sphere model, the sintering problem of mismatched shrinkage rates is analogous to the problem of thermal expansion mismatch. By adopting the elastic solution for the thermal stress problem and using the elastic–viscous analogy to transform it to the time-dependent solution for the sintering problem, we obtain the following equations for the stresses and strain rates:

$$\sigma_i = (1 - v_i) K_{cs} \dot{\varepsilon}_{fm} \tag{5.25}$$

$$\sigma_{rm} = \left[\left(\frac{a}{r} \right)^3 - v_i \right] K_{cs} \dot{\varepsilon}_{fm} \tag{5.26}$$

$$\sigma_{\theta m} = -\left[\frac{1}{2} \left(\frac{a}{r} \right)^3 + v_i \right] K_{cs} \dot{\varepsilon}_{fm} \tag{5.27}$$

where a is the inner radius of the composite sphere, v_i is the volume fraction of inclusions, and K_{cs} is given by

$$K_{cs} = \frac{1}{1/(4G_m) + v_i/(3K_m)} \tag{5.28}$$

The free strain rate (i.e., the linear densification rate of the free matrix) is given by

$$\dot{\varepsilon}_{fm} = \frac{1}{3} \left(\frac{\Sigma}{K_m} \right) \tag{5.29}$$

The hydrostatic backstress can be calculated from Equation 5.25, Equation 5.28, and Equation 5.29, giving

$$\frac{\sigma_m}{\Sigma} = -\frac{v_i}{v_i + 3K_m/4G_m} \tag{5.30}$$

According to Equation 5.30, as v_i approaches zero, so does σ_m. Furthermore, K_m and G_m are positive, so σ_m is always smaller than Σ.

The stress in the inclusion can be compared with the sintering stress by using Equation 5.25, Equation 5.28, and Equation 5.29:

$$\frac{\sigma_i}{\Sigma} = \frac{1 - v_i}{v_i + 3K_m/4G_m} \tag{5.31}$$

Using Equation 5.23 and Equation 5.24, as v_i approaches zero, the stress in an isolated inclusion is

$$\frac{\sigma_i}{\Sigma} = \frac{2(1 - v_m)}{1 + v_m} \tag{5.32}$$

Equation 5.32 shows that σ_i/Σ can be > 2 only when $v_m < 0$. Negative values of the Poisson's ratio have not been observed experimentally, so the absolute *maximum* value of σ_i predicted by Scherer's model is, therefore, 2Σ. The maximum stresses in the matrix occur at the interface between the matrix and the inclusion. The radial stress, $\sigma_{rm}(a) = \sigma_i$, so it is also predicted to be less than 2Σ. The circumferential or hoop stress is found from Equation 5.27, Equation 5.28, and Equation 5.29 to be

$$\frac{\sigma_{\theta m}(a)}{\Sigma} = -\frac{1/2 + v_i}{v_i + 3K_m/4G_m} \tag{5.33}$$

When v_i approaches zero, Equation 5.33 becomes

$$\frac{\sigma_{\theta m}(a)}{\Sigma} = -\frac{1 - 2v_m}{1 + v_m} \tag{5.34}$$

Equation 5.34 shows that $\sigma_{\theta m}(a)$ can be no greater in magnitude than Σ unless $v_m < 0$, so the viscous analysis of Scherer predicts transient stresses that are considerably smaller than those calculated by the viscoelastic models [3,23].

We can also calculate the effect of the transient stresses on the densification rate of the composite. The linear densification rate of the composite is predicted to be

$$\dot{\varepsilon}_c = \frac{(1 - v_i)K_{cs}\dot{\varepsilon}_{fm}}{4G_m} \tag{5.35}$$

Another way to express the results is to consider the densification rate of the composite relative to that predicted by the rule of mixtures. Using Equation 5.6, Equation 5.28, and Equation 5.35, this ratio is given by

$$\frac{\dot{\varepsilon}_c}{\dot{\varepsilon}_c^{rm}} = \frac{1}{1 + v_i(4G_m/3K_m)} \tag{5.36}$$

Since the denominator is always greater than 1, this equation predicts that the linear densification rate is lower than that predicted by the rule of mixtures. From Equation 5.23 and Equation 5.24, we have that

$$\frac{4G_m}{3K_m} = \frac{2(1 - 2v_m)}{1 + v_m} \tag{5.37}$$

The maximum value of this ratio is 2, which occurs when $v_m = 0$, so for v_i less than ~0.10, the composite densification rate is predicted to deviate from the rule of mixtures by not more than ~10%.

A *self-consistent model*, in which a microscopic region of the matrix is regarded as an island of sintering material in a continuum (the composite) that contracts at a slower rate, was also analyzed by Scherer [30]. The mismatch in sintering rates causes stresses that influence the densification rate of each region. It is found that the equations for the self-consistent model differ from those of the composite sphere model only in that the shear viscosity of the matrix G_m is replaced by the shear viscosity of the composite G_c. Taking Equation 5.36, the corresponding equation for the self-consistent model is, therefore,

$$\frac{\dot{\varepsilon}_c}{\dot{\varepsilon}_c^{rm}} = \frac{1}{1 + v_i(4G_c/3K_m)} \tag{5.38}$$

An approximation to G_c can be obtained from the Hashin–Shtrikman equation [31], which, for a viscous matrix, is given by

$$G_c = G_m \left[1 + \frac{15}{2} \left(\frac{v_i}{1 - v_i} \right) \left(\frac{1 - v_m}{4 - 5v_m} \right) \right] \tag{5.39}$$

In the case of viscous sintering where the matrix phase of the composite is a glass, explicit expressions can be derived for the moduli and Poisson's ratio [32]. The Poisson's ratio is given to a good approximation by

$$v_m = \frac{1}{2} \left(\frac{\rho_m}{3 - 2\rho_m} \right)^{1/2} \tag{5.40}$$

where ρ_m is the relative density of the matrix, and the uniaxial viscosity is given by

$$E_m = \frac{3\eta\rho_m}{3 - 2\rho_m} \tag{5.41}$$

where η is the viscosity of the glass.

By using Equation 5.37 and Equation 5.40, we can calculate the ratio $(4G_m)/(3K_m)$ in terms of the relative density of the matrix, and this ratio can be substituted into Equation 5.31, Equation 5.32, and Equation 5.36 to calculate the stresses and densification rates for the composite sphere model. Alternatively, for the self-consistent model, G_c can be found from Equation 5.39, and the same procedure repeated to determine the stresses and strain rates. Figure 5.19 shows the predicted values for $\dot{\varepsilon}_c/\dot{\varepsilon}_c^{rm}$ as a function of the relative density of the matrix for the composite sphere and self-consistent models. For $v_i < {\sim}20$ vol%, the predictions for the two models are almost identical, but they deviate significantly for much higher values of v_i. When $v_i < {\sim}10$–15 vol%, the predicted values of $\dot{\varepsilon}_c/\dot{\varepsilon}_c^{rm}$ are not very different from 1, and the rule of mixtures is accurate enough for these values of v_i.

An experimental test of Scherer's theory made using the data in Figure 5.15 for glass matrix composites indicates that the theory performs well for $v_i < {\sim}15$ vol% but significantly underestimates the effects of the inclusions at higher values of v_i (Figure 5.20). The agreement is less satisfactory for polycrystalline matrix composites. The results indicate that other factors must also play a role in reducing the sintering rates of the composites.

5.3.5 Percolation and Network Formation

As the volume fraction of inclusions increases, we reach a stage where the inclusions become so numerous that they form enough inclusion–inclusion contacts to produce a continuous network extending all the way across the sample. The network of inclusions is said to *percolate* through the sample, rather like water percolating through ground coffee in a coffee maker. The stage where the inclusions first form a percolating network is called the *percolation threshold*.

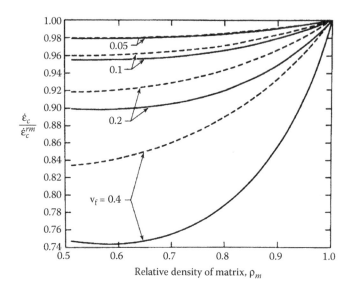

FIGURE 5.19 Comparison of the predictions of Scherer's theory of sintering with rigid inclusions with the predictions of the rule of mixtures. The linear strain rate of the composite normalized by the strain rate from the rule of mixtures (Equation 5.6) is plotted versus the relative density of the matrix for the indicated volume fraction of inclusions, v_f. The dashed curves represent the composite sphere model (Equation 5.36); solid curves represent the self-consistent model (Equation 5.38).

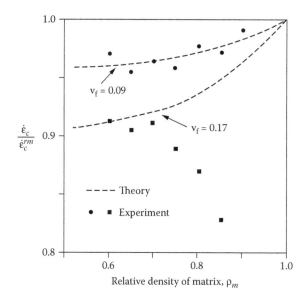

FIGURE 5.20 Comparison of the sintering data for the glass matrix composites described in Figure 5.15 with the predictions of Scherer's theory for sintering with rigid inclusions (Equation 5.36). The linear strain rate of the composite normalized by the strain rate from the rule of mixtures is plotted versus the relative density of the matrix ρ_m for the indicated volume fraction of inclusions v_f. Note the deviation starting near $\rho_m = 0.75$ for the composite with $v_f = 0.17$.

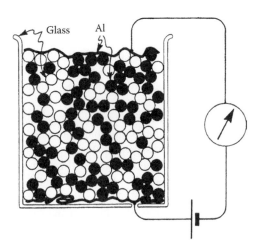

FIGURE 5.21 Sketch of an experiment to illustrate some elementary aspects of percolation in a mixture of aluminum and glass balls. When the percolation threshold for the aluminum balls is reached, the system will conduct an electrical current.

The formation of a continuous network of inclusions has important consequences for sintering as well as for other properties of the system. If the inclusions are electrically conducting, the structure can carry an electric current as soon as the percolating network forms (Figure 5.21). If the contacts between the inclusions are rigid, the structure will be mechanically rigid. This is important for sintering because the increased stiffness of the structure will retard sintering, and, if the structure is completely rigid, sintering will stop. The models considered so far neglect the formation of a percolating network of inclusions.

5.3.5.1 The Concept of Percolation

Percolation has been considered in some detail by Zallen [33]. The concept of percolation is illustrated in Figure 5.22 with a triangular, two-dimensional lattice [34]. If we place particles on sites at random, larger and larger clusters of adjoining particles will be formed as the number of particles increases. Eventually, one of these clusters, referred to as the *percolating cluster* or the

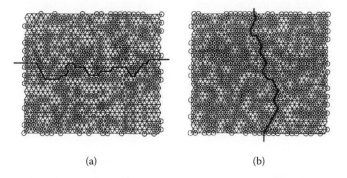

(a)   (b)

FIGURE 5.22 Illustration of percolation on a triangular lattice. When 40% of the sites are filled, no single cluster of particles extends all the way across the lattice, as indicated by the thick line in (a). When more than 50% of the sites are filled, a "spanning cluster" appears; then, as indicated by the thick line in (b), a path of occupied sites completely crosses the lattice. (From Scherer, G.W., Viscous sintering of particle-filled composites, *Ceram. Bull.*, 70, 1059, 1991. With permission.)

spanning cluster, becomes large enough to extend all the way across the lattice. The fraction of sites that must be filled before the percolating cluster appears is called the *percolation threshold, p_c*.

The percolation threshold depends on the shape and dimensionality of the lattice. For the triangular lattice shown in Figure 5.22, the percolation threshold is reached when half of the sites are occupied, i.e., $p_c = 0.5$, whereas $p_c = 0.593$ for a square, two-dimensional lattice, and $p_c = 0.311$ for a cubic lattice. For a powder system undergoing sintering, no lattice is present, but it has been shown that p_c occurs at a certain volume fraction (or area fraction in two dimensions) regardless of the nature of the lattice. In three dimensions, the percolation threshold is reached when the volume fraction of particles is ~16 vol% even when no lattice is present. If, for example, glass balls are mixed with aluminum balls as sketched in Figure 5.21, the structure will become electrically conducting when the aluminum balls occupy ~16 vol% of the space. If the glass balls and the aluminum balls are of the same size, percolation will occur when the number fraction of the aluminum balls is ~0.27. This is because the particles occupy ~64% of the total volume of the structure (the packing density for dense random packing), and the volume fraction of the aluminum balls (0.27×0.64) must equal 0.16. If glass and aluminum balls of the same size are placed at random on a simple cubic lattice, percolation will occur when the number fraction of the aluminum balls is ~0.31 because, in this case, the packing density for the simple cubic lattice is 0.52 and $0.52 \times 0.31 \approx 0.16$. The percolation threshold of ~0.16 applies to equiaxial particles in a random arrangement. This value will change if the particles agglomerate or repel one another. The value of p_c also depends on the aspect ratio and orientation of the particles. It decreases with increasing aspect ratio but increases if the particles become aligned.

5.3.5.2 Effect of Percolation on Sintering

The effect of percolation on the rigidity of the composite is an important factor, since this will influence the sintering kinetics. If the inclusions bond together on contact, the rigidity of the composite will increase dramatically near the percolation threshold, and the sintering rate will show a corresponding decrease. Even before the percolation threshold, large clusters of inclusions will form, and these will have a significant effect locally, even before the overall stiffening produced by the percolating cluster.

As the results of Figure 5.15 indicate, it is not impossible to densify composites with an inclusion volume fraction greater than the percolation threshold (~16 vol%). In fact, some practical composites with an inclusion content significantly greater than the percolation threshold can be sintered to almost full density. The reason is that the inclusions do not bond together to form a rigid network. In many cases, particularly for glass matrices and systems with a significant amount of liquid phase at the sintering temperature, the glass or the liquid phase wets the contacts between the inclusions, providing a lubricating layer. When the inclusion–inclusion contacts are wetted, the stage where the system becomes fairly rigid, referred to as the *rigidity threshold*, is expected to occur at inclusion contents significantly higher than the percolation threshold. In general, if the inclusions form strong bonds on contact (e.g., if they are not wetted by the matrix) or interlock (e.g., because of surface roughness), the densification rate will be reduced significantly near the percolation threshold. On the other hand, if the contacts between the inclusions are lubricated, higher inclusion contents can be accommodated without a dramatic reduction in the densification rate.

5.3.5.3 Numerical Simulations

Numerical simulations can be extremely valuable in exploring the effects of particle interactions on the sintering of composites, as demonstrated by the work of Scherer and Jagota [34–36], who used a finite element method to simulate the sintering of viscous matrices containing rigid inclusions. In a three-dimensional packing of spherical particles, randomly selected particles are assigned to be inclusions, and the rest are chosen to be matrix particles with the same size and surface tension as the inclusions. As sketched in Figure 5.23, the composite contains three types of contacts: inclusion–inclusion (i–i), matrix–matrix (m–m), and inclusion–matrix (i–m). The results of the finite element simulation

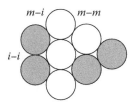

FIGURE 5.23 Contacts between inclusions (*shaded*) and matrix particles (*white*).

are found to depend on the degree of bonding between the inclusions and the ratio of the viscosity of the inclusions to the viscosity of the matrix particles, *H*. For rigid inclusions ($H = 10^6$) that form strong inclusion–inclusion bonds and for wetted inclusion–matrix contacts, Figure 5.24 shows the results of a finite element simulation, plotted in the form of the ratio $4G_c/3K_m$ versus the volume fraction of inclusions. Near the percolation threshold, the viscosity rises sharply by a factor of ~10^6, characteristic of the inclusions themselves. Also shown for comparison are the predictions of two other models: the Hashin–Shtrikman model (Equation 5.39) and the self-consistent model. These two models do not account for the rapid rise in the viscosity near the percolation threshold. They do show a rapid rise in the viscosity reminiscent of a percolation threshold, but it appears at too high a volume fraction.

5.3.6 FACTORS INFLUENCING THE SINTERING OF CERAMIC COMPOSITES

When the effects of transient stresses and network formation between the inclusions are taken into account, the models appear to provide a reasonably good description of viscous sintering with rigid inclusions. The agreement is, however, less satisfactory for the sintering of polycrystalline matrix composites. For these composites, additional factors may also contribute significantly to the reduction of the matrix densification rate. One factor is that the inclusions can seriously disrupt the packing of the matrix in the regions immediately surrounding the inclusions, so densification of these regions is slow [37].

A more important factor is the formation of a rigid skeletal network, consisting of the inclusions and the matrix phase, well before the percolation limit. Sudre and coworkers [38–40] proposed a model based on microstructural observations of the evolution of Al_2O_3 matrix composites containing

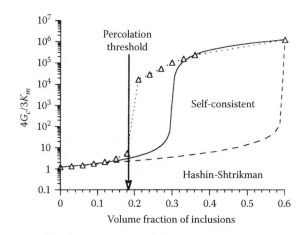

FIGURE 5.24 Shear viscosity of the composite, G_c, normalized by the bulk viscosity of the matrix, K_m, as a function of the volume fraction of inclusions (0.6 times the number fraction). Calculations by finite element method (Δ), self-consistent model (*solid line*), and Hashin–Shtrikman lower bound (*dashed line*) are shown. (From Scherer, G.W., and Jagota, A., Effect of inclusions on viscous sintering, *Ceramic Trans.*, 19, 99, 1991. With permission.)

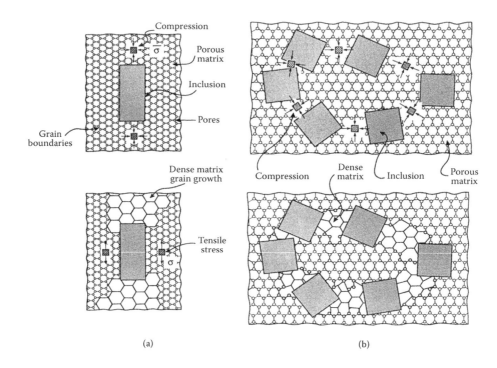

(a)   (b)

FIGURE 5.25 Schematic diagrams illustrating the sequence of events leading to the retardation of sintering and damage in bodies containing inclusions: (a) single inclusion; (b) multiple inclusions.

ZrO_2 inclusions and on numerical analysis of the transient stresses due to differential densification between the matrix and the inclusions. A sketch of the important features of the model is shown in Figure 5.25. The stress field due to differential densification leads to premature densification of certain matrix regions between the inclusions, and these regions support grain growth (Figure 5.25a). For multiple inclusions (Figure 5.25b), the premature densification and associated grain growth lead to the development of a dense annulus that resists densification, so severely reduced densification rates can occur even for a fairly sparse network of nontouching inclusions.

The importance of network formation and packing inhomogeneities has been clearly demonstrated by the work of Kapolnek and De Jonghe [41] and Hu and Rahaman [42] in sintering experiments using inclusions that were uniformly coated with the matrix phase. In the experiments by Hu and Rahaman, composites consisting of a polycrystalline ZnO matrix and inert, rigid inclusions of ZrO_2 were prepared by two separate methods. In one method, the matrix and inclusions were mixed mechanically in a ball mill; in the other, individual ZrO_2 inclusions were coated with the required thickness of ZnO powder by chemical precipitation from solution. The mechanically mixed powder and the coated powder were consolidated and sintered under nearly identical conditions. Figure 5.26 shows a sketch comparing the inclusion distribution in the matrix for the two methods. An important feature is that for the coated powder, the inclusions are separated from one another by a layer of matrix, so network formation between the inclusions is prevented. Furthermore, each inclusion is surrounded by a homogeneous layer of matrix, so packing inhomogeneities, especially in the matrix immediately surrounding the inclusions, are significantly reduced. The densification of the coated powder is very close to that for a ZnO powder containing no inclusions and, as shown in Figure 5.27, is significantly better than for the mechanically mixed powder. By severely reducing the sintering impediments through the use of coated powders, dense composites with up to 35–40 vol% of particulate, whisker, or platelet reinforcement have been prepared by conventional sintering [43,44]. Figure 5.28 shows the cross section of a composite consisting of an Al_2O_3 matrix containing 20 vol% SiC platelets that was sintered to almost full density.

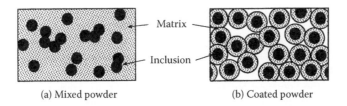

(a) Mixed powder   (b) Coated powder

FIGURE 5.26 Schematic diagrams illustrating the distribution of inclusions (cross-sectional view) in composites formed from (a) a mixture of the inclusions and powder matrix prepared by ball-milling and (b) coated inclusions prepared by chemical precipitation.

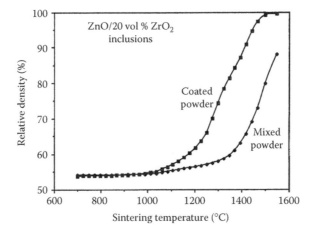

FIGURE 5.27 Data for relative density versus temperature during sintering at a constant heating rate (5°C/min) for composites formed from coated powders and from mechanically mixed powders. Both systems had the same composition, consisting of a ZnO matrix and 20 vol% ZrO_2 inclusions. Note the significantly higher density of the composite formed from the coated powders.

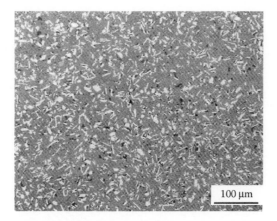

FIGURE 5.28 Optical micrograph of the polished surface of an Al_2O_3/20 vol% SiC platelet composite produced by conventional sintering of coated powders, showing a fairly uniform distribution of the platelets (*light phase*) in a dense matrix (*dark phase*). The black spots are pores produced mainly by pullout of some platelets during polishing of the cross-section.

5.4 CONSTRAINED SINTERING II: ADHERENT THIN FILMS

Many applications, particularly in the areas of electronic and optical ceramics, require the deposition of a porous thin film on a substrate by techniques such as casting, spraying, dip coating, or spin coating, after which the film is dried and sintered to produce the required microstructure. Commonly the film adheres to the substrate but is too thin to cause it to deform, so the substrate can be considered to be rigid. If the film remains attached to the substrate during sintering and does not crack, shrinkage in the plane of the substrate is inhibited, and stresses arise in the film and in the substrate. Far from the edges, all of the shrinkage occurs in the direction perpendicular to the plane of the film (Figure 5.29). The stress in the film is essentially a tensile stress that maintains strain compatibility with the substrate, subjecting the film to simultaneous creep. This is analogous to the stress in a sandwich seal caused by mismatch in the thermal expansion coefficients in the layers [45]. Compared to the free film, the stresses in the constrained film reduce the densification rate and can also lead to the growth of flaws.

5.4.1 MODELS FOR CONSTRAINED SINTERING OF THIN FILMS

Models for constrained sintering of thin adherent films on rigid substrates have been developed by Bordia and Raj [46] and by Scherer and Garino [47]. Both models consider the film to be a homogeneous continuum. The model of Bordia and Raj suffers from inconsistencies, similar in nature to those for the viscoelastic models of sintering with rigid inclusions. Scherer and Garino assumed that the sintering rate of the film could be described by a constitutive relation for a porous, viscous body. By analogy with Equation 5.22, the sintering rate along the orthogonal x, y, and z directions (Figure 5.29) can be written

$$\dot{\varepsilon}_x = \dot{\varepsilon}_f + \frac{1}{E_p}[\sigma_x - \nu_p(\sigma_y + \sigma_z)] \tag{5.42}$$

$$\dot{\varepsilon}_y = \dot{\varepsilon}_f + \frac{1}{E_p}[\sigma_y - \nu_p(\sigma_x + \sigma_z)] \tag{5.43}$$

$$\dot{\varepsilon}_z = \dot{\varepsilon}_f + \frac{1}{E_p}[\sigma_z - \nu_p(\sigma_x + \sigma_y)] \tag{5.44}$$

where $\dot{\varepsilon}_x, \dot{\varepsilon}_y, \dot{\varepsilon}_z$ and $\sigma_x, \sigma_y, \sigma_z$ are the linear densification rates and stresses in the x, y, and z directions, respectively; $\dot{\varepsilon}_f$ is the linear densification rate of the free or unconstrained film; E_p is the uniaxial

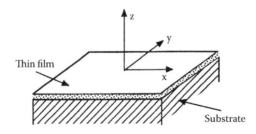

FIGURE 5.29 Geometry of a thin film attached to a rigid substrate. No shrinkage occurs in the plane of the film (xy plane); all of the shrinkage occurs in the direction perpendicular to the plane of the film (the z direction).

viscosity; and v_p is the Poisson's ratio of the porous film. As written, Equation 5.42, Equation 5.43, and Equation 5.44 are not based on any microstructural model, but a model must be chosen in order that $\dot{\varepsilon}_f$, E_p, and v_p can be specified.

Since there is no deformation of the film in the x and y directions, $\varepsilon_x = \varepsilon_y = 0$. Also, there is no constraint in the z direction, so $\sigma_z = 0$. Putting $\sigma_x = \sigma_y = \sigma$, Equation 5.42, Equation 5.43, and Equation 5.44 give

$$\sigma = -\frac{E_p \dot{\varepsilon}_f}{1 - v_p} \tag{5.45}$$

and

$$\dot{\varepsilon}_z = \left(\frac{1 + v_p}{1 - v_p}\right) \dot{\varepsilon}_f \tag{5.46}$$

The volumetric densification rate is given by

$$\frac{\dot{\rho}}{\rho} = -\frac{\dot{V}}{V} = \dot{\varepsilon}_x + \dot{\varepsilon}_y + \dot{\varepsilon}_z \tag{5.47}$$

where ρ is the relative density, V is the volume, and the dot denotes the derivative with respect to time. Using Equation 5.46 and putting $\dot{\varepsilon}_x = \dot{\varepsilon}_y = 0$, the densification rate of the constrained film is

$$\left(\frac{\dot{\rho}}{\rho}\right)_c = -\left[\frac{1 + v_p}{3(1 - v_p)}\right] 3\dot{\varepsilon}_f \tag{5.48}$$

Since the densification rate of the unconstrained film is given by

$$\left(\frac{\dot{\rho}}{\rho}\right)_u = -3\dot{\varepsilon}_f \tag{5.49}$$

the function of v_p in the square brackets of Equation 5.48 represents the amount by which the densification rate of the film is retarded by the substrate.

For amorphous films that sinter by viscous flow, we can use Scherer's model for viscous sintering (see Chapter 2) to determine the terms E_p, $\dot{\varepsilon}_f$, and v_p [32]. It is found that the term in the square brackets of Equation 5.48 is always less than 1 for $\rho < 1$, so the constrained film is predicted to sinter at a slower rate than the unconstrained film. After integration, the results can be plotted as relative density versus dimensionless time (Figure 5.30). Starting from the same initial density, ρ_o, the constrained film takes a longer time to reach a given density than does an unconstrained film of the same material. For example, for $\rho_o = 0.5$, the constrained film takes ~25% longer than the unconstrained film to reach theoretical density.

The stress in the film during sintering can be found from Equation 5.45 and the equations for E_p, $\dot{\varepsilon}_f$, and v_p. The equation can be written

$$\sigma = \frac{\gamma_{sv}}{l_o \rho_o^{1/3}} f(\rho) \tag{5.50}$$

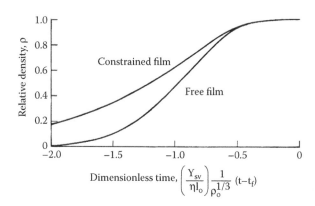

FIGURE 5.30 Predictions of the model of Scherer and Garino for the sintering of a constrained film and a free film. The constrained film is predicted to sinter more slowly than the free film. (Courtesy of G.W. Scherer.)

where γ_{sv} is the specific surface energy of the glass, l_c is the length of the unit cell for Scherer's sintering model (Figure 2.16), ρ_o is the initial density, and $f(\rho)$ is a function of the density of the film. This function reaches a maximum of ~1 at $\rho \approx 0.81$. The maximum stress is, therefore,

$$\sigma_{max} = \frac{\gamma_{sv}}{l_o \rho_o^{1/3}} \tag{5.51}$$

For a polymeric silica gel prepared by the hydrolysis of silicon tetraethoxide [1], $\gamma_{sv} \approx 0.25$ J/m², $l_c \approx 10$ nm, and $\rho_o \approx 0.50$, so $\sigma_{max} \approx 30$ MPa. This value of σ_{max} is on the order of the stresses used in hot pressing, but it is tensile in nature. If the film is not strong enough to withstand such high tensile stresses, cracking will result. For a colloidal silica gel, the corresponding values are: $\gamma_{sv} \approx 0.25$ J/m², $l_c \approx 100$ nm, $\rho_o \approx 0.25$, and $\sigma_{max} \approx 3$ MPa; that is, the maximum stress is considerably smaller.

5.4.2 Experimental Observations of Sintering of Thin Films

Garino and Bowen [48,49] studied the sintering of constrained and free films of soda-lime-silica glass, polycrystalline ZnO, and polycrystalline Al₂O₃, and compared their data with the predictions of Scherer and Garino's model. Constrained films (100–200 μm thick) were prepared by casting powder slurries onto a substrate. After drying, the shrinkage was measured using a laser reflectance apparatus, sketched in Figure 5.31. Free films were obtained by breaking off small pieces of the dried films from the substrate, and the shrinkage of these films was measured from scanning electron micrographs taken at various stages of sintering.

Figure 5.32 shows the data for the volumetric shrinkage of the free film and the constrained film of the glass powder at 650°C. The data for the free film are well fitted by Scherer's model for the viscous sintering of glass. Scherer and Garino's constrained sintering model gives a good fit to the data for the constrained film for shrinkages up to ~25% but overestimates the shrinkage at higher values. When the constrained film was sintered for times longer than those shown in Figure 5.32, the density was found to increase steadily, and it eventually reached the same final value as the free film.

The data for the sintering of the polycrystalline ZnO films are shown in Figure 5.33. In this case, the constrained film sinters considerably more slowly than the free film, and even after prolonged sintering it does not reach the same endpoint density as the free film. Instead, the

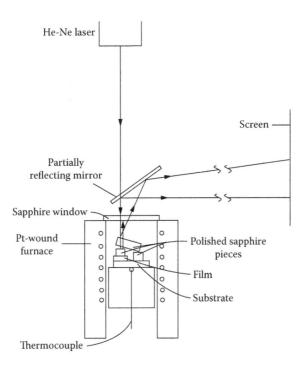

FIGURE 5.31 Schematic diagram of a laser reflectance apparatus used to measure the *in situ* shrinkage of an adherent film during sintering.

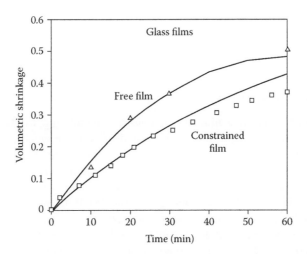

FIGURE 5.32 Isothermal shrinkage data for constrained and free glass films sintered at 650°C. The curve through the data for the free film represents the best fit to the data using Scherer's model for viscous sintering. The curve through the data for the constrained film represents the fit to the data by the thin film model of Scherer and Garino using the fitting constants derived from the fit to the free-film data. (Courtesy of T.J. Garino.)

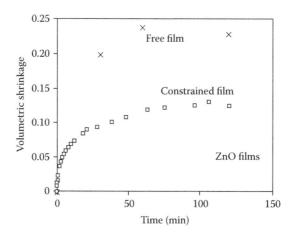

FIGURE 5.33 Shrinkage data for constrained and free films of ZnO during sintering at 778°C. (Courtesy of T.J. Garino.)

shrinkage reaches a steady value that is considerably lower than that reached by the free film. The data for the Al_2O_3 films showed trends similar to those for the ZnO films.

The data in Figure 5.32 and Figure 5.33 indicate that the constrained sintering of polycrystalline films is qualitatively different from that of glass films. The reason for this difference is not clear, but two factors require consideration. The first factor is grain growth (or coarsening), which plays no part in the sintering of glassy films. Grain growth can be accounted for in Scherer and Garino's model by assuming that the viscosity increases as G^m, where $m = 2$ for lattice diffusion and $m = 3$ for grain boundary diffusion. However, this modification still does not provide an adequate representation of the sintering data for the polycrystalline films. The second factor is differential densification. Microstructural observations of the polycrystalline films show features (e.g., large pores surrounded by dense regions) that are characteristic of differential densification. A problem is that theories indicate that differential densification in a film is possible only if the adhesion of the film to the substrate fails [50,51].

5.4.3 Crack Growth and Damage in Constrained Sintering of Films

The stresses generated in adherent films during constrained sintering, as observed earlier, can be substantial. If the film is sufficiently thin, the stress distribution normal to the plane of the film surface can be considered uniform. The interfacial adhesion or shear strength must be sufficiently high to support the force balance. For a fixed interfacial strength, a critical film thickness is found above which the interface can no longer support the tensile stress, resulting in interface failure, such as delamination. Typically, for the sintering of particulate films on rigid substrates, this critical thickness is <50 μm, but it can be much smaller when large densification strains or weak interfaces are involved [52]. For polymeric films prepared by sol-gel processing, the critical thickness is commonly <1 μm. An example of a cracked and delaminated sol-gel film after sintering is shown in Figure 5.34.

Failure by cracking can also be initiated in the film itself by the growth of preexisting flaws during sintering. These flaws are known to occur readily during the drying of sol-gel or particulate films deposited by colloidal processing. An analysis of the mechanics of sintering constrained films predicts that a preexisting crack will grow only under certain conditions that depend on the crack length, film thickness, and interfacial adhesion between the film and the substrate [51,52]. In the theory, the effects of film thickness and interfacial adhesion are treated in terms of a single parameter, referred to as a *friction parameter*. The predictions of the theory, summarized in terms

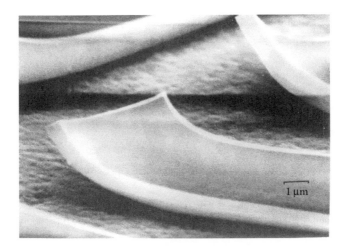

FIGURE 5.34 Delamination of a sintered zirconia thin film produced by a sol-gel method on a nanoporous alumina substrate. (Courtesy L.C. De Jonghe.)

of a crack growth map (Figure 5.35), indicate that regardless of the interfacial adhesion or film thickness, a preexisting crack will grow only if the crack length is above a critical value (*vertical line*), whereas no crack will grow if the interfacial adhesion is large enough or the film thickness small enough, such that the friction parameter is below a certain critical value (*horizontal line*). Experiments with glass films indicate that cracking occurs above a certain film thickness that is in good agreement with the theory [53].

The critical parameters are more difficult to predict for polycrystalline films, but, in general, a higher critical thickness for crack propagation results from careful drying prior to sintering, a homogeneous green structure, a high packing density, better interfacial adhesion, and low sintering rates at low sintered densities. Figure 5.36 shows an example of a thin, dense Y_2O_3-stabilized ZrO_2

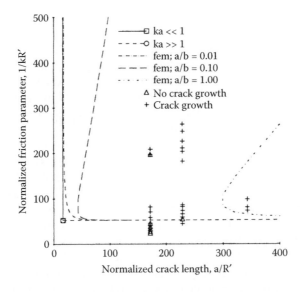

FIGURE 5.35 Crack growth map for a sintering film. The region enclosed toward the origin represents conditions under which preexisting cracks do not extend. The symbols (+ and Δ) represent experimental data from glass films on a platinum substrate. (Courtesy of R.K. Bordia.)

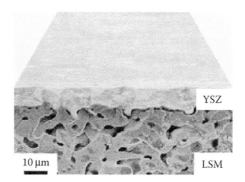

FIGURE 5.36 Dense film of Y_2O_3-stabilized ZrO_2 electrolyte, produced by colloidal deposition followed by sintering, on an invariant, porous lanthanum strontium manganate substrate. Pseudoperspective SEM micrograph. (Courtesy of L.C. De Jonghe.)

layer formed on a rigid, porous lanthanum strontium manganate substrate by colloidal deposition and sintering. Constrained sintering of a thin layer of ceramic electrolyte on a porous, rigid electrode material to produce a dense electrolyte without microstructural flaws is a key step in the manufacture of solid-oxide fuel cells for electrical power generation. In this example, the homogeneous packing and high packing density of the layer achieved by the colloidal deposition method dramatically reduced the tendency for flaw development during sintering.

5.5 CONSTRAINED SINTERING III: MULTILAYERS

Adherent multilayers, formed from thick films of conductors, resistors, and dielectrics, constitute an important segment of the microelectronics industry because they serve as interconnecting substrates for integrated circuits. An important step in the manufacture of the multilayered packages is the *cosintering* (or *cofiring*) of the metal–ceramic multilayered structure. When the different layers sinter at different rates, transient stresses are generated that hinder the densification of the individual layers and can lead to the growth of microstructural flaws (e.g., cracking and delamination), as well as to anisotropic shrinkage (e.g., warping of the multilayered structure). In practice, the problems of reduced sintering, microstructural damage, and warping are commonly minimized by matching the sintering rates of the layers through control of the powder characteristics, particle packing in the green system, and the sintering schedule. However, we can reach a clearer understanding of these problems by analyzing the development of the transient stresses and their effects on the sintering of the multilayered structure.

Figure 5.37 shows the data for the unconstrained shrinkage of thick films of gold particles and ceramic particles consisting of a mixture of cordierite and glass [54]. When the two layers are bonded together and cosintered, the sintering of the gold is constrained by the ceramic film, leading to tensile stresses in the gold film and warping toward the gold. Later, as the ceramic film tries to shrink, it is constrained by the gold. We can analyze the stresses in the films and their influence on the densification and warping of the composite structure during cosintering by using an approach similar to that of Scherer and Garino [47], outlined earlier for the constrained sintering of films. As indicated by Equation 5.48, the key parameters required for the analysis are the free sintering rates and the constitutive parameters (e.g., shear and bulk viscosities) of the individual layers. As will be described in the next section, several expressions have been proposed for the shear viscosity and the bulk viscosity of porous sintering bodies. For the bilayer system described in Figure 5.37, the use of expressions for the shear viscosity and the bulk viscosity derived by Skorokhod [55] provided a reasonable description of the constrained sintering kinetics.

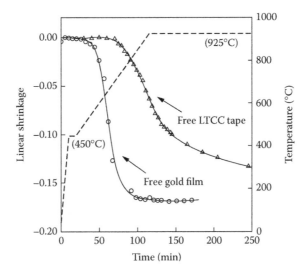

FIGURE 5.37 Shrinkage kinetics for unconstrained thick film of gold particles and LTCC tape consisting of cordierite and glass particles. The data points are fitted with smooth curves. (Courtesy of G.-Q. Lu.)

Polycrystalline films in a multilayered structure are observed to be more prone to the development of microstructural flaws than are glass films during cosintering [56], so it is particularly important to minimize the presence of processing flaws in the polycrystalline layer during forming. When it does not lead to a deterioration of the properties of the multilayered system, the incorporation of a glassy second phase improves the stress relaxation in the polycrystalline film and can serve to alleviate the development of flaws.

5.6 CONSTITUTIVE MODELS FOR POROUS SINTERING MATERIALS

It was shown earlier that for a sintering body the elastic strain is negligible, so it is sufficient to use the elastic–viscous analogy. The constitutive equation for pure shear can be written as

$$s_{ij} = 2G_p \dot{e}_{ij} \tag{5.52}$$

where s_{ij} is the shear stress, $\dot{e}_{ij}$ is the strain rate, and G_p is the shear viscosity of the porous material. The constitutive equation for the hydrostatic stress σ is given by

$$\sigma = K_p (\dot{\varepsilon} - 3\dot{\varepsilon}_f) \tag{5.53}$$

where $\dot{\varepsilon}$ is the volumetric strain rate, $\dot{\varepsilon}_f$ is the linear free strain rate, and K_p is the bulk viscosity of the porous material. The shear viscosity G_p and the bulk viscosity K_p are related to the uniaxial viscosity E_p and the Poisson's ratio v_p by Equation 5.23 and Equation 5.24. Most of the published constitutive equations for porous sintering materials are in the form of Equation 5.52 and Equation 5.53. They differ mainly in the assumed dependence of G_p and K_p on relative density ρ and grain size G. The expressions suggested by several authors are summarized in Table 5.1. In many cases, the data obtained in experiments do not have sufficient accuracy to distinguish between the expressions.

TABLE 5.1
Expressions for the Shear Viscosity G_p and the Bulk Viscosity K_p for Porous Sintering Materials

Shear Viscosity, G_p	Bulk Viscosity, K_p	Ref.
$3\eta\rho/[(6 - 4\rho) + (3\rho - 2\rho^2)^{1/2}]$	$\eta\rho/[(3 - 2\rho) - (3\rho - 2\rho^2)^{1/2}]$	32
$\eta\rho^2$	$4\eta\rho^3/3(1 - \rho)$	55
$AG^m \exp[-2\alpha(1 - \rho)]$	$AG^m \exp[-\alpha(1 - \rho)]$	58
$\eta_0\rho^n/(1 - \rho)^\lambda$	$\eta_0 b\rho/(1 - \rho)^\beta$	23
$\eta_1\rho^{5.26}$	$\eta_1\rho^{5.26}/2$	60
$\eta_1\rho^{5.26}$	$\eta_1\rho^{5.26}/c(1 - \rho)^\chi$	60

η = viscosity of the fully dense material; ρ = relative density; A = constant that depends on the matter transport mechanism; G = instantaneous grain size; m = grain size exponent that depends on the matter transport mechanism ($m = 2$ for lattice diffusion and $m = 3$ for grain boundary diffusion); α = parameter determined from creep data; η_0 and η_1 = reference viscosities that depend on the initial grain size, temperature, and initial density; n, λ, b, β, c, and χ = parameters obtained by fitting the equations to experimental data.

Scherer [32] derived expressions for G_p and K_p by using a microstructural model described in Chapter 2 for the sintering of glasses (see Section 2.6.4.4). The expressions provide an excellent description of the shear deformation and densification of glass powder compacts sintered under an applied uniaxial stress [57]. Skorokhod's expressions [55] were obtained phenomenologically. For the density range of interest for most sintering experiments ($\rho > 0.5$), the values for G_p and K_p determined from Skorokhod's expressions are not very different from those determined from Scherer's expressions [27]. Rahaman, De Jonghe, and Brook [58] developed expressions based on the analytical equations for the densification and creep rates and on data from simultaneous creep and densification experiments (see Chapter 2). The constants A and m in the expressions depend on the matter transport mechanism: $m = 2$ for lattice diffusion and $m = 3$ for grain boundary diffusion, and α is an experimentally determined parameter that depends on the dihedral angle, defined in Equation 2.93. For example, a value $\alpha = 2.0$ was obtained for CdO. The inverse of the bulk viscosity, $1/K_p$, does not go to zero as ρ goes to zero, so the expression for K_p is not applicable to the final stage of sintering. Hsueh et al. [23] assumed that the viscosity depended on the instantaneous grain size and the density, and derived expressions for G_p and K_p based on appropriate grain growth and densification equations. The expressions contain parameters that were obtained by fitting the equations to experimental data [59]. Du and Cocks [60] developed expressions suitable for use in finite element modeling of sintering, but their approach is essentially a modification of that used by Hsueh et al. [23].

5.7 MORPHOLOGICAL STABILITY OF CONTINUOUS PHASES

A continuous second phase in polycrystalline ceramics, such as a continuous pore phase or fibers in a composite, may undergo capillary-induced shape changes, leading to breakup and the formation of discrete pores or particles. Under certain conditions, thin polycrystalline films have also been observed to become unstable and break up into islands, uncovering the substrate. The morphology and spatial distribution of a phase can have important consequences for achieving the required microstructure and properties, so factors influencing the morphological stability of continuous phases can become important considerations in materials design.

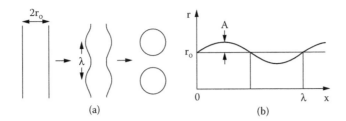

FIGURE 5.38 (a) Rayleigh instability in a cylindrical fluid (or solid) of radius r_o subject to radial fluctuations of wavelength λ. The cylinder breaks up into a line of spheres when the amplitude of the fluctuation becomes sufficiently large. (b) Geometrical parameters in the analysis of breakup phenomena in a cylindrical material.

5.7.1 Rayleigh Instability and Microstructural Evolution

An analysis by Rayleigh in 1879 [61] permits a qualitative understanding of several features of the morphological stability of continuous phases in materials. Rayleigh analyzed the morphological stability of cylindrical fluid jets under the influence of surface tension forces and found that an infinitely long cylindrical jet will eventually break up into a row of spherical droplets (Figure 5.38), in much the same way as when a narrow stream of water flows from a faucet. This type of breakup, referred to as *Rayleigh instability*, is driven by a reduction in surface energy. It is based on the fact that a sphere has a lower surface area than a cylinder of the same volume. Because mass must be redistributed during the breakup, the fluid jet does not progress directly to a single sphere but instead breaks up into multiple spheres, limited in number by the condition that there exists a maximum number of spheres with a total surface area lower than that of a cylinder with the same volume.

The Rayleigh analysis indicates that for a cylinder of radius r_o subject to infinitesimal periodic perturbations with a wavelength λ in the axial direction (Figure 5.38a), the amplitude of the perturbation will increase, due to reduction in specific surface energy, only when $\lambda > \lambda_{min} = 2\pi r_o$ (the circumference of the cylinder). Growth of the perturbations with $\lambda > \lambda_{min}$ eventually leads to the formation of one spherical particle per wavelength increment of the cylinder. The Rayleigh analysis also indicates that the rate at which the amplitude of the perturbation develops reaches a maximum value when $\lambda = 9.02r_o$. This value of λ, denoted λ_{max}, dominates the periodicity of the breakup, so the spacing of the resulting spherical particles should coincide closely with λ_{max}. For the case when $\lambda < \lambda_{min}$, the cylinder is stable and the infinitesimal perturbation readjusts itself with time.

Capillary-driven morphological instability has been studied for several different geometries. Nichols and Mullins [62,63] analyzed the stability of an infinitely long cylinder under isotropic surface energy and found that the kinetics of breakup depend on the matter transport mechanism by which the perturbation grows. For matter transport by surface diffusion, $\lambda_{max} = \sqrt{2}\,\lambda_{min}$, whereas, for lattice diffusion, $\lambda_{max} = 2.1\lambda_{min}$. Nichols [64] extended this work to cylinders with finite length and hemispherical ends. The analysis indicates that a minimum length $L_{min} = (7.2/\pi)\lambda_{min}$ is necessary for ovulation (formation of more than one discrete sphere), whereas cylinders with lengths less than L_{min} are predicted to eventually spheroidize (form a single sphere).

Stüwe and Kolednik [65] analyzed the disintegration by surface diffusion of a long cylindrical pore and found a λ_{max} value identical to that obtained by Nichols and Mullins for surface diffusion–dominated growth. They also estimated a time for disintegration of the cylindrical pore into discrete spherical pores, which, for the case when the volume of the cylinder remains constant (no densification), is given by

$$\tau \approx \frac{kT}{D_s \gamma_{sv}} \left(\frac{z^4}{z^2 - 4\pi^2} \right) \left(\frac{r_o}{\delta} \right)^4 \ln\left(\frac{r_o}{A_o} \right) \qquad (5.54)$$

where k is the Boltzmann constant; T is the absolute temperature; D_s is the surface diffusion coefficient; γ_{sv} is the specific energy of the solid–vapor interface; $z = \lambda/r_o$, where λ is the wavelength of the perturbation and r_0 is the initial radius of the cylindrical pore; δ is the diameter of an atom; and A_0 is the initial amplitude of the perturbation (Figure 5.38b). For a cylindrical pore with a given initial radius r_o, the key factors that determine the value of τ are D_s and the ratio r_o/A_o.

Complicating effects arise when the continuous second phase is located along three-grain junctions in polycrystalline materials. The dihedral angle ψ that characterizes the intersection of the second phase with the grain boundary has a strong influence on the morphological stability of the continuous second phase [66–70]. If ψ is close to 180°, the second phase will behave like a cylinder and will break up to form spheres as outlined above for the Rayleigh analysis. As ψ decreases, λ_{min} increases, and the second phase becomes increasingly stable against breakup (Figure 5.39). The value of λ_{min} tends to infinity as ψ approaches 60°, when the curvature of the second phase at the three-grain junction vanishes, so the second phase is stable against perturbations of any wavelength [69].

A wide range of microstructural phenomena involving capillary-induced shape changes has been analyzed and modeled in terms of the Rayleigh instability, including fibers in composites [71], the healing of cracks introduced as a result of thermal shock [72] or by scoring and welding of bicrystals [73], pore channels formed during wire-sintering experiments [62], and potassium-filled bubbles in tungsten wire [65]. Although the details may be different in each case, the basic principle governing the breakup is the same.

Morphological instability effects can also arise in microstructures produced by solid-state and liquid-phase sintering. In solid-state sintering, the transition from the intermediate to the final stage involves the change in pore morphology from a continuous pore network to one in which the pores

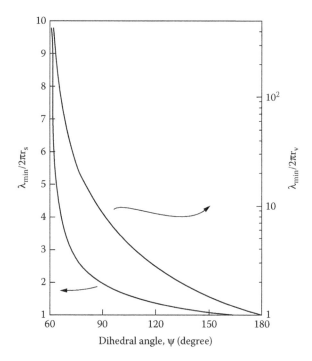

FIGURE 5.39 Dihedral angle dependence of λ_{min}, the minimum wavelength perturbation capable of increasing in magnitude, compared with values obtained assuming a cylindrical pore geometry. Cylindrical pores of radius r_v and r_s have respective volumes and surface curvatures equivalent to those of pores characterized by a dihedral angle of ψ. (Courtesy of A.M. Glaeser.)

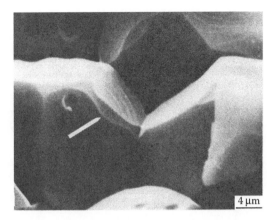

FIGURE 5.40 Scanning electron micrograph of pore channels in MgO with well-defined perturbations (arrowed). (Courtesy of A.M. Glaeser.)

are isolated or closed. The dihedral angle is commonly large (on the order of 120°), so the continuous pore channels are expected to be unstable against breakup. Rayleigh instability processes have been proposed as a potential mechanism for the transition from the intermediate to the final-stage microstructure, but the evidence for their occurrence is very limited. The most convincing evidence comes from an examination of the fractured surfaces of MgO hot pressed to ~99% of the theoretical density [74], which appears to show a few continuous pore channels resembling those that would be expected to develop during the growth of morphological perturbations (Figure 5.40). Real powder compacts contain pores with nonuniform cross sections and a distribution of dihedral angles, which would influence the pore closure conditions and sizes. Pores with smaller average radius, larger amplitude of perturbation (Equation 5.54), and larger dihedral angle at the grain boundary are expected to close more rapidly and at a lower density, influencing the development of microstructural inhomogeneities. In liquid-phase-sintered materials, a continuous glassy phase is often present along the grain boundaries. However, liquid-phase sintering requires a low dihedral angle (zero in the ideal case), so the continuous glassy phase at the grain boundaries is expected to be stable against breakup and to remain interconnected.

5.7.2 Morphological Stability of Thin Films

A factor that must be considered for thin films is their stability against shape changes, particularly those shape changes that tend to break up the continuous film into a group of islands. As an example, Figure 5.41 shows the breakup of an Y_2O_3-stabilized ZrO_2 film (8 mol% Y_2O_3) that was prepared by spin coating of mixed solutions of zirconium acetate and yttrium nitrate onto single-crystal Al_2O_3 substrates [75]. After pyrolysis at 1000°C to decompose the metal salts, sintering at 1400°C produced nearly fully dense polycrystalline films that, on further heating, started to break up into islands, uncovering the substrate. This instability is driven by capillarity (surface tension) due to the thermodynamic requirement of minimization of the interfacial energy. A necessary condition is that the film–substrate interface has a higher specific energy than the substrate surface.

Films are stable against small perturbations, but they can become unstable against large-amplitude perturbations such as pinholes. Holes with radii smaller than a critical value shrink, whereas holes with radii larger than the critical value grow, eventually leading to the breakup of the film into islands with a size determined by the surface energies [76]. A polycrystalline microstructure is particularly susceptible to breakup. Grain boundaries and three-grain junctions lead to the development of large perturbations that may eventually intersect the substrate and cause the film to break up into islands.

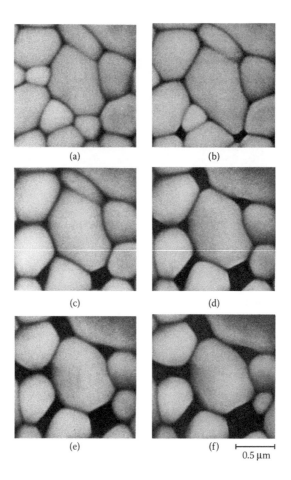

(a)   (b)

(c)   (d)

(e)   (f)

0.5 μm

FIGURE 5.41 Microstructural evolution of a ZrO_2 (8 mol% Y_2O_3) thin film during heat treatment at 1400°C for the following times (in hours): (a) 1, (b) 2, (c) 3, (d) 4, (e) 5, and (f) 6. (From Miller, K.T., Lange, F.F., and Marshall, D.B., The instability of polycrystalline thin films: experiment and theory, *J. Mater. Res.*, 5, 151, 1990. With permission.)

To see how these perturbations originate, we start with the well-known observation that a groove forms where the grain boundary intercepts the free surface, with the equilibrium angle at the root of the groove being the dihedral angle ψ given by $\cos(\psi/2) = \gamma_{gb}/2\gamma_{sv}$ (see Figure 2.4). Assuming that the film consists of uniform grains with an initial size $G = 2a$ and thickness h, and that the minimum energy surface shape is a spherical cap meeting the grain boundary at the equilibrium angle (Figure 5.42), then the requirement that the film volume be conserved yields an equilibrium groove depth, d, measured relative to the surface of the flat film, given by

$$d = a\,\frac{2 - 3\cos\theta + \cos^3\theta}{3\sin^3\theta} \tag{5.55}$$

where $\theta = \pi/2 - \psi/2$. This result shows that the grain boundary groove will go to a finite depth after an infinite time of annealing. We can determine the conditions under which the groove will intercept the substrate by putting $d \geq h$, giving

$$\frac{a}{h} \geq \frac{3\sin^3\theta}{2 - 3\cos\theta + \cos^3\theta} \tag{5.56}$$

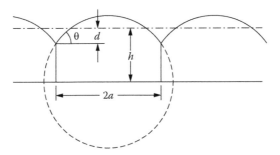

FIGURE 5.42 The equilibrium grain boundary groove configuration for a grain with a circular cross-section. The shape of the surface is a spherical cap that intersects the grain boundary at an angle $\theta = 90° - \psi/2$, where ψ is the equilibrium dihedral angle. The grain diameter is $2a$, the film thickness is h, and the equilibrium groove depth, measured with respect to the flat film, is d.

The critical value of the grain radius to the film thickness $(a/h)_c$ above which the grain boundary groove intercepts the substrate is plotted in Figure 5.43. As θ goes to zero, the value of $(a/h)_c$ tends to infinity, so the grain boundary groove will never intercept the substrate for finite a and h. On the other hand, as θ goes to its maximum value of $\pi/2$, $(a/h)_c$ tends toward 3/2. In particular, for $\theta = 30°$ (corresponding to $\psi = 120°$), $(a/h)_c$ is on the order of 7. Thus, for grain sizes larger than the film thickness, grain boundary grooves can reach the substrate, resulting eventually in island formation.

Miller, Lange, and Marshall [75] analyzed the breakup phenomena observed in Figure 5.41 by considering the free energy changes associated with the spheroidization of the grains and the uncovering of the substrate. The total energy of the surfaces and interfaces can be expressed by

$$E = A_{sv}\gamma_{sv} + A_{gb}\gamma_{gb} + A_i\gamma_i + A_{sub}\gamma_{sub} \tag{5.57}$$

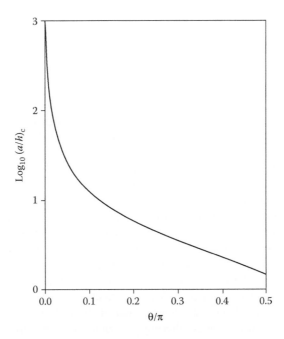

FIGURE 5.43 The critical ratio of the grain radius to the film thickness $(a/h)_c$, above which grain boundary grooving will lead to film rupture, versus the equilibrium groove angle, θ. The parameters a, h, and θ are defined in Figure 5.42.

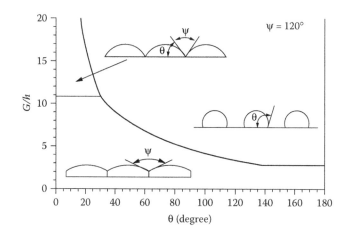

FIGURE 5.44 Two-dimensional equilibrium configuration diagram for the sintering of a thin film. The ratio of the grain diameter G to the film thickness h is plotted versus the contact angle θ for an equilibrium dihedral angle $\psi = 120°$. (From Miller, K.T., Lange, F.F., and Marshall, D.B., The instability of polycrystalline thin films: experiment and theory, *J. Mater. Res.*, 5, 151, 1990. With permission.)

where A_{sv} is the solid–vapor interfacial area of the film, A_{gb} is the grain boundary area, A_i is the film–substrate interfacial area, A_{sub} is the substrate–vapor interfacial area, and γ_{sv}, γ_{gb}, γ_i, and γ_{sub} are the corresponding interfacial energies. By evaluating E to determine its minimum value as a function of the configuration of the film, we can find the most stable configuration. The results of such calculations can be represented in terms of an equilibrium configuration diagram, showing the minimum energy configuration for any desired values of G/h, ψ, and θ. As an example, the results of the minimum energy configuration are shown in Figure 5.44, where G/h is plotted as a function of θ for $\psi = 120°$. The diagram is divided into three regions, showing the conditions for the completely covered substrate, the uncovered substrate, and the partially connected film (corresponding to the situation where the grain boundary just intercepts the substrate).

5.8 SOLID SOLUTION ADDITIVES AND THE SINTERING OF CERAMICS

As outlined in Chapter 3, the use of solid solution additives (or dopants) provides a very effective approach for the fabrication by sintering of ceramics with high density and controlled grain size. The most widely recognized example of this approach was reported by Coble [77], who showed in 1961 that small additions of MgO (0.25 wt%) to Al_2O_3 allowed the production of polycrystalline translucent alumina with theoretical density (Lucalox). Figure 5.45 illustrates the effect of MgO additive on the microstructure of Al_2O_3. Since Coble's work, many examples of the effectiveness of the solid solution approach can be found in the ceramic literature (see Table 3.1 for a selected list).

A problem with the solid solution approach is that, apart from a few recognized systems, the role of the dopant is not understood very well. The main reason for this gap in the understanding is that dopants can often display a variety of functions, which makes an understanding of the dopant's role difficult. As a result, the selection of additives has remained largely empirical. An additive can influence both the kinetic and thermodynamic factors in sintering. An additive can alter the defect chemistry of the host (see Chapter 1) and so change the diffusion coefficient for transport of ions through the lattice, D_1. Segregation of the additive can alter the structure and composition of surfaces and interfaces, altering the grain boundary diffusion coefficient D_b, the surface diffusion coefficient D_s, and the diffusion coefficient for the vapor phase D_g (i.e., the evaporation–condensation process). Segregation can also alter the interfacial energies, so the additive can also act thermodynamically to change the surface energy γ_{sv} and the grain boundary energy γ_{gb}.

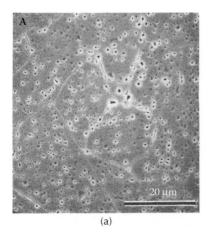 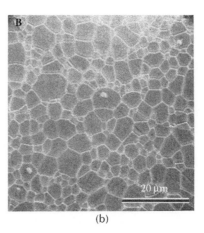

FIGURE 5.45 Microstructures of sintered Al_2O_3. (a) Undoped material showing pore–grain boundary separation and abnormal grain growth; (b) MgO-doped material showing high density and equiaxial grain structure. (Courtesy M.P. Harmer.)

Another consequence of segregation, described in Chapter 3, is that the additive can alter the intrinsic grain boundary mobility, M_b.

In principle, an additive will influence each of these factors to a certain extent. In a general sense, an effective additive is one that alters many phenomena in a favorable way but few phenomena in an unfavorable way. In practice, it is commonly found that when a dopant is effective for sintering, its major role is the ability to reduce M_b significantly.

5.8.1 Effect of Additives on Kinetic Factors

Although an additive can alter each of the diffusion coefficients for matter transport (D_l, D_{gb}, D_s, and D_g), historically the major emphasis has been placed on the ability of the additive to alter D_l through its effect on the defect chemistry of the host. To determine how an additive will influence D_l, the defect chemistry of the host must be known. Specifically, we must know the nature of the rate-controlling species (anion or cation), the type of defect (vacancy or interstitial), and the state of charge of the defect. In practice, this information is known in only a few cases. To illustrate the approach, let us consider Al_2O_3, a system that has been widely studied. According to Kroger [78], the intrinsic defect structure consists of cationic Frenkel defects:

$$Al_{Al}^x \rightleftharpoons Al_i^{\bullet\bullet\bullet} + V_{Al}^{///} \tag{5.58}$$

It has also been suggested that the diffusion of aluminum ions controls the rate of densification, with the faster diffusion of oxygen ions occurring along the grain boundary. Based on Equation 5.58, we may assume that the rate-controlling process is the diffusion of triply charged aluminum interstitial ions. A possible defect reaction for the incorporation of MgO into the Al_2O_3 lattice is

$$3MgO + Al_{Al}^x \xrightarrow{Al_2O_3} 3Mg_{Al}' + Al_i^{\bullet\bullet\bullet} + 3O_o^x \tag{5.59}$$

According to this equation, the effect of MgO would be to increase the densification rate through an increase in the concentration of aluminum interstitial ions.

Let us now consider an alternative situation, by supposing that the densification rate is controlled by the diffusion of triply charged aluminum vacancies. In this case, the creation of aluminum interstitials according to Equation 5.59 would lead to a reduction in the concentration of aluminum vacancies as described by Equation 5.58 and a corresponding reduction in the densification rate.

If instead of MgO we use TiO_2 as a dopant, a possible defect reaction for the incorporation into Al_2O_3 is

$$3TiO_2 \xrightarrow{Al_2O_3} 3Ti_{Al}^{\bullet} + V_{Al}^{///} + 6O_O^x \qquad (5.60)$$

According to Equation 5.60, if the rate-controlling process is the diffusion of aluminum vacancies, the addition of TiO_2 would increase the densification rate through an increase in the aluminum vacancy concentration. On the other hand, if the rate-controlling process is the diffusion of aluminum interstitials, then the addition of TiO_2 would lead to a decrease in the densification rate.

As this example of Al_2O_3 shows, if we know the intrinsic defect structure of the host, it is possible to select dopants to increase D_l. However, this single factor of an increase in D_l is insufficient to guarantee an effective additive. There are cases, such as TiO_2-doped Al_2O_3, where the dopant has been reported to produce an increase in D_l but the attainment of theoretical density is not realized. Other possible functions of the additive must be examined.

An additive can also act favorably to decrease the surface diffusion coefficient D_s, but this by itself is also insufficient to guarantee the achievement of theoretical density. The reason why we require a decrease (and not an increase) in D_s for favorably influencing the sintering can be argued as follows: A decrease in D_s reduces the rate of coarsening, thereby increasing the densification rate. Therefore, the pore size r will be smaller at any given stage of microstructural development (e.g., grain size). Because of its strong dependence on r (see Equation 3.86), the pore mobility M_p increases considerably and prevents separation of the pores from the boundary, so abnormal grain growth is prevented. Furthermore, the diffusion distance for matter transport into the pores is kept small, improving the probability of achieving high density.

5.8.2 EFFECT OF ADDITIVES ON THERMODYNAMIC FACTORS

The effects of dopants on γ_{sv} and γ_{gb} have been investigated in only a few instances. In one of the most detailed studies, Handwerker et al. [79,80] measured the effect of MgO dopant on the distribution of dihedral angles in Al_2O_3 from grain boundary grooves on polished and thermally etched samples. We recall that the dihedral angle provides a measure of the ratio γ_{gb}/γ_{sv}. As Figure 5.46

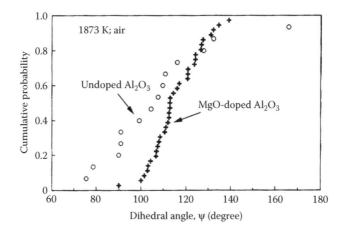

FIGURE 5.46 Cumulative distribution of dihedral angles in undoped Al_2O_3 and MgO-doped Al_2O_3. (Courtesy of C.W. Handwerker.)

shows, the MgO dopant reduces the width of the distribution of dihedral angles without significantly altering the mean value (117°). The reduction in the spread of the dihedral angles can be interpreted qualitatively to mean that local variations in the driving force for sintering and coarsening are reduced, so a more homogeneous microstructure is favored. An improvement in the microstructural homogeneity, we will recall, favors the attainment of a higher density at a given stage of microstructural development and reduces the potential for initiating abnormal grain growth. The dihedral angle results, therefore, indicate that a function of MgO dopant in Al_2O_3 is to reduce the consequences of microstructural inhomogeneity.

5.8.3 SEGREGATION OF ADDITIVES

The solute drag mechanism proposed by Cahn [81] was discussed in Chapter 3 as an important approach for reducing the intrinsic grain boundary mobility M_b, and it was recognized that the effectiveness of the method depends critically on the ability of the solute to segregate at the grain boundaries. An extensive review of grain boundary segregation in ceramics was published in 1974 by Kingery [82,83]. The major driving forces leading to segregation of an equilibrium concentration of solute at the grain boundaries in ceramics are the reduction in the *elastic strain energy* due to a difference in size between the solute and the host atom for which it substitutes and the *electrostatic potential* of interaction between aliovalent solutes and charged grain boundaries. Segregation driven by elastic strain energy is limited to the core of the grain boundary, typically to a width of <1 nm. In contrast, for segregation driven by the electrostatic interaction, a compensating *space charge* layer adjacent to the charged grain boundaries produces a greater width of segregation, typically 1 nm to a few tens of nanometers. Coupling of the elastic strain and electrostatic effects, as well as dipolar effects, can also influence segregation [84], but we shall consider only the major driving forces: elastic strain and electrostatic potential.

5.8.3.1 Elastic Strain Energy

When the radius of the solute ion a_s is different from that of the host a, substitution leads to a misfit in the lattice Δa defined by

$$\Delta a / a = (a_s - a)/a \tag{5.61}$$

The misfit produces an increase in the elastic strain energy of the lattice given by [85]

$$U_o = \frac{6\pi a^3 K(\Delta a/a)^2}{1 + 3K/4G} \tag{5.62}$$

where K is the bulk modulus of the solute and G is the shear modulus of the host crystal. According to this equation, the strain energy is proportional to the square of the misfit. Appendix D provides a list of ionic radii that can be used to estimate the misfit.

If the grain boundary is regarded as a thin region of disorder between the crystalline grains, it may be expected that there will be some fraction of sites at the boundary where the energy associated with the addition of solute ions is small. Segregation of the solute ions to the grain boundary region, therefore, provides partial relaxation of the elastic strain energy in the lattice of the grains. We may expect that the extent of segregation would depend on the difference in energy $\Delta G_a = E_l - E_b$, for the solute in the lattice E_l and at the boundary E_b. It is evident that segregation is increasingly favored for larger values of ΔG_a, since this represents a greater reduction in the free energy of the system. Furthermore, we might expect that ΔG_a will have some spatial dependence, because the structural disorder varies with distance from the center of the boundary. Details of the spatial dependence are not clear, but it is not believed to extend more than a few lattice spacings from the center of the boundary.

McLean [86] showed that the segregated solute concentration at the grain boundary C_{gb} is related to the solute concentration in the lattice C_o by

$$C_{gb} = C_o \exp\left(\frac{\Delta G_a}{RT}\right) \tag{5.63}$$

where R is the gas constant and T is the temperature. By annealing Al_2O_3 samples at several temperatures, quenching the samples, and using Auger electron spectroscopy to measure the Ca content at the grain boundaries, Johnson [87] found that Equation 5.63 provides a good description of Ca segregation in Al_2O_3 when $\Delta G_a \approx 117$ kJ/mol. A simple model in which ΔG_a is assumed to be proportional to U_o and, hence, to $(\Delta a/a)^2$ provides a qualitative description of the segregation of isovalent solutes.

5.8.3.2   Electrostatic Interaction and the Space Charge Concept

Frenkel [88] first proposed that a surface charge and a compensating space charge could develop in ionic crystals in which the formation energy for cation defects differs from that for anion defects. The results have since been generalized to include charge distribution at other lattice discontinuities, such as grain boundaries and dislocations [89–91]. We can see how this charge (or potential) develops by considering the formation of a Schottky defect in an ionic solid such as NaCl (see Chapter 1). The energy of formation of a Schottky defect Δg_S can be separated into energies of formation of the Na cation vacancy $\Delta g_{V_{Na}}$ and the Cl anion vacancy $\Delta g_{V_{Cl}}$, defined as the energy required to bring the ion to the lattice discontinuity (e.g., the grain boundary). We can express this as

$$\Delta g_S = \Delta g_{V_{Na}} + \Delta g_{V_{Cl}} \tag{5.64}$$

The energy of formation of the cation defect is likely to be different from that for the anion defect, and, on heating, this will determine the defect concentration at the grain boundary. On the other hand, the bulk of the crystal must be electrically neutral, and the defect concentration is determined by the principle of electroneutrality. Thus, differences in the formation energies of the individual defects cause the defect concentration at the grain boundary to be different from that in the bulk. The grain boundary can carry a net charge, and, to retain electroneutrality of the system, this net charge is balanced by an adjacent space charge layer that decays with distance, typically on the order of up to a few tens of nanometers. At equilibrium, there exists an electrical potential difference between the grain boundary and the bulk, and its interaction with charged solute ions can cause segregation at the grain boundary.

To get an idea of the variation of the potential and defect concentration in the space charge layer, let us consider NaCl, for which such effects have been examined in some detail [82,83]. In the simplest formulation, it is assumed that the grain boundaries act as perfect sources and sinks for vacancies, so vacancies can be readily created or destroyed. If Schottky defects dominate, following the treatment of defect chemistry in Chapter 1, the equilibration of the cation and anion with the grain boundary can be represented by the following defect reactions:

$$Na_{Na}^x \rightleftharpoons Na_{gb}^\bullet + V_{Na}' \qquad K_{SC} = [Na_{gb}^\bullet][V_{Na}'] \tag{5.65}$$

$$Cl_{Cl}^x \rightleftharpoons Cl_{gb}' + V_{Cl}^\bullet \qquad K_{SA} = [Cl_{gb}'][V_{Cl}^\bullet] \tag{5.66}$$

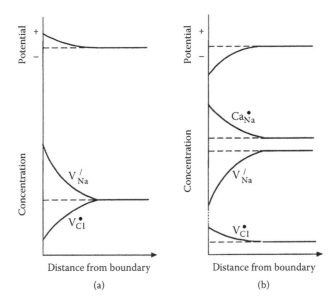

FIGURE 5.47 Schematic variations of the electrostatic potential and defect concentration as a function of distance from the grain boundary for (a) undoped NaCl and (b) NaCl doped with CaCl$_2$.

where K_{SC} and K_{SA} are the equilibrium constants for the Schottky cation and anion formation reactions, respectively. If the ions transported to the grain boundary are assumed to simply extend the perfect lattice, then the Schottky reaction is recovered:

$$\text{null} \rightleftharpoons V_{Na}' + V_{Cl}^{\bullet} \qquad K_S = [V_{Na}'][V_{Cl}^{\bullet}] \tag{5.67}$$

where K_S is the equilibrium constant for the Schottky reaction. In NaCl, $\Delta g_{V_{Na}}$ is estimated to be one-half $\Delta g_{V_{Cl}}$, so the grain boundary is enriched in the cation $[Na^{\bullet}_{gb}]$ and, through the Schottky equilibrium, Equation 5.67, is depleted in $[Cl'_{gb}]$ at the same time. The grain boundary has a positive charge due to the excess cations, and this is compensated by a space charge that has an excess negative charge. The variation of the electrical potential and the defect concentration is summarized in Figure 5.47a. The defect concentration varies with distance in the space charge layer according to the spatially varying potential $V(x)$ and is given by

$$\left[V_{Na}' \right] = \exp\left(\frac{-\Delta g_{V_{Na}} + eV(x)}{kT} \right) \tag{5.68}$$

$$\left[V_{Cl}^{\bullet} \right] = \exp\left(\frac{-\Delta g_{V_{Cl}} - eV(x)}{kT} \right) \tag{5.69}$$

where e is the electronic charge, k is the Boltzmann constant, and T is the absolute temperature. It has been estimated that $\Delta g_{V_{Na}} = 0.65$ eV and $\Delta g_{V_{Cl}} = 1.21$ eV. Since the electroneutrality condition $[V_{Na}'] = [V_{Cl}^{\bullet}]$ holds far from the grain boundary, we find from Equation 5.68 and Equation 5.69 that the electrostatic potential between the grain boundary and the bulk of the crystal is

$$V_{\infty} = \frac{1}{2e}(\Delta g_{V_{na}} - \Delta g_{V_{Cl}}) = -0.28 \text{ volt} \tag{5.70}$$

This estimate shows that the potential is not trivial in magnitude.

The space charge potential changes when the bulk concentrations of the defects are altered by solutes or nonstoichiometry. Taking the case where an aliovalent dopant such as $CaCl_2$ is present in NaCl, the defect reaction for the incorporation of the solute can be written

$$CaCl_2 \xrightarrow{\text{NaCl}} Ca_{Na}^{\bullet} + V_{Na}^{/} + 2Cl_{Cl}^{x} \qquad (5.71)$$

Additional cation vacancies are generated in the bulk of the lattice, and, on equilibration, this leads to a reduction in the concentration of the Na^+ ions at the grain boundary, resulting in an excess of Cl^- at the boundary. We recall from Chapter 1 that the defect concentrations produced by impurities and dopants are normally much larger than those for intrinsic defects, so the overall effect is a reversal in the sign of the electrostatic potential. The resulting negative electrostatic potential of the boundary will cause the Ca ion with the positive effective charge to segregate at the grain boundary as a space charge to maintain the electroneutrality of the bulk of the crystal. The variation of the electrostatic potential and the defect concentration for Ca-doped NaCl is summarized in Figure 5.47b. The cation vacancy concentration still varies according to Equation 5.68, and, far from the boundary, it can be written as

$$\left[V_{Na}^{/}\right] = \left[Ca_{Na}^{\bullet}\right] = \exp\left(\frac{-\Delta g_{V_{Na}} + eV_{\infty}}{kT}\right) \qquad (5.72)$$

Hence, the potential difference between the grain boundary and the bulk is given by

$$V_{\infty} = \frac{1}{e}\left(\Delta g_{V_{Na}} + kT\ln\left[Ca_{Na}^{\bullet}\right]\right) \qquad (5.73)$$

This voltage is positive for reasonable values of the dopant cation concentration.

Using a similar argument, it can be shown that the addition of extrinsic chlorine vacancies to the lattice will result in a reduction of anions from the grain boundary, giving rise to a larger negative potential V_{∞} than for the intrinsic case. Thus, at a given temperature, the potential can vary from negative to positive depending on the defect concentration. At some defect concentration, the potential is zero, and this point is referred to as the *isoelectric point,* which is analogous to the isoelectric point for aqueous colloidal suspensions [1]. The isoelectric point depends on the defect concentration as well the temperature, as can be deduced from Equation 5.73 for NaCl. The existence of the isoelectric point has been confirmed by direct observation of the migration of low-angle grain boundaries in NaCl bicrystals maintained in an electric field [92]. The results in Figure 5.48 for NaCl bicrystals with a bulk impurity content of ~2 ppm show that the boundary migration is reversed in experiments at 600°C and at 640°C, indicating an isoelectric point between these two temperatures.

In oxides, observations appear to support the space charge concept in several systems, including Al_2O_3 [93], MgO [94], ZrO_2 [95], TiO_2 [96,97], $BaTiO_3$ [98], and $MgAl_2O_4$ [99]. The effects are particularly clear in the case of TiO_2. A detailed study of the solute segregation indicates that the electrostatic potential at the grain boundary and, hence, the segregation of aliovalent solutes depend on the doping level, the temperature, and the oxygen partial pressure. As illustrated in Figure 5.49, for TiO_2 doped simultaneously with an acceptor (Al) and a donor (Nb), the Al-rich samples show Al segregation and Nb depletion, consistent with a positive boundary and a negative space charge layer (negative V_{∞}). On the other hand, Nb-rich samples show Nb segregation and Al depletion. No detectable segregation or depletion is found for a slightly Nb-rich sample (Figure 5.49b), and this composition appears to be near the isoelectric point at 1350°C. Above this temperature Al segregation was observed, whereas below this temperature Nb segregation was observed. Determination of the

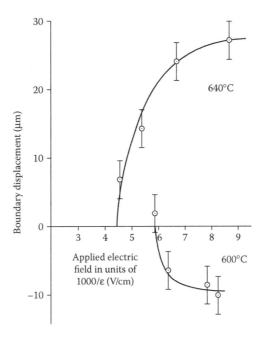

FIGURE 5.48 Migration of pure tilt low-angle grain boundary in NaCl as a function of applied electric field. The change in the direction of motion between 600°C and 640°C reflects a change in the sign of the boundary electrical potential. (From Schwensfeir, R.J., Jr. and Elbaum, C., Electric charge on dislocation arrays in sodium chloride, *J. Phys. Chem. Solids*, 28, 597, 1967. With permission.)

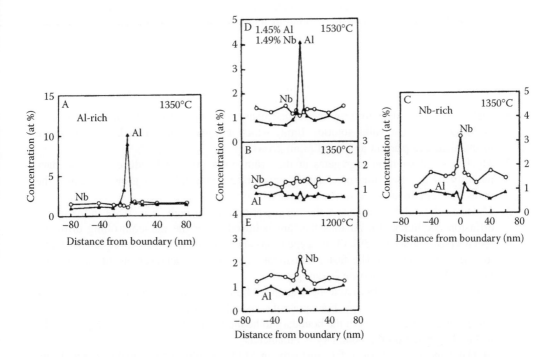

FIGURE 5.49 Scanning transmission electron microscope measurements of the solute distribution across grain boundaries in TiO$_2$ doped simultaneously with varying concentrations of Al and Nb. *(B)* represents a sample very close to the isoelectric point. (Courtesy of Y.-M. Chiang.)

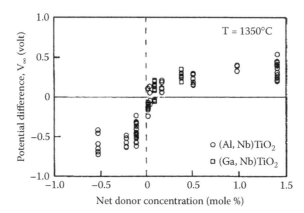

FIGURE 5.50 Potential difference V_∞ between the grain boundary and the bulk in TiO_2. (Courtesy of Y.-M. Chiang.)

potential difference V_∞ between the grain boundary and the bulk (Figure 5.50) indicates that the isoelectric point is slightly to the donor-doped side. This is because TiO_2 is somewhat nonstoichiometric and, at high temperatures, there is sufficient intrinsic reduction even in an oxidizing atmosphere, so some donor doping is required to compensate for the defects introduced by reduction.

5.8.4 THE ROLE OF MgO IN Al₂O₃

Since the work of Coble in 1961 [77], the MgO-doped Al_2O_3 system has represented one of the most celebrated examples of the effectiveness of the solid solution approach for microstructure control. Although the dramatic effect of MgO on the sintered microstructure of Al_2O_3 was easily reproduced (Figure 5.45), understanding the mechanism by which MgO acts proved to be more difficult, despite numerous investigations. One of the main obstacles in the earlier work was the use of powders containing impurities that tended to mask the true effect of MgO. As described in a thorough review by Bennison and Harmer [100], the role of MgO in the sintering of Al_2O_3 is now well understood, at least at a phenomenological level.

The specific mechanisms by which MgO has been suggested to act in Al_2O_3 are as follows:

1. MgO reacts with Al_2O_3 to form fine second-phase particles of $MgAl_2O_4$, due to an excess of MgO above the solid solution limit, which pin the boundaries by a Zener-type mechanism and prevent abnormal grain growth.
2. MgO in solid solution segregates at the grain boundaries and reduces the boundary mobility by a solute-drag mechanism, preventing abnormal grain growth because the pores can remain attached to the boundary.
3. MgO in solid solution enhances the densification rate through an increase in the lattice diffusion coefficient for Al ions that are believed to be the rate-controlling diffusing species (see Equation 5.59). (The oxygen ions are assumed to diffuse more rapidly along the grain boundaries.) The faster densification allows the achievement of full density prior to the onset of abnormal grain growth.
4. MgO lowers the dihedral angle by changing the relative values of the grain boundary energy and the surface energy. For a pore of constant volume, the reduction in the dihedral angle causes a greater area of the boundary to be intersected by the pore (see Figure 2.31), so the drag force on the boundary is enhanced.
5. MgO enhances the rate of surface diffusion in Al_2O_3, thereby enhancing the pore mobility through its dependence on the surface diffusion coefficient (Equation 3.86), allowing the pores to migrate with the boundary and avoiding abnormal grain growth.

In a classic experiment, Johnson and Coble [101] proved that mechanism 1 was not necessary for the achievement of theoretical density, so this mechanism can be ruled out. In the experiment, an undoped Al_2O_3 powder compact was sintered in close proximity to a pre-equilibrated two-phase mixture of $MgAl_2O_4$ spinel and Al_2O_3. The experiment was designed to allow transfer of MgO from the spinel–alumina compact to the undoped Al_2O_3 compact in concentrations below the solid solubility limit. It produced an Al_2O_3 pellet with a fully dense, fine-grained, MgO-doped outer surface surrounding a core devoid of MgO, which consisted of abnormal grains and entrapped porosity. Mechanism 5 by itself cannot lead to an enhancement of densification. An increase in the surface diffusion coefficient leads to enhanced microstructural coarsening (i.e., increase of the pore size and grain size), so the pore mobility and the densification rate actually decrease at any given density.

The current understanding is that MgO additions affect all the key parameters controlling the sintering of Al_2O_3 to some extent (i.e., D_l, D_{gb}, D_s, γ_{gb}/γ_{sv}, and M_b), so mechanisms 2 through 5 are all influenced to some degree [102–104]. The extent to which Al_2O_3 responds to MgO doping appears to depend on the concentration and composition of the background impurities in the powder. Surface diffusion appears to increase slightly, whereas lattice diffusion or grain boundary diffusion can be increased or decreased slightly. However, the single most important effect of MgO doping is the ability to significantly reduce the grain boundary mobility through a solute drag mechanism (mechanism 2). The consequence of this significantly reduced mobility is a reduced tendency for abnormal grain growth to occur (Figure 3.54).

The important role of MgO as a microstructural homogenizer, to reduce anisotropies in the surface and grain boundary energies and mobilities, and to stabilize the microstructure against the consequences of inhomogeneous densification has also been recognized. The observed effect of MgO to narrow the distribution of dihedral angles without significantly altering the mean value in Al_2O_3 (Figure 5.46) can be interpreted as effectively reducing local variations in the driving force for sintering and grain growth, promoting a more uniform development of the microstructure.

Baik, White, and Moon [105–107] studied the segregation of Ca and Mg at free surfaces of sapphire (i.e., single-crystal Al_2O_3), and at the grain boundaries of polycrystalline Al_2O_3. Strong Mg segregation occurs at several surfaces of sapphire, but the segregation of Ca is strongly dependent on the surface orientation. In the experiments with polycrystalline Al_2O_3, powder compacts were doped with controlled amounts of CaO (100 ppm) or MgO (300 ppm) or with a combination of CaO (100 ppm) and MgO (300 ppm). With CaO alone, segregation is inhomogeneous, with only some grain boundaries enriched with a high concentration of Ca. On the other hand, for the sample doped with both CaO and MgO, the distribution of the Ca in the grain boundaries is more homogeneous. Assuming that MgO segregation at grain boundaries is as effective as that observed at the surfaces of sapphire, these observations may explain the effectiveness of MgO (solute drag mechanism) and the ineffectiveness of CaO (inhomogeneous segregation) as a grain growth inhibitor in Al_2O_3.

Variations in the packing density of the green body, we will recall, give rise to differential densification in which the densely packed regions undergo local densification, resulting in the development of dense, pore-free regions in an otherwise porous microstructure. These dense regions are better able to support grain growth and commonly form the sites for the initiation of abnormal grain growth. Consistent with its ability to reduce the grain boundary mobility in Al_2O_3, MgO is found to suppress grain growth within the dense regions, allowing the porous regions to densify without the occurrence of abnormal grain growth in the already dense regions [108,109]. The role of MgO can be interpreted as stabilizing the microstructure against the consequences of inhomogeneous densification.

Another feature related to microstructural homogenization is the sintering of Al_2O_3 in the presence of a small amount of liquid phase. Many Al_2O_3 powders contain small amounts of SiO_2

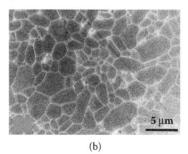

(a)   (b)

FIGURE 5.51 Microstructures of (a) dense undoped Al_2O_3 and (b) dense MgO-doped Al_2O_3. Note that the MgO addition leads to a more uniform grain structure and a smaller grain size. (Courtesy of M.P. Harmer.)

and other impurities that form a silicate liquid phase, leading to the growth of anisotropic (faceted) abnormal grains (see Section 3.6). MgO additions suppress the development of such abnormal grains and produce a more uniform microstructure (Figure 5.51). Kaysser et al. [110] performed experiments in which controlled amounts of anorthite glass were added to Al_2O_3 powders containing a small amount of SiO_2 impurity, and large sapphire spheres were included to seed abnormal grain growth. Rapid growth of anisotropic grains occurred for the undoped Al_2O_3, and any anisotropy of the seed crystal was hard to detect. With 0.1 wt% MgO, the matrix grains were fine and equiaxial in shape, whereas the seed crystal grew abnormally but in an isotropic manner. With 0.9 wt% anorthite, the seed crystal grew abnormally, with faceting along the basal planes, and the addition of 0.1 wt% MgO was unable to suppress the faceting. The results indicate that abnormal grain growth in systems containing a liquid can be caused by anisotropic growth of boundaries containing intergranular liquid films.

5.9 SINTERING WITH CHEMICAL REACTION: REACTION SINTERING

Reaction sintering, sometimes referred to as reactive sintering, is a particular type of sintering process in which the chemical reaction of the starting materials and the densification of the powder compact are both achieved in a single heat treatment step. These systems can be divided into two main classes, depending on whether single-phase solids or composites are produced. For a powder compact consisting of a mixture of two reactant powders, the simplest example of reaction sintering is shown in Equation 5.74 and in Figure 5.52a. During sintering, reaction between two starting powders A (e.g., ZnO) and B (e.g., Fe_2O_3) and densification occur to produce a polycrystalline, single-phase solid C (e.g., $ZnFe_2O_4$):

$$ZnO + Fe_2O_3 \rightarrow ZnFe_2O_4 \tag{5.74}$$

A more complex example of reaction sintering is shown in Equation 5.75 and in Figure 5.50b. Reaction between two starting powders D (e.g., Al_2O_3) and E (e.g., zircon, $ZrSiO_4$) and densification occur to produce a composite solid consisting of two phases F (mullite, $3Al_2O_3 \cdot 2SiO_2$) and G (ZrO_2):

$$3Al_2O_3 + 2(ZrO_2 \cdot SiO_2) \rightarrow 3Al_2O_3 \cdot 2SiO_2 + 2ZrO_2 \tag{5.75}$$

A variation of this second class of reaction-sintering system is obtained in systems for which one of the product phases is a liquid. This occurs when the sintering temperature is above the eutectic temperature and corresponds to the process of liquid-phase sintering discussed in detail in

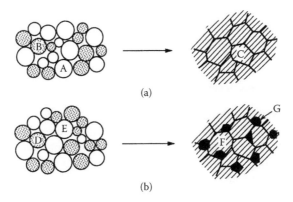

(a)

(b)

FIGURE 5.52 Schematic diagram illustrating the production of: (a) a single phase solid C by reaction sintering of a compacted mixture of two powders, A and B; (b) a composite solid, consisting of a matrix F and inclusions G, by reaction sintering of a compacted mixture of two powders, D and E.

Chapter 4. Here, we will confine our discussion of reaction sintering mainly to systems in which the products are solids at the sintering temperature.

Since the energy changes for chemical reactions are much larger than those for surface area changes (see Chapter 1), it would be very desirable if the free energy of the reaction could be used to drive the densification process. Unfortunately, there is no evidence that the energy of the reaction can act directly as a driving force for densification. Reaction sintering has the benefit of eliminating the pre-reaction (or calcination) step in the formation of solids with complex composition. Taking $ZnFe_2O_4$ as an example, the conventional route for producing a dense polycrystalline solid involves calcination of a loose mixture of ZnO and Fe_2O_3 powders to form a single-phase $ZnFe_2O_4$ powder and milling the calcined powder to break down agglomerates, followed by powder compaction and sintering. In reaction sintering, the reaction and densification occur in the same heating cycle, so the calcination and subsequent milling steps in the conventional route are eliminated.

In practice, reaction sintering has several shortcomings, so the process finds little use in the production of single-phase solids. The large energy changes associated with the reaction can lead to the development of microstructures that inhibit densification. An example of such a microstructure is shown in Figure 5.53 for the $CaO–Al_2O_3$ system [111]. The plate-like calcium hexa-aluminate

FIGURE 5.53 Scanning electron micrograph of calcium hexa-aluminate (CA_6 arrowed) formed during the densification of Al_2O_3 powder mixed with 5 wt% CaO powder, after 20 h at 133°C. The plate-shape CA_6 phase replicates the porous Al_2O_3 matrix during solid-state reaction/grain growth and inhibits densification.

phase formed during the reaction replicates the porous Al_2O_3 matrix and inhibits densification. Other shortcomings include the risk of chemically inhomogeneous products due to incomplete reaction and difficulties in controlling the microstructure as a result of the added complexity introduced by the reaction.

5.9.1 Influence of Process Variables

Depending on the processing conditions, such as particle size, temperature, heating rate, and applied pressure, reaction and densification can occur in sequence, concurrently, or in some combination of the two. Qualitatively, it is convenient to consider the reaction and the densification as two separate processes and to develop principles that allow predictions of the influence of key process variables that are useful for the design of reaction-sintering processes [112,113].

5.9.1.1 Particle Size of Reacting Powders

Herring's scaling law predicts the effect of particle size on the rate of matter transport during sintering (see Chapter 2). The densification rate varies with particle (or grain) size G according to $1/G^4$ for grain boundary diffusion and $1/G^3$ for lattice diffusion. For the reaction, if the product forms coherently on the surfaces of the particles, the reaction rate varies as $1/G^2$, but if the product does not form coherently, then the rate is expected to vary as $1/G$. Both the densification rate and the reaction rate increase with decreasing particle size (Figure 5.54), but, because of the stronger size dependence, the densification mechanism is more strongly influenced than the reaction mechanism. Therefore, a reduction in the particle size increases the densification rate relative to the reaction rate.

5.9.1.2 Sintering Temperature

Although densification and reaction will most likely occur by different mechanisms, they both involve diffusion and are, therefore, expected to be thermally activated, so the rates have an

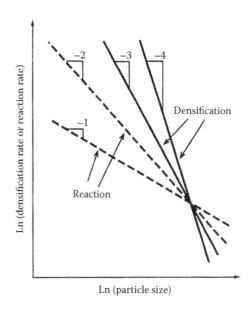

FIGURE 5.54 Predicted effect of particle size on the rates of densification and reaction during reaction sintering of a powder mixture. Smaller particle size enhances the rate of densification relative to the rate of reaction.

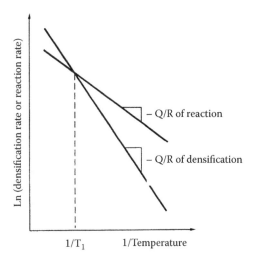

FIGURE 5.55 Predicted effect of sintering temperature on the rates of densification and reaction during reaction sintering of a powder mixture when the activation energy for densification is greater than that for the reaction. Higher sintering temperature enhances the densification rate relative to the reaction rate.

Arrhenius dependence on temperature. The dependence of the densification rate and the reaction rate on temperature is sketched in Figure 5.55, where it is assumed that the activation energy for densification is greater than that for the reaction. In this case, increasing the temperature leads to an increase in the densification rate relative to the reaction rate.

5.9.1.3 Applied Pressure

The application of an external pressure p_a increases the driving force for densification (see Chapter 2). The effect of applied pressure on the reaction rate is difficult to predict, but, because of the strong effect on the densification process, the role of applied pressure is likely to lie in its ability to accelerate the densification process, rather than in any effect on the reaction. In practice, hot pressing provides one of the surest ways to adjust the relative rates of densification and reaction in the desired direction.

5.9.1.4 Processing Trajectories in Reaction Sintering

Three different trajectories in reaction sintering are sketched in Figure 5.56, depending on the relative rates of densification and reaction. When the reaction rate is much faster than the densification rate (curve A), the reaction occurs predominantly before any significant densification, and, thus, densification must be achieved for a microstructure consisting of the fully reacted powder. Curve C shows the trajectory when the reaction rate is much slower than the densification rate; in this case, densification occurs without any significant reaction, so the reaction process must be carried out in a fully dense microstructure. Curve B represents the trajectory for a system where the reaction rate is comparable to the densification rate.

The conceptual arguments for the effect of the process variables lead to the conclusion that the best trajectory for reaction sintering is one where densification is completed before the reaction can significantly interfere with the microstructural development (curve C), assuming that the difference in molar volumes between the reactants and the product is not large enough to cause cracking of the body. It is often not possible to achieve this processing trajectory simply by reducing

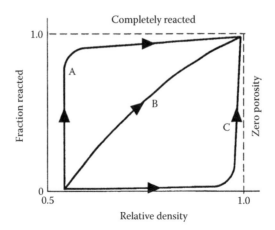

FIGURE 5.56 Sketch showing three different processing trajectories in reaction sintering. Curve A is the trajectory for a system in which the reaction rate is much higher than the densification rate; curve B is for a system where the reaction rate and densification rate are of roughly the same magnitude; for the system following trajectory C, the densification rate is much higher than the reaction rate.

the particle size or optimizing the sintering temperature, so the use of an applied pressure, such as in hot pressing, is warranted.

5.9.2 Experimental Observations of Reaction Sintering

Systems in which two simple oxides react to form a complex oxide form useful models for investigating the production of single-phase ceramics by reaction sintering because of the available data on the reaction mechanisms [114] and the relative simplicity of the reactants and product. Two examples are the formation of zinc ferrite by Equation 5.74 and the formation of zinc aluminate spinel by the reaction between ZnO and Al_2O_3 powders:

$$ZnO + Al_2O_3 \rightarrow ZnAl_2O_4 \qquad (5.76)$$

These two systems have important similarities and differences, and they serve to illustrate the importance of the reaction process in the densification and microstructural evolution. For both systems, the molar volume of the reactants is almost equal to that of the product. The reaction is rapid, being completed prior to any significant densification, so the processing trajectory follows curve A, instead of the more favorable curve C in Figure 5.56.

The observed reaction-sintering results for these two systems show significant differences. For the $ZnO-Fe_2O_3$ reaction sintering system [115], the reaction produces little change in the volume of the powder compact. Despite the less favorable processing trajectory, high final densities (>95% of the theoretical) are achieved, which are comparable to the final densities achieved in the sintering of single-phase $ZnFe_2O_4$ powder compacts (Figure 5.57). For the $ZnO-Al_2O_3$ system, the compact expands by 25–30 vol% during the reaction (Figure 5.58), and densification is significantly lower than that for a single-phase $ZnAl_2O_4$ powder compact [116].

The origins of the significantly different sintering behavior observed for the $ZnO-Fe_2O_3$ and $ZnO-Al_2O_3$ systems appear to reside in the microstructural changes produced as a result of the reaction process. These microstructural changes may, in turn, be related to the different solid-state reaction mechanisms in the two systems. At the temperatures used in sintering, the reaction between

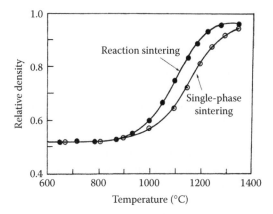

FIGURE 5.57 Relative density versus temperature for the reaction sintering of ZnO–Fe$_2$O$_3$ powder mixtures and for the sintering of a single-phase ZnFe$_2$O$_4$ powder. The samples were heated at a constant rate of 5°C/min.

ZnO and Fe$_2$O$_3$ occurs by a Wagner counterdiffusion mechanism where the cations migrate in opposite directions and the oxygen ions remain essentially stationary. The product is expected to form on both the ZnO and the Fe$_2$O$_3$ particles, causing little disruption of the particle packing, so the subsequent densification is not significantly affected by the reaction. The reaction between ZnO

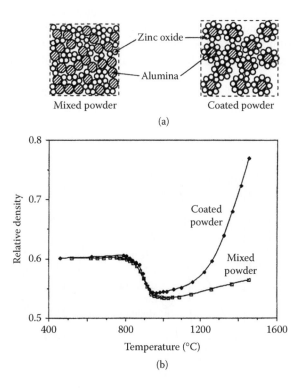

FIGURE 5.58 Effect of reactant powder distribution on densification during reaction sintering of ZnO–Al$_2$O$_3$ powders. (a) Sketch illustrating the distribution (cross-sectional view) of the ZnO and Al$_2$O$_3$ powders in compacts formed from mechanically mixed powders and from coated powders; (b) sintering data showing significantly enhanced densification for the coated powders.

and Al_2O_3 is not as clear, but it is believed to occur either by a mechanism in which the one-way diffusion of Zn^{2+} ions through the $ZnAl_2O_4$ product layer is rate controlling or by a gas–solid reaction between ZnO vapor and Al_2O_3 [114]. In either case, the product is expected to form predominantly on the percolating network of original Al_2O_3 particles, which inhibits subsequent densification. The sintering can, however, be improved. One way is to apply pressure during the subsequent sintering stage to break down the network and improve the particle packing. Another way is to modify the particle packing of the initial powder compact to reduce the disruption caused by the reaction. Coating the Al_2O_3 particles with ZnO (as opposed to the normal mechanical mixing of the powders) leads to enhanced densification rates (Figure 5.58), which can approach the densification rate for the single-phase $ZnAl_2O_4$ powder.

Because of the difficulties commonly encountered in the production of single-phase ceramics with high density by reaction sintering, it appears that the process might be better applied to the fabrication of composites, since this class of materials requires inhomogeneity as a microstructural characteristic. Displacement reactions such as that described by Equation 5.75 have been the subject of research for the fabrication of ZrO_2-toughened ceramics. Different accounts of the reaction and densification sequence have been reported for this system. The work of Claussen and Jahn [117] provides an example of reaction sintering in which the reaction is delayed until after the densification is completed. Using a two-stage heating schedule (Figure 5.59), it was possible to achieve nearly full densification at ~1450°C, and the reaction between zircon and Al_2O_3 subsequently initiated and proceeded to completion at ~1600°C. The resulting composite consists of a fine dispersion of monoclinic and tetragonal ZrO_2 in a mullite matrix (Figure 5.60). The data of Di Rupo et al. [118] show that the densification and reaction processes occur simultaneously during hot pressing and sintering at 1450°C, whereas Yangyun and Brook [113] report a type of behavior that is intermediate between those observed by Claussen and Jahn and by Di Rupo et al. These different results are most probably a consequence of differences in the composition and particle size of the starting powders. The starting powders have different impurity levels and have been milled to varying extent. The powders that densify first appear to have a finer grain size and a higher level of impurity.

Reaction sintering involving displacement reactions has also been investigated as a processing route for the production of ceramic–metal composites in which metal phase serves to improve the

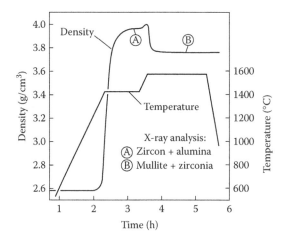

FIGURE 5.59 Sintering and reaction schedule for reaction sintering of zircon–Al_2O_3 powder mixtures. At A, only zircon and Al_2O_3 are detected by X-ray analysis; at B, the reaction to produce mullite and ZrO_2 is completed.

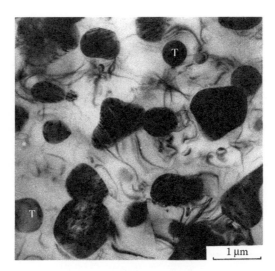

FIGURE 5.60 Bright field transmission electron micrograph of a ZrO$_2$–mullite material produced by reaction sintering of zircon–Al$_2$O$_3$ powder mixtures. The material was annealed for 1 h. Larger twinned particles are monoclinic ZrO$_2$; smaller particles (T) are tetragonal ZrO$_2$. Note the considerable stress contours in the mullite matrix. (Courtesy of N. Claussen.)

fracture toughness of the ceramic. An example is the production of Ni-reinforced Al$_2$O$_3$ from mixtures of Al and NiO powders, where the following reaction takes place:

$$2Al + 3NiO \rightarrow Al_2O_3 + 3Ni \tag{5.77}$$

Densities of 96% to 98% of the theoretical have been achieved for this system at sintering temperatures of 1600°C to 1700°C [119].

An interesting exploitation of the reaction-sintering concept is the formation of single-phase mullite and SiC-reinforced mullite composites from Al$_2$O$_3$ and SiC particles coated with an amorphous silica layer (Figure 5.61). This process, developed by Sacks, Bozkurt, and Scheiffele [120] and referred to as transient viscous sintering, involves rapid densification due to viscous flow of the interparticle silica glass, whereas the subsequent reaction at a higher temperature between the Al$_2$O$_3$ cores and the silica glass produces the required mullite phase for SiC-reinforced mullite.

5.10 VISCOUS SINTERING WITH CRYSTALLIZATION

Sol-gel techniques can be used to prepare porous gels in the form of a film, monolithic solid, or fine powders, which are commonly amorphous [1]. If the composition is crystallizable, then during the conversion of the dried gel or powder compact to a dense ceramic by viscous sintering, it is possible that crystallization and densification can occur in sequence, concurrently, or in some combination of the two. For the same chemical compound, the crystalline phase has a considerably higher viscosity than the amorphous phase, so the sintering of a polycrystalline material is significantly more difficult than the amorphous phase. This argument suggests that the most favorable processing trajectory for amorphous materials is the achievement of full density prior to any significant crystallization. In Figure 5.56, if we replace reaction with crystallization, the most favorable processing trajectory is path C.

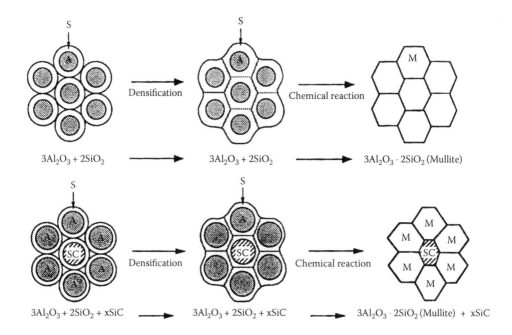

FIGURE 5.61 Schematic illustration of transient viscous sintering for the formation of (*top*) single-phase mullite and (*bottom*) mullite–SiC composite.

5.10.1 Effect of Process Variables

By analogy with our earlier discussion of reaction sintering, it is convenient to consider the crystallization and the densification as two separate processes and to develop principles that allow qualitative predictions of the influence of key process variables on the relative rates of densification and crystallization.

5.10.1.1 Viscosity

The viscosity can be altered by changes in the sintering temperature or in the chemical composition of the glass, but it is more meaningful to consider the influence of the viscosity on sintering and crystallization instead of the separate variables of temperature or composition. The rate of viscous sintering is inversely proportional to the viscosity of the glass η (see Chapter 2). If η is constant, as for example in sintering at a fixed temperature, the extent of densification after a time t is proportional to t/η. When η changes with time, as for example in sintering at a constant heating rate, the extent of densification is proportional to $\int dt/\eta$. The effect of viscosity on the kinetics of crystallization has been considered by Zarzycki [121]. For a fixed number of nuclei formed, and when the thermodynamic barrier to formation of a nucleus is constant, Zarzycki showed that the extent of crystallization also depends on the quantity t/η or, if η changes with time, on $\int dt/\eta$. Since the extent of densification and crystallization depends on the same function of the viscosity, any factor that changes the viscosity influences both processes in such a way that the volume fraction of crystals is exactly the same by the time a given density is reached. Changes in the viscosity cannot, therefore, be used to vary the extent of densification relative to crystallization.

5.10.1.2 Pore Size

The crystallization process does not depend on the microstructure of the glass, but, according to viscous sintering theories (see Chapter 2), the densification rate is inversely proportional to the pore size r. Small pores favor rapid sintering without affecting the driving force for crystallization. The reduction in the pore size provides one of the most effective methods for increasing the ratio of the densification rate to the crystallization rate. In practice, the pores cannot be made too small (e.g., less than approximately 5 nm), because the sintering rate may become so rapid that full density is achieved prior to complete burnout of the organic constituents present in materials prepared, for example, by sol–gel processing. The impurities remaining in the dense solid can lead to a substantial deterioration of the properties.

5.10.1.3 Applied Pressure

Applied pressure can be very effective for increasing the densification rate if it is significantly higher than the sintering stress, Σ (see Chapter 2). For pores with a radius r, the sintering stress can be taken to be $\approx \gamma_{sv}/r$, where γ_{sv} is the specific surface energy of the solid–vapor interface. Taking $\gamma_{sv} \approx 0.25$ J/m^2 and $r \approx 10$ nm gives $\Sigma \approx 25$ MPa, a value that is comparable to the pressures available in hot pressing. It is, therefore, expected that hot pressing would have a significantly greater effect for larger pore sizes. The applied pressure generally has little effect on crystallization, but in some cases the effect of pressure on the crystallization process cannot be ignored. Crystallization often leads to a reduction in specific volume, and the application of an applied pressure can accelerate the process.

5.10.1.4 Heating Rate

Densification depends on viscous flow, but crystallization depends on two kinetic steps, nucleation and growth. The nucleation depends on the viscosity but also on the undercooling, so increasing the heating rate at lower temperatures, where the nucleation rate of the crystals is fast, can delay the onset of crystallization. In this way, higher density can be achieved for a given amount of crystallization.

5.10.2 ANALYSIS OF VISCOUS SINTERING WITH CRYSTALLIZATION

There are two main approaches for analyzing how the occurrence of crystallization can influence the densification process. In one approach, put forward by Uhlmann et al. [122], the results are represented in terms of *time–temperature–transformation* (TTT) diagrams to show the conditions of temperature and time under which crystallization and densification will occur. TTT diagrams have long been used by metallurgists to show the influence of thermal history on phase transformations, particularly in steels. In the other approach, developed by Scherer [123], the crystals formed during sintering are considered to be rigid inclusions in a viscous sintering matrix, and their effects on densification are analyzed by adapting the theory of sintering with rigid inclusions discussed earlier.

5.10.2.1 TTT Diagrams

An example of a TTT diagram for the crystallization of a glass is sketched in Figure 5.62. The curve, typically C-shaped, shows the times for the beginning (and sometimes the end) of crystallization at a given temperature. If sintering is to be completed prior to crystallization, the conditions must be chosen to remain to the left of the curve C_S, representing the onset of crystallization. The nose of the curve, denoted by T_{max}, represents the maximum crystallization rate.

In constructing the TTT diagram [124], if it is assumed that the nucleation and growth rates are constant, which is applicable when the volume crystallized is small, the volume fraction of crystals produced by a given thermal history can be calculated by Avrami's equation [125–127],

$$v = 1 - \exp\left(-\frac{\pi I_v u^3 t^4}{3}\right) \approx \frac{\pi I_v u^3 t^4}{3} \tag{5.78}$$

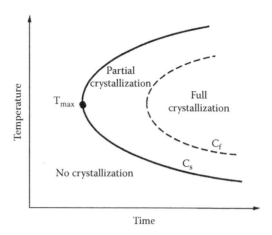

FIGURE 5.62 Sketch illustrating the regions in a time–temperature–transformation (TTT) diagram for the crystallization of an amorphous material.

where I_v is the rate of crystal nucleation, u is the rate of crystal growth, and t is the time. The determination of v from Equation 5.78 relies on theories or measured values for I_v and u. According to the classical theory, the nucleation rate, defined as the number of stable nuclei produced per unit volume of untransformed material per unit time, varies according to [128]

$$I_v \sim \exp\left[-\frac{\Delta G^* + \Delta G_n}{kT}\right] \tag{5.79}$$

where ΔG^* is the free energy for the formation of a nucleus, ΔG_n is the free energy of activation for atomic migration across the interface between the nucleus and the untransformed material, k is the Boltzmann constant, and T is the absolute temperature. As the temperature decreases below the liquidus, the nucleation rate increases rapidly because the thermodynamic driving force increases. The rate eventually decreases again because of the decreasing atomic mobility. This variation leads to the C shape of the TTT curve (Figure 5.62), with a maximum crystallization rate at T_{max}. Below T_{max}, the crystal growth rate can be approximated by

$$u \sim T \exp(-\Delta G_c / kT) \tag{5.80}$$

where ΔG_c is the free energy of activation for atomic migration across the interface between the crystal and the uncrystallized material. It is generally assumed that $\Delta G_n \approx \Delta G_c$ and that both are equal to the activation energy, Q, for viscous flow defined by

$$\eta = \eta_o \exp(Q / RT) \tag{5.81}$$

where η_o is a constant and R is the gas constant. Substituting for Q in Equation 5.79 and Equation 5.80 to determine I_v and u, followed by substitution of these quantities into Equation 5.78, gives

$$v \sim (t/\eta)^4 T^3 \exp(-\Delta G^* / kT) \tag{5.82}$$

For a spherical nucleus, ΔG^* is given by

$$\Delta G^* = (16\pi/3)\gamma_{sl}^3 / (\Delta G_v)^2 \tag{5.83}$$

where γ_{sl} is the specific energy of the crystal–liquid interface and ΔG_v is the free energy change per unit volume crystallized.

Considering now the densification process, the time to reach full density during sintering by viscous flow is given approximately by (see Equation 2.68)

$$t_f \approx \frac{\eta}{\gamma_{sv} N^{1/3}} \approx \left(\frac{4\pi}{3}\right)^{1/3} \left(\frac{\eta r}{\gamma_{sv}}\right) \tag{5.84}$$

Thus, t_f can be found if the pore radius r, the specific surface energy of the glass–vapor interface γ_{sv}, and η are known.

Figure 5.63 shows an example of a TTT curve for silica gel, which was constructed by using viscosity data obtained from sintering studies by Sacks and Tseng [129], as well as viscosity data for conventional dry silica, to find I_v and u and substituting into Equation 5.78 to find the condition for a given volume fraction v of crystals. The value $v = 10^{-6}$ was taken to represent the onset of crystallization. The figure shows two TTT curves, one for "wet" silica with a high hydroxyl content and the other for dry silica. The viscosity of the "wet" silica is so much lower than that of the dry silica that the TTT curve for the wet silica is shifted significantly to lower times at any temperature. The curves show reasonable agreement with the crystallization data of Sacks and Tseng.

Using Equation 5.84 to determine t_f, the dashed curves in Figure 5.63 represent the time to reach full density at each temperature for two assumed values of the pore size of 5 nm and 500 nm. If the sintering curve does not cross the crystallization curve, then densification can be completed before detectable crystallization occurs. Consistent with the earlier discussion, gels with

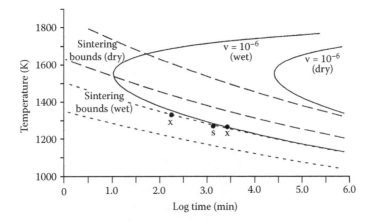

FIGURE 5.63 Sintering (*dashed line*) and crystallization (*solid line*) curves for dry silica and for "wet" silica with a higher hydroxyl content. The lower viscosity of the "wet" silica leads to a shift of the curves to lower times at any temperature. x marks treatments that resulted in crystallization with little sintering; s marks treatments that produced sintered glasses free of crystallinity. (From Uhlmann, D.R., Zelinski, B.J., Silverman, L., Warner, S.B., Fabes, B.D., and Doyle, W.F., Kinetic processes in sol–gel processing, in *Science of Ceramic Chemical Processing*, Hench, L.L., and Ulrich, D.R., Eds., Wiley, New York, 1986, p. 173. With permission.)

smaller pores are seen to sinter faster, so they are less likely to crystallize before the completion of sintering.

5.10.2.2 Analysis in Terms of Sintering with Rigid Inclusions

The approach based on TTT diagrams is valid only if the volume fraction of crystals is small. For higher volume fraction, the presence of the crystals influences the sintering rate, and this must be taken into account. Scherer [123] analyzed the effect of crystallization on the sintering of an amorphous matrix as a case of sintering with rigid inclusions, discussed earlier in Section 5.3. Compared to the earlier discussion, the size and volume fraction of the crystals (i.e., the inclusions) increase during sintering and must be included in the analysis.

The effect of crystallization on sintering depends on where the crystals are formed and on the nature of the crystallization process. Scherer considered the case where the nucleation rate is low and the growth rate is high. As sketched in Figure 5.64, the crystals grow to a size that is large compared to the scale of the gel microstructure, so the volume fraction of the crystals formed can easily be calculated. It is readily recognized from Figure 5.64 that if crystallization occurs prior to significant densification, considerable porosity will be trapped within the crystalline zone and the achievement of high density will be severely limited.

Because the effects are less difficult to predict in isothermal sintering than in the more realistic case of constant heating rate sintering, we shall limit our discussion to isothermal sintering. The linear densification rate of the free matrix (i.e., the glass) can be found from the models for viscous sintering described in Chapter 2. The Mackenzie and Shuttleworth equation is used because of its simple form, giving

$$\dot{\varepsilon}_{fm} = \frac{1}{2}\left(\frac{4\pi}{3}\right)^{1/3}\left(\frac{\gamma_{sv}N^{1/3}}{\eta}\right)\left(\frac{1}{\rho}-1\right)^{2/3} \tag{5.85}$$

where γ_{sv} is the specific energy of the solid–vapor interface, N is the number of pores per unit volume of the solid phase, and ρ is the relative density of the glass. The linear densification rate of the composite containing a volume fraction v of crystals is assumed to follow Equation 5.38 for the self-consistent model:

$$\dot{\varepsilon}_c = \frac{(1-v)\dot{\varepsilon}_{fm}}{1+v(4G_c/3K_m)} \tag{5.86}$$

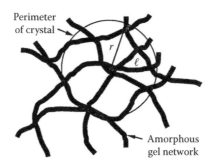

FIGURE 5.64 Schematic diagram illustrating the crystallization of an amorphous gel network when the nucleation rate is low and the growth is fast. Individual crystals are large compared to the scale of the gel structure. For a crystal to grow to a radius of r, it must trace a path of length l.

and since the shrinkage of the body results entirely from the densification of the matrix, we also have that

$$\dot{\varepsilon}_c = (1-v)\frac{1}{3\rho}\left(\frac{d\rho}{dt}\right) \tag{5.87}$$

Equating the right-hand sides of Equation 5.86 and Equation 5.87, and substituting for $\dot{\varepsilon}_{fm}$ from Equation 5.85, we have that

$$\frac{d\rho}{d\theta} = \frac{3}{2}\left(\frac{4\pi}{3}\right)^{1/3}\left(\frac{\rho^{1/3}(1-\rho)^{2/3}}{1+v(4G_c/3K_m)}\right) \tag{5.88}$$

where

$$\theta = \left(\frac{\gamma_{sv}N^{1/3}}{\eta}\right)t \tag{5.89}$$

The evaluation of Equation 5.88 requires making some assumptions about the rheology of the system. Scherer assumed a relationship that follows from the self-consistent theory:

$$1+v\left(\frac{4G_c}{3K_m}\right) = \frac{(1+v_c)(1-v)}{1+v_c-3(1-v_c)v} \tag{5.90}$$

where v_c is the Poisson's ratio of the composite, given approximately by

$$v_c = \frac{v}{5}+\left(\frac{1-v}{2}\right)\left(\frac{\rho}{3-2\rho}\right)^{1/2} \tag{5.91}$$

Assuming homogeneous nucleation and growth where I_v and u are constant, Scherer derived the following equation for v in terms of a parameter C and a function of the density:

$$v = C\rho\int_0^\theta \exp\left[-C\int_{\theta'}^\theta \rho(\theta'')\theta''^{1/3}d\theta''\right]\theta'^3 d\theta' \tag{5.92}$$

where

$$C = \frac{(4\pi/3)I_v u^3}{(\gamma_{sv}N^{1/3}/\eta)4} \tag{5.93}$$

Using Equation 5.84 for t_f, we find that C is also given by

$$C = \frac{4\pi}{3}I_v u^3 t_f^4 \tag{5.94}$$

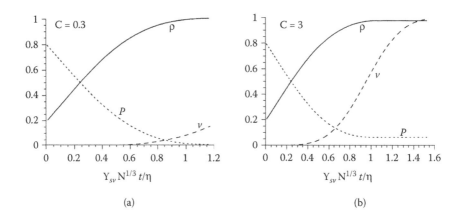

FIGURE 5.65 Predicted effect of crystallization on the sintering of an amorphous gel. The density of the matrix ρ, volume fraction of crystals v, and porosity P are plotted versus reduced time for a gel with initial relative density of 0.2. Calculations are shown for two values of the parameter C: (a) C = 0.3; (b) C = 3.0. (From Scherer, G.W., Effect of inclusions on shrinkage, *Mater. Res. Soc. Symp. Proc.*, 180, 503, 1990. With permission.)

Comparison of Equation 5.94 with Equation 5.78 shows that C is roughly the volume fraction of crystals that appear during the time required for complete densification of the free matrix.

Taking fixed values of C, we can find the parameters v and ρ by numerically evaluating Equation 5.88 and Equation 5.78. The results show that the densification behavior is highly dependent on C. For low values of C, the porosity can be eliminated (Figure 5.65a). However, for larger values of C, substantial porosity becomes trapped within the crystals (Figure 5.65b). When v reaches ~50 vol%, densification of the matrix virtually stops because the glass is held in the interstices of the network of crystals. The matrix crystallizes completely, trapping residual porosity of ~10 vol%. Densification is predicted to stop when $v \approx 50$ vol%, because Equation 5.90 gives a sort of percolation threshold at ~50 vol%. However, we know that the threshold occurs earlier, so densification is arrested at a smaller volume fraction of crystals.

5.10.3 EXPERIMENTAL OBSERVATIONS OF SINTERING WITH CRYSTALLIZATION

The inhibiting effects of crystallization prior to full densification have been observed for many compositions [124]. In the case of alkali silicate gels, the crystallization rate is so fast that it is easier to prepare dense glasses by the more traditional melting route. Panda, Mobley, and Raj [130,131] found that a cordierite-type glass can be sintered to full density during constant heating rate sintering at two different heating rates, 0.2°C/min and 2.0°C/min, but later the shear viscosity reached a steady value as the glass crystallized. On the other hand, an aluminosilicate glass shows only limited sinterability when heated at 0.2°C/min, because crystallization occurs prior to the glass reaching full density. Increasing the heating rate to 2.0°C/min delays the onset of crystallization, and the glass can be sintered to full density.

Mullite aerogels prepared by supercritical drying have a low initial relative density $\rho_o \approx 0.05$, and these gels are difficult to densify, reaching a final relative density ρ_f of only ~0.50 after sintering at 5°C/min to 1250°C [132]. If the supercritically dried gel is first compacted to an initial density $\rho_o \approx 0.50$, then a final density $\rho_f \approx 0.97$ is obtained under the same sintering conditions. The difficulty in densifying the low-density aerogel is attributed to the relatively large pores that lower the densification rate relative to the crystallization rate. Appreciable crystallization occurs and prevents the sample from reaching a high density.

Amorphous TiO_2 films prepared by sol–gel processing reached a final density $\rho_f \approx 0.70$ after sintering at 1°C/min to 750°C [133,134]. The low final density is attributed to the inhibiting effects of crystallization. When the film is heated rapidly to avoid excessive nucleation at lower temperatures, densification is completed prior to extensive crystallization, and a dense material is obtained, which can subsequently be crystallized. Microstructural analysis reveals that the films stop densifying when the volume fraction of the crystals reaches a certain value, most likely corresponding to the percolation threshold. High heating rates (up to 60°C/min) delay the onset of crystallization, resulting in the production of films with higher density.

5.11 CONCLUDING REMARKS

In this chapter, we considered several important issues that are not taken into account in the simple sintering models. Inhomogeneities in a powder compact are a common source of problems for sintering, leading to differential densification that can severely retard the sintering rate and cause the growth of microstructural flaws. Although a few techniques may reverse the deterioration in the microstructure, the most practical approach is to minimize the extent of inhomogeneities in the green compact by careful processing (e.g., by colloidal methods).

In composites, rigid inclusions such as reinforcing particles, platelets, or fibers can severely retard the sintering rate of a powder matrix, particularly in the case of a polycrystalline matrix. Realistic models predict that the transient stresses due to differential shrinkage between the inclusions and the sintering matrix are small. The major impediment to densification appears to be the formation of a rigid percolating network of the inclusion phase. The sintering of thin films attached to a rigid substrate is impeded by transient stresses caused by the constrained sintering in the plane of the film. Continuum models provide an adequate explanation of the sintering of amorphous films. The sintering of polycrystalline films is qualitatively different from the sintering of amorphous films and is not adequately described by the models. Constrained sintering is also important in the cosintering (or cofiring) of multilayered structures for several industrial applications, such as the manufacture of multilayered packages for the electronics industry.

Rayleigh instability effects can arise and influence the morphological stability of continuous second phases in microstructures produced by solid-state and liquid-phase sintering. Thin adherent polycrystalline films are also susceptible to breakup, forming islands and uncovering the substrate. Grain boundaries and three-grain junctions form the main sources of perturbations that may eventually intersect the substrate and cause the film to break up.

We discussed the present understanding of the role of solid solution additives (dopants) in the sintering of ceramics, using MgO-doped Al_2O_3 as a special example. Although an effective dopant can potentially influence all of the kinetic and thermodynamic factors in sintering, its most dominant role appears to be a reduction in the grain boundary mobility by a solute drag mechanism. This mechanism is critically dependent on solute segregation at the grain boundaries, which, in turn, is controlled both by electrostatic effects due to a space charge layer and by elastic strain energy effects.

The principal processing parameters that control reaction sintering and viscous sintering with crystallization were considered. For such systems, successful fabrication generally depends on the ability to achieve a processing path in which full densification occurs prior to the reaction (reaction sintering) or in which full densification occurs prior to crystallization (viscous sintering).

PROBLEMS

5.1 Is grain growth beneficial for sintering? Explain.

5.2 Compare the sintering of a powder with a broad particle size distribution with that for a compact with a narrow size distribution when the average particle size of the two powders is the same.

5.3 Derive Equation 5.2 for the volume fraction of rigid inclusions in a porous powder matrix. If the volume fraction of inclusions in the fully dense matrix $v_f = 0.20$, plot the inclusion volume fraction of the porous composite v_i as a function of the relative density of the matrix ρ_m for values of ρ_m in the range of 0.5–1.0.

5.4 Derive Equation 5.3, Equation 5.4, and Equation 5.5 relating the matrix and composite density and densification rates.

5.5 The sintering of an Al_2O_3 powder compact formed from a mixture of coarse and fine powders follows approximately the rule of mixtures [9]. However, the sintering of compacts consisting of a fine Al_2O_3 powder matrix and coarse, rigid inclusions of ZrO_2 is significantly lower than the predictions according to the rule of mixtures [38]. Discuss the reasons for the difference in sintering behavior between these two systems.

5.6 Derive Equation 5.6 giving a relationship for the rule of mixtures in terms of the linear densification rates.

5.7 Explain why a long, narrow cylinder of any material or a long, narrow cylindrical pore is unstable and will tend to break up into a row of spheres. Show that there exists a maximum number of spheres into which the cylindrical material or pore will break up. If the radius of the cylinder is r_o, determine the radius of the sphere for this maximum number of spheres.

5.8 Pinholes can often occur in thin films deposited on a substrate by techniques such as evaporation, solution-based methods, and sputtering. Assuming that the pinholes have a circular cross-section and the film is amorphous, determine the conditions under which the pinholes will grow or shrink, in terms of the relative values of the interfacial energies, if the film is annealed at a high enough temperature for mass transport to occur.

5.9 Derive Equation 5.55 for the equilibrium groove depth of an adherent polycrystalline film. For a dihedral angle of 120°, use Equation 5.55 to determine the critical value of the grain diameter to the film thickness above which the grain boundary groove intercepts the substrate. How does this value compare with the results of the minimum energy configuration calculations in Figure 5.44?

5.10 For ZrO_2, assuming that anion vacancies are the dominant point defects and that the formation energy of an anion vacancy is smaller than that for a cation vacancy, sketch schematically the variation of the electrostatic potential and the defect concentrations as functions of the distance from the grain boundary. Repeat the procedure for CaO-doped ZrO_2, assuming that the defect concentrations produced by the dopant are significantly larger than those generated thermally.

5.11 An early explanation for the role of MgO in the sintering of Al_2O_3 was that MgO enhances the rate of surface diffusion in Al_2O_3, thereby enhancing the pore mobility, allowing the pores to migrate with the boundary and avoiding abnormal grain growth. Explain the flaw in this explanation.

Briefly explain what you believe to be the current understanding of the role of MgO in the sintering of Al_2O_3.

5.12 Consider the reaction between equimolar mixtures of ZnO and Al_2O_3 to form $ZnAl_2O_4$ spinel. The molar volume of the reactants is approximately equal to that of the product. Experimentally, when equimolar mixtures of ZnO and Al_2O_3 powders with approximately the same particle size are reaction sintered, a volume expansion of 20% to 30% is observed (Figure 5.58). Suggest possible explanations for this volume expansion.

REFERENCES

1. Rahaman, M.N., *Ceramic Processing*, CRC Press, Boca Raton, FL, 2006.
2. Evans, A.G., Considerations of inhomogeneity effects in sintering, *J. Am. Ceram. Soc.*, 65, 497, 1982.
3. Raj, R., and Bordia, R.K., Sintering behavior of bimodal powder compacts, *Acta Metall.*, 32, 1003, 1984.

4. Lange, F.F., Densification of powder rings constrained by dense cylindrical cores, *Acta Metall.*, 37, 697, 1989.
5. Exner, H.E., Principles of single phase sintering, *Rev. Powder Metall. Physical Ceram.*, 1, 1, 1979.
6. Weiser, M.W., and De Jonghe, L.C., Rearrangement during sintering in two-dimensional arrays, *J. Am. Ceram. Soc.*, 69, 822, 1986.
7. Rhodes, W.H., Agglomerate and particle size effects on sintering of yttria-stabilized zirconia, *J. Am. Ceram. Soc.*, 64, 19, 1981.
8. Yeh, T.-S., and Sacks, M.D., Effect of green microstructure on sintering of alumina, *Ceramic Trans.*, 7, 309, 1990.
9. Smith, J.P., and Messing, G.L., Sintering of bimodally distributed alumina powders, *J. Am. Ceram. Soc.*, 67, 238, 1984.
10. Kingery, W.D., and Francois, B., Sintering of crystalline oxides, I. Interactions between grain boundaries and pores, in *Sintering and Related Phenomena*, Kuczynski, G.C., Hooton, N.A., and Gibbon, G.F., Eds., Gordon & Breach, New York, 1967, p. 471.
11. Lange, F.F., Sinterability of agglomerated powders, *J. Am. Ceram. Soc.*, 67, 83, 1984.
12. Harmer, M.P., and Zhao, J., Effect of pores on microstructure development, in *Ceramic Microstructures '86: Role of Interfaces, Mater. Sci. Res.*, Vol. 21, Pask, J.A., and Evans, A.G., Eds., Plenum Press, New York, 1987, p. 455.
13. Lin, M., Rahaman, M.N., and De Jonghe, L.C., Creep-sintering and microstructure development of heterogeneous MgO powder compacts, *J. Am. Ceram. Soc.*, 70, 360, 1987.
14. Chu, M.-Y., De Jonghe, L.C., Lin, M.K.F., and Lin, F.J.T., Pre-coarsening to improve microstructure and sintering of powder compacts, *J. Am. Ceram. Soc.*, 74, 2902, 1991.
15. Lin, F.J.T., De Jonghe, L.C., and Rahaman, M.N., Microstructure refinement of sintered alumina by a two-step sintering technique, *J. Am. Ceram. Soc.*, 80, 2269, 1997.
16. Lin, F.J.T., De Jonghe, L.C., and Rahaman, M.N., Initial coarsening and microstructural evolution of fast-fired and MgO-doped alumina, *J. Am. Ceram. Soc.*, 80, 2891, 1997.
17. Evans, A.G., and Hsueh, C.H., Behavior of large pores during sintering and hot isostatic pressing, *J. Am. Ceram. Soc.*, 69, 444, 1986.
18. Pan, J., Ch'ng, H.N., and Cocks, A.C.F., Sintering kinetics of large pores, *Mech. Mater.*, 37, 705, 2004.
19. Flinn, D.D., Bordia, R.J., Zimmermann, A., and Rödel, J., Evolution of defect size and strength of porous alumina during sintering, *J. Europ. Ceram. Soc.*, 20, 2561, 2000.
20. De Jonghe, L.C., Rahaman, M.N., and Hsueh, C.H., Transient stresses in bimodal compacts during sintering, *Acta Metall.* 34, 1467, 1986.
21. Rahaman, M.N., and De Jonghe, L.C., Effect of rigid inclusions on the sintering of glass powder compacts, *J. Am. Ceram. Soc.*, 70, C348, 1987.
22. Timoshenko, S.P., and Goodier, J.N., *Theory of Elasticity*, 3rd ed., McGraw Hill, New York, 1970.
23. Hsueh, C.H., Evans, A.G., Cannon, R.M., and Brook, R.J., Viscoelastic stresses and sintering damage in heterogeneous powder compacts, *Acta Metall.*, 34, 927, 1986.
24. Scherer, G.W., *Relaxation in Glass and Composites*, Wiley-Interscience, New York, 1986.
25. Christiansen, R.M., *Theory of Viscoelasticity, An Introduction*, Academic Press, New York, 1982.
26. Bordia, R.K., and Scherer, G.W., On constrained sintering — I. Constitutive model for a sintering body, *Acta Metall.*, 36, 2393, 1988.
27. Bordia, R.K., and Scherer, G.W., On constrained sintering — II. Comparison of constitutive models, *Acta Metall.*, 36, 2399, 1988.
28. Bordia, R.K., and Scherer, G.W., On constrained sintering — III. Rigid inclusions, *Acta Metall.*, 36, 2411, 1988.
29. Scherer, G.W., Sintering with rigid inclusions, *J. Am. Ceram. Soc.*, 70, 719, 1987.
30. Scherer, G.W., Viscous sintering of a bimodal pore-size distribution, *J. Am. Ceram. Soc.*, 67, 709, 1984.
31. Hashin, Z., and Shtrikman, S., A variational approach to the theory of the elastic behavior of multiphase materials, *J. Mech. Phys. Solids*, 11, 127, 1963.
32. Scherer, G.W., Sintering of inhomogeneous glasses: application to optical waveguides, *J. Non-Cryst. Solids*, 34, 239, 1979.
33. Zallen, R., *The Physics of Amorphous Solids*, Wiley, New York, 1983, Chap. 4.
34. Scherer, G.W., Viscous sintering of particle-filled composites, *Ceramic Bulletin*, 70, 1059, 1991.
35. Scherer, G.W., and Jagota, A., Effect of inclusions on viscous sintering, *Ceramic Trans.*, 19, 99, 1991.

36. Jagota, A., and Scherer, G.W., Viscosities and sintering rates of a two-dimensional granular composite, *J. Am. Ceram. Soc.*, 76, 3123, 1993.

37. Fan, C.-L., and Rahaman, M.N., Factors controlling the sintering of ceramic particulate composites: I, conventional processing, *J. Am. Ceram. Soc.*, 75, 2056, 1992.

38. Sudre, O., and Lange, F.F., Effect of inclusions on densification: I, microstructural development in an Al_2O_3 matrix containing a high volume fraction of ZrO_2 inclusions, *J. Am. Ceram. Soc.*, 75, 519, 1992.

39. Sudre, O., Bao, G., Fan, B., Lange, F.F., and Evans, A.G., Effect of inclusions on densification: II, numerical model, *J. Am. Ceram. Soc.*, 75, 525, 1992.

40. Sudre, O., and Lange, F.F., Effect of inclusions on densification: III, the desintering phenomenon, *J. Am. Ceram. Soc.*, 75, 3241, 1992.

41. Kapolnek, S., and De Jonghe, L.C., Particulate composites from coated powders, *J. Europ. Ceram. Soc.*, 7, 345, 1991.

42. Hu, C.-L., and Rahaman, M.N., Factors controlling the sintering of ceramic particulate composites: II, coated inclusion particles, *J. Am. Ceram. Soc.*, 75, 2066, 1992.

43. Hu, C.-L., and Rahaman, M.N., SiC whisker/Al_2O_3 composites by sintering of coated powders, *J. Am. Ceram. Soc.*, 76, 2549, 1993.

44. Hu, C.-L., and Rahaman, M.N., Dense ZrO_2/Al_2O_3 composites by sintering of coated particles *J. Am. Ceram. Soc.*, 77, 815, 1994.

45. Hagy, H.E., and Smith, A.F., The sandwich seal in the development and control of sealing glasses, *J. Can. Ceram. Soc.*, 38, 63, 1969.

46. Bordia, R.K., and Raj, R., Sintering behavior of ceramic films constrained by a rigid substrate, *J. Am. Ceram. Soc.*, 68, 287, 1985.

47. Scherer, G.W., and Garino, T.J., Viscous sintering on a rigid substrate, *J. Am. Ceram. Soc.*, 68, 216, 1985.

48. Garino, T.J., and Bowen, H.K., Deposition and sintering of particle films on a rigid substrate, *J. Am. Ceram. Soc.*, 70, C315, 1987.

49. Garino, T.J., and Bowen, H.K., Kinetics of constrained-film sintering, *J. Am. Ceram. Soc.*, 73, 251, 1990.

50. Jagota, A., and Hui, C.Y., Mechanics of sintering thin films — I. Formulation and analytical results, *Mech. Mater.*, 9, 107, 1990.

51. Jagota, A., and Hui, C.Y., Mechanics of sintering thin films — II. Cracking due to self-stress, *Mech. Mater.*, 11, 221, 1991.

52. Hu, M., Thouless, M., and Evans, A.G., The decohesion of thin films from brittle substrates, *Acta Metall.*, 36, 1301, 1988.

53. Bordia, R.K., and Jagota, A., Crack growth and damage in constrained sintering films, *J. Am. Ceram. Soc.*, 76, 2475, 1993.

54. Lu, G.Q., Sutterlin, R.C., and Gupta, T.K., Effect of mismatched sintering kinetics on camber in a low temperature cofired ceramic package, *J. Am. Ceram. Soc.*, 76, 1907, 1993.

55. Skorokhod, V.V., On the phenomenological theory of densification or the sintering of porous bodies, *Poroshk. Metall.*, 1, 14, 1961.

56. Cheng, T., and Raj, R., Flaw generation during constrained sintering of metal–ceramic and metal–glass multilayer films, *J. Am. Ceram. Soc.*, 72, 1649, 1989.

57. Rahaman, M.N., and De Jonghe, L.C., Sintering of spherical glass powder under a uniaxial stress, *J. Am. Ceram. Soc.*, 73, 707, 1990.

58. Rahaman, M.N., De Jonghe, L.C., and Brook, R.J., Effect of shear stress on sintering, *J. Am. Ceram. Soc.*, 69, 53, 1986.

59. Coble, R.L., and Kingery, W.D., Effect of porosity on physical properties of sintered alumina, *J. Am. Ceram. Soc.*, 39, 377, 1956.

60. Du, Z.Z., and Cocks, A.C.F., Constitutive models for the sintering of ceramic components — I. Material models, *Acta Metall. Mater.*, 40, 1969, 1992.

61. Lord Rayleigh, On the instability of jets, *Proc. London Math. Soc.*, 10, 4, 1879.

62. Nichols, F.A., and Mullins, W.W., Surface (interface) and volume diffusion contributions to morphological changes driven by capillarity, *Trans. Metall. Soc. AIME*, 233, 1840, 1965.

63. Nichols, F.A., and Mullins, W.W, Morphological changes of a surface of revolution due to capillary-induced surface diffusion, *J. Appl. Phys.*, 36, 1826, 1965.

64. Nichols, F.A., On the spheroidization of rod-shaped particles of finite length, *J. Mater. Sci.*, 11, 1077, 1976.

65. Stüwe, H.P., and Kolednik, O., Shape instability of thin cylinders, *Acta Metall.*, 36, 1705, 1988.

66. Beeré, W., A unifying theory of the stability of penetrating liquid phases and sintering pores, *Acta Metall.*, 23, 131, 1975.

67. Beeré, W., The sintering and morphology of interconnected porosity in UO_2 powder compacts, *J. Mater. Sci.*, 8, 1717, 1973.

68. Tucker, M.O., and Turnbull, J.A., The morphology of interlinked porosity in nuclear fuels, *Proc. Roy. Soc. Lond.* A, 343, 299, 1975.

69. Carter, W.C., and Glaeser, A.M., Dihedral angle effects on the stability of pore channels, *J. Am. Ceram. Soc.*, 67, C124, 1984.

70. Carter, W.C., and Glaeser, A.M., The morphological stability of continuous intergranular phases: thermodynamic considerations, *Acta Metall.*, 35, 237, 1987.

71. Stapley, A.J., and Beevers, C.J., Stability of sapphire whiskers in nickel at elevated temperatures: II, *J. Mater. Sci.*, 8, 1296, 1973.

72. Gupta, T.K., Instability of cylindrical voids in alumina, *J. Am. Ceram. Soc.*, 61, 191, 1978.

73. Yen, C.F., and Coble, R.L., Spheroidization of tubular voids in Al_2O_3 crystals at high temperatures, *J. Am. Ceram. Soc.*, 55, 507, 1972.

74. Drory, M.D., and Glaeser, A.M., The stability of pore channels: experimental observations, *J. Am. Ceram. Soc.*, 68, C14, 1985.

75. Miller, K.T., Lange, F.F., and Marshall, D.B., The instability of polycrystalline thin films: experiment and theory, *J. Mater. Res.*, 5, 151, 1990.

76. Srolovitz, D.J., and Safran, S.A., Capillary instabilities in thin films. I. Energetics, *J. Appl. Phys.*, 60, 247, 1986.

77. Coble, R.L., (a) Sintering crystalline solids. II Experimental test of diffusion models in powder compacts, *J. Appl. Phys.* 32, 793, 1961. (b) Transparent alumina and method of preparation, *U.S. Patent No. 3,026,210,* 1962.

78. Kroger, F.A., Oxidation-reduction and the major type of ionic disorder in α-Al_2O_3, *J. Am. Ceram. Soc.*, 66, 730, 1983.

79. Handwerker, C.A., Dynys, J.M., Cannon, R.M., and Coble, R.L., Metal reference line technique for obtaining dihedral angles from surface thermal grooves, *J. Am. Ceram. Soc.*, 73, 1365, 1990.

80. Handwerker, C.A., Dynys, J.M., Cannon, R.M., and Coble, R.L., Dihedral angles in magnesia and alumina: distributions from surface thermal grooves, *J. Am. Ceram. Soc.*, 73, 1371, 1990.

81. Cahn, J.W., The impurity drag effect in grain boundary motion, *Acta Metall.*, 10, 789, 1962.

82. Kingery, W.D., Plausible concepts necessary and sufficient for interpretation of grain boundary phenomena: I, grain boundary characteristics, structure, and electrostatic potential, *J. Am. Ceram. Soc.*, 57, 1, 1974.

83. Kingery, W.D., Plausible concepts necessary and sufficient for interpretation of grain boundary phenomena: II, solute segregation, grain boundary diffusion, and general discussion, *J. Am. Ceram. Soc.*, 57, 74, 1974.

84. Yan, M.F., Cannon, R.M., and Bowen, H.K., Space charge, elastic field, and dipole contributions to equilibrium solute segregation at interfaces, *J. Appl. Phys.*, 54, 764, 1983.

85. Eshelby, J.D., in *Solid State Physics*, Vol. 3, Seitz, F., and Turnbull, D., Eds., Academic Press, New York, 1956, p. 79.

86. McLean, D., *Grain Boundaries in Metals*, Clarendon Press, Oxford, 1957.

87. Johnson, W.C., Grain boundary segregation in ceramics, *Metall. Trans.*, 8A, 1413, 1977.

88. Frenkel, J., *Kinetic Theory of Liquids*, Oxford University Press, New York, 1946.

89. Lehovec, K., Space charge layer and distribution of lattice defects at the surface of ionic crystals, *J. Chem. Phys.*, 21, 1123, 1953.

90. Eshelby, J.D., Newey, C.W.A., Pratt, P.L., and Lidiard, A.B., Charged dislocations and the strength of ionic crystals, *Philos. Mag.*, Vol. 3 (Series 8), 75, 1958.

91. Kliewer, K.L., and Koehler, J.S., Space charge in ionic crystals, I. General approach with application to NaCl, *Phys. Rev. A*, 140, 1226, 1965.

92. Schwensfeir, R.J., Jr. and Elbaum, C., Electric charge on dislocation arrays in sodium chloride, *J. Phys. Chem. Solids*, 28, 597, 1967.

93. Li, C.W., and Kingery, W.D., Solute segregation at grain boundaries in polycrystalline Al_2O_3, in *Advances in Ceramics*, Vol. 10, Kingery, W.D., Ed., American Ceramic Society, Columbus, OH, 1984, p. 368.

94. Chiang, Y.M., Hendrickson, A.F., Kingery, W.D., and Finello, D., Characterization of grain boundary segregation in MgO, *J. Am. Ceram. Soc.*, 64, 385, 1981.

95. Hwang, S.L., and Chen, I.W., Grain size control of tetragonal zirconia polycrystals using the space charge concept, *J. Am. Ceram. Soc.*, 73, 3269, 1990.

96. Ikeda, J.A.S., and Chiang, Y.M., Space charge segregation at grain boundaries in titanium dioxide: I, relationship between lattice defect chemistry and space charge potential, *J. Am. Ceram. Soc.*, 76, 2437, 1993.

97. Ikeda, J.A.S., Chiang, Y.M., Garratt-Reed, A.J., and Vander Sande, J.B., Space charge segregation at grain boundaries in titanium dioxide: II, model experiments, *J. Am. Ceram. Soc.*, 76, 2447, 1993.

98. Chiang, Y.M., and Takagi, T., Grain boundary chemistry of barium titanate: I, high temperature equilibrium space charge, *J. Am. Ceram. Soc.*, 73, 3278, 1990.

99. Chiang, Y.M., and Kingery, W.D., Grain boundary migration in nonstoichiometric solid solutions of magnesium aluminate spinel: II, effects of grain boundary nonstoichiometry, *J. Am. Ceram. Soc.*, 73, 1153, 1990.

100. Bennison, S.J., and Harmer, M.P., A history of the role of MgO in the sintering of α-Al_2O_3, *Ceramic Trans.*, 7, 13, 1990.

101. Johnson, W.C., and Coble, R.L., A test of the second phase and impurity segregation models for MgO enhanced densification of sintered alumina, *J. Am. Ceram. Soc.*, 61, 110, 1978.

102. Bennison, S.J., and Harmer, M.P., Grain growth kinetics for alumina in the absence of a liquid phase, *J. Am. Ceram. Soc.*, 68, C22, 1985.

103. Berry, K.A., and Harmer, M.P., Effect of MgO solute on the microstructure development in Al_2O_3, *J. Am. Ceram. Soc.*, 69, 143, 1986.

104. Bennison, S.J., and Harmer, M.P., Effect of magnesia solute on surface diffusion in sapphire and the role of magnesia in the sintering of alumina, *J. Am. Ceram. Soc.*, 73, 833, 1990.

105. Baik, S., Segregation of Mg to the (0001) surface of single-crystal alumina: quantification of AES results, *J. Am. Ceram. Soc.*, 69, C101, 1986.

106. Baik, S., and White, C.L., Anisotropic calcium segregation to the surface of Al_2O_3, *J. Am. Ceram. Soc.*, 70, 682, 1987.

107. Baik, S., and Moon, J.H., Effect of magnesium oxide on grain boundary segregation of calcium during sintering of alumina, *J. Am. Ceram. Soc.*, 74, 819, 1991.

108. Harmer, M.P., Bennison, S.J., and Narayan, C., Microstructural characterization of abnormal grain growth development in Al_2O_3, *Mater. Sci. Res.*, 15, 309, 1983.

109. Shaw, N.J., and Brook, R.J., Structure and grain coarsening during the sintering of alumina, *J. Am. Ceram. Soc.*, 69, 107, 1986.

110. Kaysser, W.A., Sprissler, M., Handwerker, C.A., and Blendell, J.E., Effect of a liquid phase on the morphology of grain growth in alumina, *J. Am. Ceram. Soc.*, 70, 339, 1987.

111. Wu, S.J., De Jonghe, L.C., and Rahaman, M.N., Subeutectic densification and second-phase formation in Al_2O_3–CaO, *J. Am. Ceram. Soc.*, 68, 385, 1985.

112. Kolar, D., Microstructure development during sintering in multicomponent systems, *Sci. Ceram.*, 11, 199, 1981.

113. Yangyun, S., and Brook, R.J., Preparation of zirconia-toughened ceramics by reaction sintering, *Sci. Sintering*, 17, 35, 1985.

114. Schmalzried, H., *Solid State Reactions*, Academic Press, New York, 1974.

115. Rahaman, M.N., and De Jonghe, L.C., Reaction sintering of zinc ferrite during constant rates of heating, *J. Am. Ceram. Soc.*, 76, 1739, 1993.

116. Hong, W.S., De Jonghe, L.C., Yang, X., and Rahaman, M.N., Reaction sintering of ZnO–Al_2O_3, *J. Am. Ceram. Soc.*, 78, 3217, 1995.

117. Claussen, N., and Jahn, J., Mechanical properties of sintered, in situ-reacted mullite-zirconia composites, *J. Am. Ceram. Soc.*, 63, 228, 1980.

118. Di Rupo, E., Gilbart, E., Carruthers, T.G., and Brook, R.J., Reaction hot-pressing of zircon-alumina mixtures, *J. Mater. Sci.*, 14, 705, 1979.

119. Tuan, W.H., and Brook, R.J., Processing of alumina/nickel composites, *J. Europ. Ceram. Soc.*, 10, 95, 1992.

120. Sacks, M.D., Bozkurt, N., and Scheiffele, G.W., Fabrication of mullite and mullite-matrix composites by transient viscous sintering of composite powders, *J. Am. Ceram. Soc.*, 74, 2428, 1991.

121. Zarzycki, J., Crystallization of gel-produced glasses, in *Nucleation and Crystallization in Glasses*, Advances in Ceramics, Vol. 4, Simmons, J.H., Uhlmann, D.R., and Beall, G.H., Eds., The American Ceramic Society, Columbus, OH, 1982, p. 204.

122. Uhlmann, D.R., Zelinski, B.J., Silverman, L., Warner, S.B., Fabes, B.D., and Doyle, W.F., Kinetic processes in sol–gel processing, in *Science of Ceramic Chemical Processing*, Hench, L.L., and Ulrich, D.R., Eds., Wiley, New York, 1986, p. 173.

123. Scherer, G.W., Effect of inclusions on shrinkage, *Mater. Res. Soc. Symp. Proc.*, 180, 503, 1990.

124. Brinker, C.J., and Scherer, G.W., *Sol–Gel Science*, Academic Press, New York, 1990, Chap. 11.

125. Avrami, M., Kinetics of phase change. I, general theory, *J. Chem. Phys.*, 7, 1103, 1939.

126. Avrami, M., Kinetics of phase change. II, transformation-time relations for random distribution of nuclei, *J. Chem. Phys.*, 8, 212, 1940.

127. Avrami, M., Kinetics of phase change. III, granulation, phase change, and microstructure, *J. Chem. Phys.*, 9, 177, 1941.

128. Christian, J.W., *The Theory of Phase Transformations in Metals and Alloys*, 2nd ed., Pergamon Press, New York, 1975.

129. Sacks, M.D., and Tseng, T.-Y., Preparation of SiO_2 glass from model powder compacts: II, sintering, *J. Am. Ceram. Soc.*, 69, 532, 1984.

130. Panda, P.C., and Raj, R., Sintering and crystallization of glass at constant heating rates, *J. Am. Ceram. Soc.*, 72, 1564, 1989.

131. Panda, P.C., Mobley, W.M., and Raj, R., Effect of heating rate on the relative rates of sintering and crystallization in glass, *J. Am. Ceram. Soc.*, 72, 2361, 1989.

132. Rahaman, M.N., De Jonghe, L.C., Shinde, S.L., and Tewari, P.H., Sintering and microstructure of mullite aerogels, *J. Am. Ceram. Soc.*, 71, C338, 1988.

133. Keddie, J.L., and Giannelis, E.P., Effect of heating rate on the sintering of titanium dioxide thin films: competition between densification and crystallization, *J. Am. Ceram. Soc.*, 74, 2669, 1991.

134. Keddie, J.L., Braun, P.V., and Giannelis, E.P., Interrelationship between densification, crystallization, and chemical evolution in sol–gel titania thin films, *J. Am. Ceram. Soc.*, 77, 1592, 1994.

6 Sintering Process Variables and Sintering Practice

6.1 INTRODUCTION

The common requirements for the final sintered product are high density and small grain size, because these microstructural characteristics enhance most engineering properties. It is well recognized that the processing steps prior to sintering have a significant effect on the microstructural development of the fabricated ceramic [1]. In particular, the sintering rate and the ability to achieve a high final density are strongly dependent on the particle size and particle packing of the green body. However, assuming that proper powder preparation and consolidation procedures are in effect, successful fabrication remains dependent on the ability to control the microstructure through manipulation of the process variables in the sintering stage. A wide variety of sintering techniques have been developed to obtain ceramics with the required density, microstructure, and composition. In general, these methods involve the manipulation of some combination of the heating schedule, atmosphere, and applied pressure.

The reader will recall from Chapter 1 the four categories of sintering (solid-state sintering, liquid-phase sintering, pressure-assisted sintering, and vitrification) and the simple heating schedules (isothermal sintering and constant heating rate sintering). Sintering in which no external pressure is applied to the body is sometimes referred to as conventional sintering, free sintering, or pressureless sintering. Because it is more economical, conventional sintering is the preferred method, but the additional driving force available in pressure-assisted sintering often guarantees the attainment of high density and a fine-grained microstructure. Pressure-assisted sintering is commonly used for the production of high-cost ceramics and prototype ceramics where high density must be guaranteed.

Heating of the powder compact is commonly achieved with electrical resistance furnaces. The heating schedule can be simple, as in isothermal sintering, or it may have a complex temperature–time relationship, as in some industrial sintering. The use of microwave energy for sintering ceramics has attracted attention within the last twenty years, but this heating method sees only limited use in industrial sintering. Microwave heating is fundamentally different from that in conventional resistance furnaces: the heat is generated internally by interaction of the microwaves with the material. It is effective for heating ceramic bodies rapidly, and some studies have shown considerable enhancement of sintering with microwave heating. However, the achievement of sufficiently uniform heating can be difficult.

Control of the sintering atmosphere is also important, and precise control of the oxygen or nitrogen partial pressure as a function of temperature may in some cases be beneficial or essential. The atmosphere has a strong effect on processes such as decomposition, evaporation of volatile constituents, and vapor transport. It can also influence the oxidation number of atoms (particularly those of the transition elements) and the stoichiometry of the solid. Insoluble gases trapped in closed pores may hinder the final stages of densification or lead to swelling of the compact after densification, and in these cases a change of the sintering atmosphere or the use of vacuum is indicated.

6.2 SINTERING MEASUREMENT TECHNIQUES

It is often required to monitor the progress of sintering, to provide data for understanding how the microstructure develops and for optimizing the process variables to achieve the desired microstructure. Although several techniques are available to monitor sintering, only a few selected

TABLE 6.1
Heating Elements Commonly Used in Electrical
Resistance Furnaces

Material	Maximum Temperature (°C)	Furnace Atmosphere
Nichrome	1200	Oxidizing; inert; reducing
Pt	1400	Oxidizing
SiC	1450	Oxidizing
MoSi$_2$	1600	Oxidizing
LaCrO$_3$	1750	Oxidizing
Mo, W, Ta	2000	Vacuum; inert; reducing[a]
C (graphite)	2800	Inert; reducing[b]

[a] Atmospheres containing nitrogen or carbon must be avoided above ~1500°C.
[b] Vacuum may be used below ~1500°C.

measurements prove to be very useful in practice. Measurement of the shrinkage or density as a function of time or temperature, coupled with an examination of the final microstructure using microscopy, is performed routinely. It is also useful to examine the microstructure at several intervals during the heating cycle to characterize the grain growth and microstructural evolution. Measurement of the average pore size and the pore size distribution of the partially sintered samples using mercury porosimetry can provide additional information about the characteristics of the open pores and the homogeneity of the microstructure.

6.2.1 FURNACES

A variety of furnaces are available commercially to provide the heating schedule and atmosphere used in conventional sintering. Sizes range from those for small, research-type powder compacts to large furnaces that can accommodate parts several feet in diameter for industrial production. The most common types of sintering furnaces are electrical resistance furnaces, in which a current-carrying resistor, called the furnace element or winding, serves as the source of heat. In addition to size and cost, important considerations in the selection of a furnace are the maximum temperature capability and the atmosphere in which it can be operated for extended periods. Table 6.1 provides a selected list of common furnace elements. Several metal alloys (e.g., nichrome) can be used as furnace elements for temperatures up to ~1200°C in both oxidizing and reducing atmospheres. For extended use in air, other furnace elements can deliver higher temperatures, e.g., Pt (up to ~1400°C); SiC (1450°C); MoSi$_2$ (1600°C); and LaCrO$_3$ (1750°C). For higher temperatures, furnace elements consisting of various refractory metals (e.g., Mo, W, and Ta) can be used in vacuum, inert, or reducing atmospheres up to ~2000°C. At these high temperatures, the metallic elements must not be exposed to atmospheres containing carbon or nitrogen because they readily form carbides and nitrides, which reduce the life of the element considerably. Graphite elements, heated electrically or by induction, can provide temperatures up to ~2800°C in inert or reducing atmospheres. For extended use of graphite elements above ~2000°C, high-purity He gas from which the trace impurities of oxygen have been removed (e.g., by passing the gas through an oxygen getter prior to its entering the furnace) provides a useful atmosphere.

A temperature controller is an integral part of most furnaces to take the system through the required heating cycle. In many cases, it is also necessary to have fairly precise control of the sintering atmosphere around the powder compact or to use a sintering atmosphere that is not compatible with the furnace element. For laboratory-scale experiments, a tube furnace (Figure 6.1) can often provide the desired temperature and atmosphere control. Silica tubes can be used for

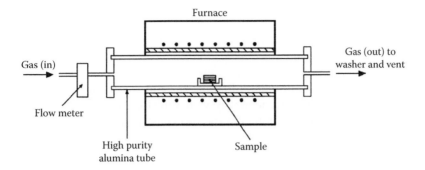

FIGURE 6.1 Schematic diagram of a tube furnace used for controlled atmosphere sintering.

temperatures up to ~1100°C, whereas high-purity Al_2O_3 tubes can be used up to temperatures in the range of 1600°C–1800°C, but the low thermal shock resistance of Al_2O_3 limits its maximum heating or cooling rate to 10°C/min to 20°C/min, depending on the thickness of the tube.

6.2.2 SHRINKAGE AND DENSITY

The shrinkage can be determined directly by measuring the initial dimensions (e.g., length and diameter for a cylindrical pellet) and the dimensions after a given time t (or temperature T). The linear shrinkage is defined as $\Delta L/L_o$, where L_o is the initial length and $\Delta L = L - L_o$, where L is the length after time t. In this definition, $\Delta L/L_o$ is negative for sintering, but the negative sign is often ignored and the magnitude is reported. Comparison of the shrinkages in the axial and radial directions provides a measure of the shrinkage anisotropy during sintering. The density is determined from the mass and dimensions (volume) of a sample with a regular shape, whereas the Archimedes method may be used to measure the density of a sample with a regular or irregular shape.

To determine the sintering kinetics, the density or shrinkage at several times or temperatures is measured, giving a plot of density or shrinkage (y-axis) versus sintering time or temperature (x-axis). This can be very time consuming, because a different sample must be used for each given time (or temperature). For laboratory-scale experiments, the technique of dilatometry provides continuous monitoring of the linear shrinkage of the powder compact over the complete sintering schedule. A variety of dilatometers are available commercially for providing the required accuracy in shrinkage measurement, as well as control of the temperature and atmosphere. The equipment is similar to that commonly used to measure the thermal expansion coefficient of solids (Figure 6.2).

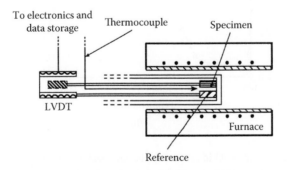

FIGURE 6.2 Schematic diagram of a dual pushrod dilatometer for continuous monitoring of shrinkage kinetics.

The data are plotted as shrinkage $\Delta L/L_o$ versus time (or temperature). If the mass of the powder compact remains constant and the shrinkage is isotropic, the relative density ρ can be determined from the shrinkage data by using Equation 1.2.

Analysis of the sintering kinetics is often performed by determining the densification rate $\dot{\rho}$, defined as $(1/\rho)d\rho/dt$. The procedure involves fitting a smooth curve through the density data and differentiating with respect to time to find the slope at any instant. As an example of the information that can be obtained economically with a dilatometer, Figure 6.3a shows the data for $\Delta L/L_o$ and ρ versus temperature T for a ZnO powder compact (initial relative density $\rho_o = 0.52$) that was sintered at a constant heating rate of 5°C/min to ~1100°C. For this powder compact,

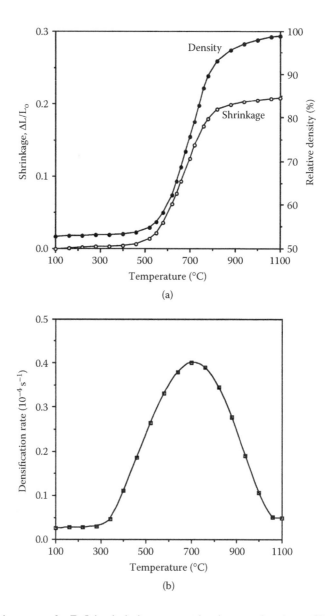

FIGURE 6.3 Sintering curves for ZnO in air during constant heating rate sintering at 5°C/min obtained from the dilatometric shrinkage curve, showing (a) the shrinkage and density as a function of temperature and (b) the densification rate as a function of temperature.

the shrinkage (or densification) starts at ~500°C and is almost completed at ~1000°C. The maximum in the densification rate occurs at ~800°C (Figure 6.3b), corresponding to a compact density of $\rho \approx 0.80$. A convenient starting point for selecting an isothermal sintering temperature is often a temperature slightly above the maximum of the densification rate versus temperature curve, which would be expected to provide a reasonable densification rate without excessive grain growth.

6.2.3 GRAIN SIZE

Grain size can be estimated from scanning electron microscopy (SEM) observations of fractured surfaces of the sample, but planar cross-sections taken from the three-dimensional sample should be used for more accurate data. The average grain size and the standard deviation can be determined in a straightforward manner from SEM micrographs with lineal analysis [2,3], but imaging software is now commonly used. In lineal analysis, a line of known length X is drawn randomly on the micrograph, and the number of intersections N_L that the line makes with the grain boundaries is found. The intercept length is determined from the equation

$$L = \frac{X}{MN_L} \tag{6.1}$$

where M is the magnification of the micrograph. The procedure is repeated many times until more than 200 grain boundary intersections are counted, and the mean intercept length $\overline{L}$ and its standard deviation are calculated. The average grain size G is related to $\overline{L}$ by a geometrical parameter that depends on the characteristics of the grains. For a log-normal distribution of grain sizes and tetrakaidecahedral grain shapes, the relationship is [4]

$$G = 1.56\overline{L} \tag{6.2}$$

When analyzing samples containing a second phase, such as porosity, the length of the test line falling on the pores must be excluded. The intercept length is found from a modified form of Equation 6.1 [5]:

$$L = \frac{X_{eff}}{MN_{eff}} \tag{6.3}$$

where X_{eff} is the length of the test line falling on the grains and $N_{eff} = N_{ss} + N_{sv}/2$, where N_{ss} is the number of intercepts with the grain boundaries (solid–solid boundaries) and N_{sv} is the number of intercepts with the solid–vapor interfaces.

6.3 CONVENTIONAL SINTERING

Although the sintering behavior of real powders is considerably more complex than that assumed in the models, the sintering theories clearly indicate the key process variables that must be controlled to optimize sintering. The particle size and particle packing of the green body, the heating schedule, and the sintering atmosphere have the strongest effects on sintering, but other factors, such as the size distribution, shape, and structure of the particles, can also have a significant influence.

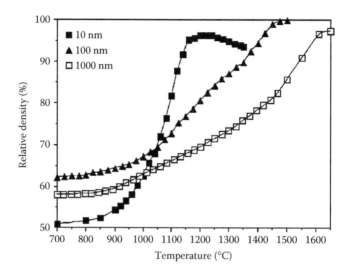

FIGURE 6.4 Effect of particle size on the sintering of CeO$_2$ powder compacts during constant heating rate sintering (10°C/min) in an oxygen atmosphere.

6.3.1 Particle and Green Compact Effects

6.3.1.1 Particle Size

The densification rate of polycrystalline powder compacts depends strongly on particle size (Equation 2.51 and Equation 2.53), and the reduction of the particle size provides an important method for speeding up sintering. This is illustrated by the data shown in Figure 6.4 for CeO$_2$ powder compacts with three different particle sizes [6]. Because of packing difficulties, the observed sintering rate is lower than that predicted by the models, but the enhancement of the rate with decreasing particle size is still considerable. A consequence of the enhanced sintering rate is the use of lower sintering temperatures. Nanoscale particles (<50–100 nm in size) commonly exhibit large reductions in the sintering temperature. The data in Figure 6.4 indicate that isothermal sintering of 10-nm CeO$_2$ particles can be achieved at less than ~1150°C, compared to 1500°C–1600°C for 1-µm particles. Unfortunately, finer particles are more prone to agglomeration, so the green density is often lower than that for coarser particles, and the shrinkage during sintering is larger. Contamination can also be a problem due to the large surface area of the powder. The removal of surface impurities, such as hydroxyl groups, from the surface of nanoscale particles can also be difficult. Decomposition during sintering can lead to trapped gases in the pores that can limit the final density, as seen in Figure 6.4 for the 10-nm particles.

6.3.1.2 Particle Size Distribution

The solid-state sintering models generally assume that the particles are monodisperse, but a particle size distribution can have a significant effect on sintering. Coble [7] modeled the initial-stage sintering of a linear array of particles with different sizes and found that an equivalent particle size, considerably smaller than the average size, can be used to account for the shrinkage. The sintering rates of binary mixtures of particles were intermediate between those for the end-member sizes. A simple rule of mixtures has in fact been proposed to describe the sintering data of binary mixtures of alumina powders where the coarser phase is relatively inactive [8,9]. Commonly, differential densification between the coarser phase and the finer phase, coupled with interactions between the particles in the coarser phase, can severely inhibit densification (see Chapter 5).

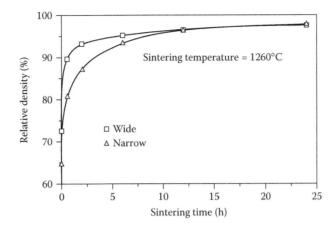

FIGURE 6.5 Effect of particle size distribution on the sintering of slip-cast Al$_2$O$_3$ powder compacts prepared from wide and narrow size distribution powders having the same median particle size. (From Yeh, T.S., and Sacks, M.D., Effect of green microstructure on sintering of alumina, *Ceramic Trans.*, 7, 309, 1990.)

The use of mixtures of discrete particle sizes or a wide distribution of particle size can result in an increase in the packing density (because the finer particles fit within the interstices of the larger particles), so the shrinkage required for complete densification is reduced. This reduction in the shrinkage is important in industrial sintering of large objects. Figure 6.5 shows that, for the same average particle size, an increase in the width of the particle size distribution leads to an increase in the sintering rate in the early stages of sintering [10], presumably due to the presence of the fine particles and the fine pores. The behavior in the later stages of sintering depends critically on the particle packing of the green body. For heterogeneous packing, an increase in the width of the particle size distribution is expected to enhance the detrimental effects of differential densification and grain growth (due to the enhanced driving force arising from the size difference), so the attainment of a high final density may be difficult. On the other hand, if the particle packing is homogeneous, with small pores of narrow size distribution, a high final density can be achieved regardless of the width of the initial particle size distribution (Figure 6.5), indicating the importance of the pore characteristics in sintering.

6.3.1.3 Particle Shape and Particle Structure

Particle shape influences sintering primarily through its effect on the packing of the green body. Deviation from the spherical or equiaxial shape leads to a reduction in the packing density and packing homogeneity, resulting in a reduction of densification. Compacts of acicular (or elongated) particles can, however, be sintered to high density if the particles are aligned and the packing is homogeneous, as has been demonstrated for acicular Fe$_2$O$_3$ particles [11]. Some solids maintain or develop faceted shapes during sintering, but the effects of shape are often complicated by other factors, such as packing and composition.

The solid-state sintering models generally assume that the particles are dense, discrete single-crystals units. Although many powders used in practical sintering consist of dense, single-crystal particles, some powder synthesis routes may result in the production of particles that are polycrystalline at some stage. For example, spherical, monodisperse particles synthesized by the Stober process are actually porous agglomerates of much finer primary particles [1]. The influence of this particle substructure on sintering of TiO$_2$ has been studied by Edelson and Glaeser [12]. Heating a powder compact of TiO$_2$ to 700°C leads to rapid sintering and grain growth of the fine primary particles in the spheres, as well as their conversion to rutile, so each sphere consists of only one to three grains by the time the spheres start to sinter at ~1000°C.

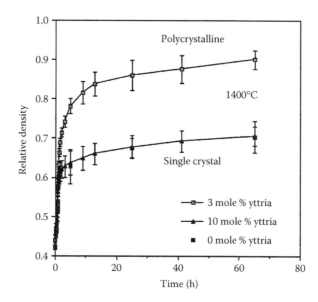

FIGURE 6.6 Densification of ZrO_2 and Y_2O_3-stabilized ZrO_2 powder compacts at 1400°C. The polycrystalline powder compact sinters faster than the compact of single-crystal particles. (From Slamovich, E.B., and Lange, F.F., Densification behavior of single-crystal and polycrystalline spherical particles of zirconia, *J. Am. Ceram. Soc.,*73, 3368, 1990. With permission.)

Slamovich and Lange [13] synthesized single-crystal spherical particles of Y_2O_3-stabilized ZrO_2 (0 mol% and 10 mol% Y_2O_3) and polycrystalline spherical particles (3 mol% Y_2O_3) and studied the sintering behavior of powder compacts, as well as clusters or pairs of particles. Compacts of the polycrystalline particles were observed to sinter faster than compacts of single-crystal particles (Figure 6.6), and this was attributed to the larger number of grain boundaries contributing to mass transport. Microstructural observations showed that while necks between the single-crystal particles reached a stable size, necks between the polycrystalline particles continued to grow, and the polycrystalline particles continued to densify as a result of the neck growth (Figure 6.7).

6.3.1.4 Particle Packing

The effect of particle packing on sintering was considered in detail in Chapter 5 when we discussed microstructural inhomogeneities and their effects on sintering. The reader will recall that for enhanced sintering rates and the attainment of high density, the particles must be homogeneously packed with a high packing density. These packing characteristics result in fine, uniform pores with a low pore coordination number. Furthermore, the number of particle contacts is maximized, providing many grain boundaries and short diffusion paths for rapid matter transport into the fine pores. In Chapter 5, we described the work of Yeh and Sacks [14] on the effect of particle packing on the sintering of Al_2O_3 powders. Prior to the work of Yeh and Sacks, Rhodes [15] provided a clear demonstration of these principles in a study of the sintering of Y_2O_3-stabilized ZrO_2. Using a fine powder (crystallite size of approximately 10 nm), he prepared a suspension and allowed the agglomerates to settle. The fine particles in the supernatant were then used to prepare compacts by gravitational settling in a centrifuge. After drying, the compact had a density of 0.74 of the theoretical density. Sintering for 1 h at 1100°C produced almost complete densification, whereas compacts prepared by die-pressing of the as-received, agglomerated powder reached a relative density of only 0.95 after sintering for 1 h at 1500°C.

Barringer and Bowen [16,17] developed a processing approach based on the uniform packing of monodisperse, spherical particles. Although rapid sintering was achieved at considerably lower temperatures than for randomly packed particles, the compacts could not be sintered to full density

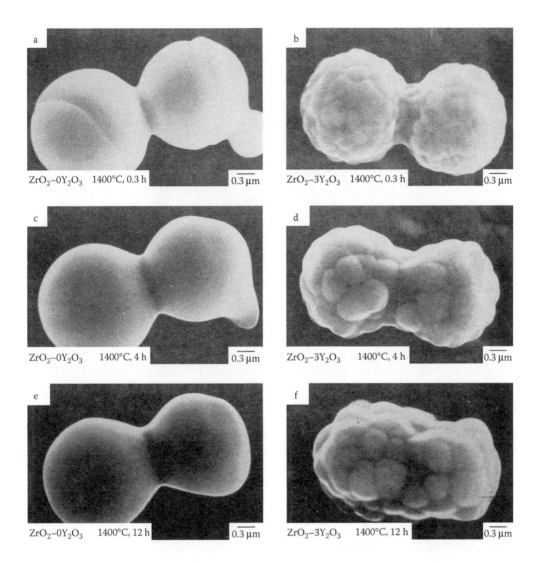

FIGURE 6.7 Pairs of particles sintered on a sapphire substrate at 1400°C for (a, b) 0.3 h; (c, d) 4 h; (e, f) 12 h. Necks between single-crystal particles reach a stable size, whereas neck growth between polycrystalline particles continues, resulting in a spheroidizing group of crystals. (From Slamovich, E.B., and Lange, F.F., Densification behavior of single-crystal and polycrystalline spherical particles of zirconia, *J. Am. Ceram. Soc.*, 73, 3368, 1990. With permission.)

(the residual porosity was approximately 5%). A problem is that the uniform consolidation of monodisperse particles leads to the formation of small regions with three-dimensional, ordered packing (typical of the packing in crystals), referred to as domains, which are separated by packing flaws (voids) at the domain boundaries. The faster sintering of the ordered regions, when compared to the domain boundaries, leads to differential stresses that cause enlargement of the voids (Figure 6.8). These large voids cannot easily be removed during sintering, and they form the residual porosity in the final body. Based on this difficulty of removing the large pores associated with the domain boundaries, Liniger and Raj [18] suggested that dense, random packing of particles with a distribution of sizes may provide an alternative, ideal packing geometry for sintering. When compared to the ordered packing of monodisperse particles, the random structure would be less densely packed than the domains, but the density fluctuations should be less severe, leading to more homogeneous sintering

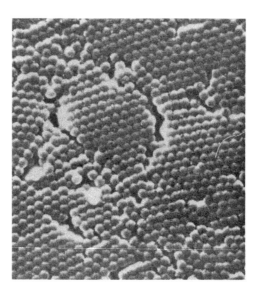

FIGURE 6.8 Partial densification of a periodically packed, multilayered arrangement of polymer spheres, showing the opening displacements at the domain boundaries.

and a reduction in the number of large, residual pores because of the reduced tendency for differential sintering. Homogeneously packed Al_2O_3 green bodies formed by colloidal processing of fine powders have in fact been sintered to almost full density at temperatures as low as ~1200°C [19].

It is important to appreciate the implications of deviating from the ideal structure, because most practical sintering, particularly for industrial applications, is not performed with an ideal system of homogeneous, densely packed, fine particles. A key problem is differential densification, which may lead to enlargement of the voids or to the generation of crack-like voids in the less dense regions. The large voids limit densification. Another problem is that the denser regions can support grain growth. Enhanced grain growth starts at an earlier point during sintering, increasing the probability for the initiation of abnormal grain growth.

6.3.1.5 Effect of Green Density

Higher green density implies that less shrinkage is required to reach a given sintered density. Several studies have shown a correlation between the sintered density and the green density for the same sintering conditions. The sintered density is observed to decrease with decreasing green density for green density values below 55%–60% of the theoretical [20,21]. Compacts with green densities lower than 40%–45% of the theoretical are often difficult to sinter to high density. Increasing green density is found to delay the onset of enhanced grain growth during the later stages of sintering [22]. These trends in the data can be explained in terms of the higher probability, with decreasing green density, of the occurrence of pores with large coordination numbers, coupled with the enhancement of differential densification. In general, the benefits of higher green density for sintering are clear, so optimization of the green density is a useful approach. For some powders, particularly some nanoscale powders, a low green density does not necessarily limit the ability to reach a high sintered density. Figure 6.9 shows the effect of green density on the sintering of a submicron ZnO powder during constant heating rate sintering [20]. Although the sintered density at any temperature increases with the green density, the compact with an initial relative density ρ_o = 0.39 still reaches a final density that is greater than 95% of the theoretical value after sintering to 1100°C. Compacts of nanoscale CeO_2 powders with a green density as low as 20% have been sintered to almost full density [23].

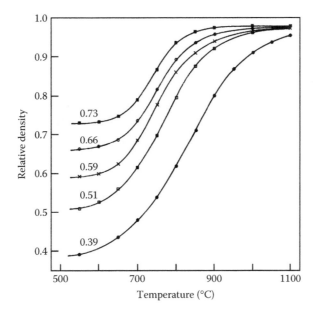

FIGURE 6.9 Effect of green density on the sintering of ZnO powder compacts during constant heating rate sintering (5°C/min to 1100°C), showing that for this powder, the final density is almost independent of the green density (shown on the curves).

6.3.2 ANISOTROPIC SINTERING SHRINKAGE

Pronounced differences are often observed in the shrinkages along different directions of some powder compacts during sintering. Anisotropic shrinkage makes dimensional control of the final article difficult, and it is a particularly serious problem in the manufacture of some industrial components, such as multilayer ceramic capacitors, where exact dimensional tolerance must be achieved. The powder characteristics and the method used to form the powder into the green body are the major causes of anisotropic sintering shrinkage. They may lead to density variations in the green body, to anisotropy in the pore shape, or to alignment of nonequiaxial particles in the green compact. The sintering conditions, such as temperature gradients in the furnace, friction between the sintering body and the substrate, and nonuniform application of the pressure during pressure-assisted sintering, can also lead to anisotropic shrinkage. With proper sintering conditions, the sintering process itself is capable of producing very uniform final dimensions. Thus, when anisotropic sintering occurs, the most useful approach is to identify the causes and, whenever possible, reduce or eliminate them. In many cases, it is not possible to eliminate anisotropic shrinkage, so the magnitude of the sintering shrinkage anisotropy should be determined and compensated for in the final dimensions.

6.3.2.1 Pore Shape Anisotropy

Even when the particles have equiaxial shapes, compacts formed by methods such as uniaxial pressing can exhibit anisotropic shrinkage. Cylindrical compacts formed by uniaxial pressing of spherical glass powders have been observed to show pronounced differences in the axial and radial shrinkage during sintering [24,25]. Exner and Giess [24] found that the shrinkage anisotropy factor, defined as the ratio of the axial shrinkage to the radial shrinkage, increased from a value of 0.3 to 0.7 with increasing density (Figure 6.10a). Furthermore, as Figure 6.10b shows, the shrinkage anisotropy factor was independent of the size and shape of the particles (crushed,

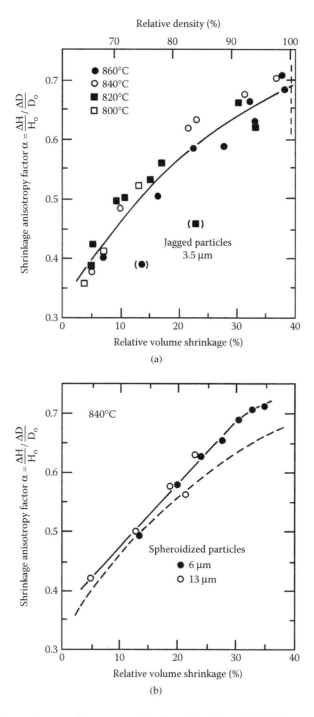

FIGURE 6.10 Shrinkage anisotropy factor α, equal to the ratio of the axial shrinkage to the radial shrinkage $(\Delta H/H_o)/(\Delta D/D_o)$, as a function of relative volume shrinkage for (a) jagged glass powder compacted uniaxially at 150 MPa and (b) two fractions of spheroidized glass powder compacted uniaxially at 75 MPa. For comparison, the data for jagged particles in (a) are shown by the dashed curve. (From Exner, H.E., and Giess, E.A., Anisotropic shrinkage of cordierite-type glass powder cylindrical compacts, *J. Mater. Res.,* 3, 122, 1988. With permission.)

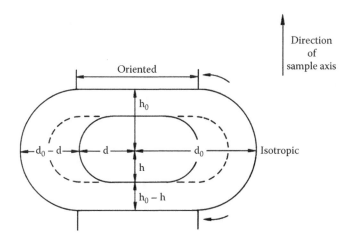

FIGURE 6.11 Schematic representation of the vertical cross-section through a partially oriented pore surface. The orientation is planar, with the isotropic part and the oriented part represented by the semicircular portions and the straight portions of the circumference, respectively.

jagged particles and spherical particles). A detailed observation of polished cross-sections of the sintered compacts revealed significant anisotropy in the pore shape. The pore shape anisotropy factor, taken as the ratio of the mean pore intercept length in the axial and radial directions, remained almost constant during sintering, having values of ~0.5 for the spherical particles and ~0.75 for the jagged particles. The shrinkage anisotropy can be qualitatively explained in terms of the pore anisotropy produced as a result of the uniaxial pressing of the green compacts (Figure 6.11). Assuming that the pore shape anisotropy can be represented by a planar cross-section, the isotropic part of the pore surface, represented by the semicircular portions, gives rise to isotropic shrinkage, while the oriented part, the straight portions, leads to additional shrinkage in the radial direction. However, the continuous decrease in the shrinkage anisotropy observed by Exner and Giess cannot be quantitatively explained in terms of the nearly constant pore anisotropy factor.

Isostatically pressed powders show considerably smaller shrinkage anisotropy during sintering than do uniaxially pressed powders. Because of the nature of the pressure application during compaction, uniaxially pressed powders should have a more elongated pore shape (with a smaller linear dimension in the direction of pressing) when compared to isostatically pressed powders.

The effect of pore shape on the shrinkage anisotropy during sintering was modeled by Olevsky and Skorokod [26], who considered the powder compact to be a linear viscous continuum containing uniformly distributed elliptical pores with the same size (Figure 6.12). This continuum model would be more appropriate for glass that has no grain boundaries. The driving force for sintering is not the same in all directions, as a result of the varying radius of curvature of the pore. For an elliptical pore with long and short axes equal to a and c, respectively, the radii of curvature in the directions of the long axis (r_l) and the short axis (r_s) are

$$r_l = \frac{c^2}{a}; \quad r_s = \frac{a^2}{c} \tag{6.4}$$

According to Equation 6.4, the driving force for sintering will be larger in the direction of the long axis because of the smaller value of the radius of curvature in this direction. The body should, therefore, shrink more in the direction of the long axis. However, anisotropy in the material response

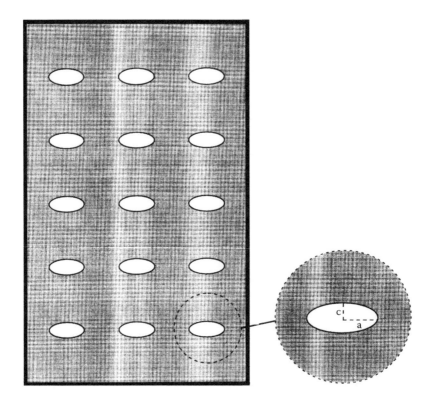

FIGURE 6.12 Illustration of an anisotropic porous body consisting of a linear viscous continuum and elliptical pores with long axis and short axis equal to a and c, respectively.

of the porous sintering body should also be taken into account. The viscosity of the body is also larger in the direction of the long axis of the pore, reducing the overall shrinkage anisotropy due to the pore curvature alone. In general, the shrinkage anisotropy depends on the interplay between the sintering driving force due to the pore curvature and the anisotropy of the material viscosity. A model for the sintering of uniformly distributed elliptical pores with the same size in a polycrystalline material also predicts that elongated pores lead to anisotropic shrinkage and that the shrinkage is greater in the direction of the long axis of the pores [27].

6.3.2.2 Particle Alignment

Anisometric or nonequiaxial particles can lead to anisotropic shrinkage during sintering if they become aligned or oriented during the forming of the green body. There are many reports of anisotropic shrinkage caused by alignment of anisometric particles in green compacts formed by uniaxial pressing in a die, injection molding, extrusion, and tape casting [28–32]. It is generally observed that the shrinkage perpendicular to the direction of alignment is greater than that parallel to the alignment direction. This can be taken to result from faster or preferential matter transport into the pores from the grain boundaries oriented in the direction of the particle alignment versus from the grain boundaries oriented perpendicular to direction of alignment.

Numerical models have been developed to simulate the anisotropic shrinkage of a polycrystalline system of elongated particles in two dimensions. For a system of uniformly packed, elliptical particles, the shrinkage was found to be higher in the direction perpendicular to the long axis of the particles [33]. Another model, which takes into account densification by grain boundary diffusion,

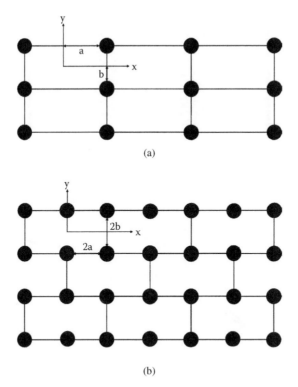

(a)

(b)

FIGURE 6.13 Model initial microstructures used for analyzing anisotropic shrinkage in two dimensions. Monosized elongated particles in (a) simple packing and (b) more closely packed arrangement.

grain growth driven by differences in grain boundary curvature, and pore migration by surface diffusion, suggests that the direction of higher (or lower) shrinkage may depend on the particle packing arrangement, as well as on the particle and pore characteristics [34]. Assuming isotropic kinetic and thermodynamic properties of the system, for an idealized model with a simple packing arrangement of monosized elongated particles (Figure 6.13a), shrinkage is higher in the direction of the orientation (the x-direction). On the other hand, for the more closely packed arrangement shown in Figure 6.13b, shrinkage is higher in the direction perpendicular to the elongation (the y-direction). However, the direction of higher (or lower) shrinkage also depends on the particle and pore characteristics. For example, if the particles in Figure 6.13b were sufficiently long ($a > 2b$), then shrinkage in the x-direction would exceed that in the y-direction. For a more realistic system of aligned, elongated particles of random size and shape (Figure 6.14a), the shrinkage was higher in the direction perpendicular to the elongation (Figure 6.14b). The shrinkage rate and the anisotropy in the shrinkage rates were found to decrease with time as the grains grew and the microstructure became more isotropic.

Powder matrices containing anisometric inclusions, such as short fibers, whiskers, or platelets, can also suffer from significant anisotropic shrinkage during sintering if the inclusion phase becomes oriented during forming of the green body [35–37]. Generally, the shrinkage in the direction of orientation is lower than that perpendicular to the direction of orientation. The presence of rigid inclusions leads to an increase in the viscosity of the body (see Chapter 5). For oriented inclusions, the rigidity of the body is higher in the direction of orientation, so the shrinkage is lower. Anisotropic sintering shrinkage is also significant in systems fabricated by the templated grain growth procedure (see Chapter 3), in which large elongated seed crystals are aligned in a powder matrix to serve as templates to pattern the grain growth [38].

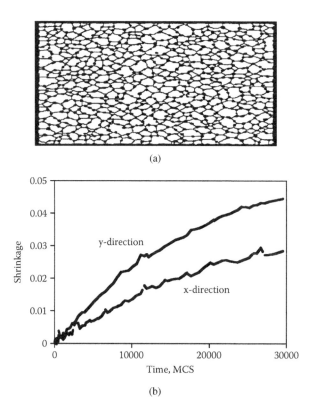

(a)

(b)

FIGURE 6.14 (a) Initial, two-dimensional model microstructure of randomly packed elongated particles with random size and shape, oriented in the *x*-direction. (b) Shrinkage in the *x*-direction (parallel to orientation) and *y*-direction (perpendicular to orientation) obtained from numerical simulation of sintering. (From Tikare, V., Braginsky, M., Olevsky, E., and Johnson, D.L., Numerical simulation of anisotropic shrinkage in a 2D compact of elongated particles, *J. Am. Ceram. Soc.*, 88, 59, 2005. With permission.)

6.3.3 HEATING SCHEDULE

Heating schedules can be simple, as in isothermal sintering or constant heating rate sintering of single-phase powders in model laboratory-scale experiments (Figure 6.15), or they can have a more complex temperature–time relationship, as in the sintering of ceramics for industrial applications. A general heating schedule is shown in Figure 6.16. Such a schedule typically consists of six stages, outlined below.

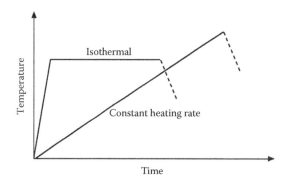

FIGURE 6.15 Sketch of simple temperature-versus-time schedules for isothermal and constant heating rate sintering.

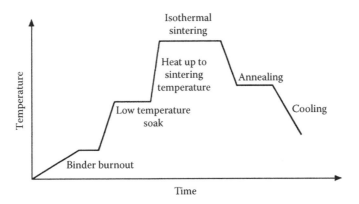

FIGURE 6.16 Sketch of a general heating schedule.

6.3.3.1 Stage 1: Binder Burnout

In this stage, the removal of volatile materials such as adsorbed water and the conversion of additives such as metal organic compounds or organic binders take place. The heat-up rate is often slow and carefully controlled, often <2°C/min, since rapid heating may cause boiling and evaporation of organic additives, leading to swelling and even cracking of the specimen. Typically, the hold temperature in this stage is, at most, 400°C–500°C.

6.3.3.2 Stage 2: Low-Temperature Soak

Stage 2 may be included to promote chemical homogenization or the reaction of powder components. The hold temperature is commonly below that for the onset of measurable sintering. Chemical homogenization may, for example, involve a solid-state reaction in which a small amount of dopant is incorporated into the powder, or it may involve a reaction leading to the formation of a liquid phase.

6.3.3.3 Stage 3: Heat-Up to the Sintering Temperature

This stage involves heat-up to the isothermal sintering temperature. The heating rate is limited by the sample size and the thermal characteristics of the furnace. For large bodies, the heat-up times can stretch for many hours to avoid temperature gradients that could lead to cracking or to avoid the formation of a dense outer layer on an incompletely densified core, as would result from differential densification. In laboratory-scale experiments with small specimens, it is often observed that a faster heating rate in this stage enhances the densification in the subsequent isothermal sintering stage. A possible explanation for this observation is that the coarsening of the powder during the heat-up is reduced due to the shorter time taken to reach the isothermal sintering temperature, resulting in a finer microstructure at the start of isothermal sintering.

6.3.3.4 Stage 4: Isothermal Sintering

The isothermal sintering temperature is chosen to be as low as possible yet compatible with the requirement that densification be achieved within a reasonable time (typically less than 24 hours). Higher sintering temperatures lead to faster densification, but the rate of coarsening also increases. The increased coarsening rate may lead to abnormal grain growth where pores are trapped inside large grains. Although densification proceeds faster, the final density may be limited. Figure 6.17 shows qualitatively some possible sintering profiles for different isothermal sintering temperatures.

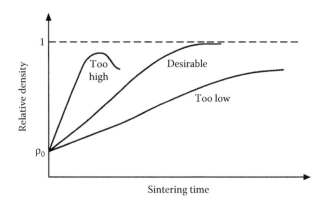

FIGURE 6.17 Sketch of sintering curves to illustrate the selection of an optimum isothermal sintering temperature.

6.3.3.5 Stage 5: Annealing

An additional hold stage may be included prior to final cool-down of the material to relieve thermal stresses, to allow for precipitation of second phases, or to modify the chemical composition or the microstructure. Annealing to reduce thermal stresses is common in systems that contain a glassy matrix or that undergo a crystallographic transformation with an accompanying volume change. Modification of the chemical composition and the microstructure is common in many functional ceramics, such as ferrites. In structural ceramics, a good example of the use of an annealing step to modify the microstructure is the crystallization of the glassy grain boundary phase in Si_3N_4 to improve the high-temperature creep resistance (see Chapter 4).

6.3.3.6 Stage 6: Cool-Down to Room Temperature

The cooling rate can be fairly fast for relatively small articles but is much slower for large articles to prevent large temperature gradients that can lead to cracking. When compositional or micro-structural modification must be achieved, the cooling rate needs to be carefully controlled. As will be discussed later, some functional ceramics, such as ferrites, also require that the cooling be carried out in a controlled atmosphere to control the stoichiometry and microstructure of the final material. In some ceramics, the cooling rate can influence the precipitation of second phases and their distribution in the fabricated article. In Sr-doped TiO_2 [39], the solubility limit of Sr in TiO_2 is ~0.5 at% at 1400°C. Even for concentrations below the solubility limit, cooling from 1400°C leads to the precipitation of $SrTiO_3$ at the TiO_2 grain boundaries. This precipitation is probably due to the segregation of Sr ions that have a large size misfit in the TiO_2 lattice. The morphology of the precipitates depends on the cooling rate. For an Sr concentration of 0.2 at%, well below the solubility limit, slow cooling from 1400°C leads to the precipitation of fine discrete particles. On the other hand, if the same composition is cooled rapidly to room temperature, no second phase is formed. Near the solubility limit (0.5 at%), slow cooling leads to the precipitation of a continuous second phase between 1400°C and 1250°C, but this breaks up on further cooling below 1200°C to produce discrete particles.

6.3.3.7 Isothermal Sintering

In isothermal sintering (Figure 6.15), the temperature is increased monotonically to a fixed sintering temperature, maintained at this temperature for the required time, and finally lowered to room temperature. The holding time is often long compared to the heat-up time, and, generally, the isothermal sintering temperature should be well above the intended service temperature of the article. As outlined earlier, the heat-up times are limited by the sample size and by the characteristics of the furnace. Considerable densification and microstructural changes can occur during the heat-up stage of isothermal sintering, but these changes are often ignored in sintering studies. The data for this nonisothermal stage are unsuitable for comparison with the predictions of sintering models that assume idealized isothermal conditions.

6.3.3.8 Constant Heating Rate Sintering

In constant heating rate sintering, the sample is heated at a constant rate to a specified temperature and immediately cooled (Figure 6.15). Provided that the sintering mechanisms do not change over the temperature range and the activation energies for densification and coarsening are available, constant heating rate experiments may be simpler to analyze theoretically than the isothermal experiments because ideal isothermal sintering is not possible [40,41]. As outlined earlier, the heating rate is limited by the sample size and the thermal characteristics of the furnace. High heating rates are commonly most useful in laboratory-scale experiments.

6.3.3.9 Effect of Heating Rate on Sintering

Manipulation of the heating rate can be useful in cases where a reaction or crystallization occurs during sintering (see Chapter 5). The heating rate can also influence the sintering and microstructural evolution of single-phase ceramics that do not undergo reaction or crystallization.

6.3.3.9.1 Amorphous Ceramics

In the case of amorphous ceramics, coarsening is absent, and viscous flow is the dominant densification mechanism. The influence of heating rate depends on the structure of the amorphous material. For powders prepared from a melted glass [42], the observed sintering kinetics are in good agreement with the predictions of the theoretical models for viscous sintering (see Chapter 2) when the viscosity η of the bulk glass is described by a single-valued function of temperature. For many oxide glasses, the dependence of the glass viscosity η on the absolute temperature T is well represented by the *Fulcher equation*:

$$\eta = A \exp\left(\frac{B}{T - T_o}\right) \tag{6.5}$$

where A, B, and T_o are constants for a given glass composition. Another useful equation is

$$\eta = \eta_o \exp\frac{Q}{RT} \tag{6.6}$$

where η_o is a constant, Q is the activation energy, and R is the gas constant. With increasing heating rate, the shrinkage or densification curves shift to higher temperatures because the glass has less time for densification at a given temperature.

The densification of gels during constant heating rate sintering depends on the gel structure [43,44]. Particulate gels show the same type of sintering behavior as powder compacts formed from a melted glass [1]. The sintering data are well represented by the theories for viscous sintering when the viscosity is described by a single-valued function of temperature. On the other hand, the sintering kinetics of polymeric gels do not show good agreement with the theoretical predictions, regardless of the geometrical model used. Figure 6.18 shows a comparison of the shrinkage data with the predictions of Scherer's model for viscous sintering (see Chapter 2)

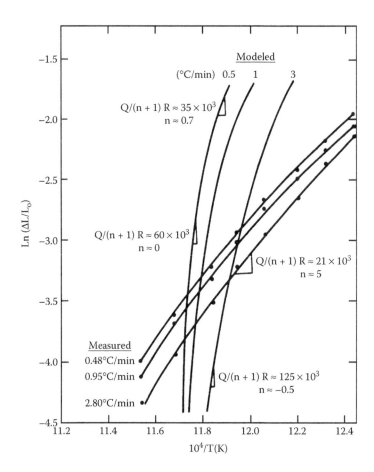

FIGURE 6.18 Natural logarithm of the shrinkage versus the reciprocal of the temperature for a multicomponent gel during constant heating rate sintering: calculated curves (MODELED) and measured data (MEASURED). (From Brinker, C.J., Scherer, G.W., and Roth, E.P., Sol → Gel → Glass: II. Physical and structural evolution during constant heating rate experiments, *J. Non-Cryst. Solids*, 72, 345, 1985. With permission.)

for a polymeric gel sintered at heating rates of 0.5°C/min, 1°C/min, and 3°C/min. The predicted values were calculated by assuming a constant activation energy (Equation 6.6) for viscous flow (~500 kJ/mol) throughout the densification process. The approach fails because the viscosity of the gels is not a single-valued function of the temperature. The apparent activation energy depends on the hydroxyl content of the gel, which varies with temperature and time, so the viscosity changes with time and temperature. For polymeric gels, the densification kinetics depend on the thermal history and not just the current temperature.

The effect of heating rate is of considerable interest and importance for the sintering of gels. Particulate gels, as outlined earlier, show normal behavior where the shrinkage curves shift to higher temperatures with increasing heating rate. For polymeric gels, the behavior depends on the interplay between the hydroxyl content of the gel (which influences the viscosity) and the time available for sintering. It is commonly observed that above a certain value of the heating rate, the shrinkage curves shift to lower temperatures with increasing heating rate. This is because the more rapidly heated gel retains more hydroxyl groups, which lower the viscosity and offset the shorter time available for sintering. Isothermal sintering of polymeric gels leads to a rapid increase in the viscosity, so the densification rate decreases drastically. It is, therefore, advantageous to increase the temperature continuously during sintering, so that the rising temperature can compensate for the loss of hydroxyl groups and the corresponding increase in the viscosity. The faster the heating rate, the lower the temperature at which densification is completed. However, this applies when crystallization does not interfere with the densification. In practice, very fast heating rates can lead to undesirable effects such as incomplete burnout of the organics and trapped gases that expand and cause cracking or bloating of the article. The maximum heating rate must be determined empirically for each gel.

6.3.3.9.2 Polycrystalline Ceramics

The constant heating rate sintering of polycrystalline materials is complicated, even for the initial stage, by the occurrence of multiple mechanisms [45]. For a wider range of density, the simultaneous occurrence of densification and coarsening must be taken into account. Lange [46] studied the sintering of Al_2O_3 powder compacts at constant heating rates of 2.5 to 20°C/min up to 1500°C and found that for each heating rate, the densification strain rate increased to a maximum at a relative density of ~0.8, followed by a decrease at higher density. He suggested that densification processes dominated at relative densities below ~0.8, whereas coarsening dominated above this relative density.

The sintering and microstructural evolution of ZnO powder compacts was studied by Chu et al. [47] over a wide range of constant heating rates (0.5 to 15°C/min). The data for the relative density ρ versus temperature T show that the curves are clustered within a fairly narrow band (Figure 6.19). An interesting observation is that the derivative of the densification strain with respect to T, that is $(1/\rho)d\rho/dT$, for this range of constant heating rates falls on the same master curve (Figure 6.20). In constant heating rate sintering, the sintering temperature T and the sintering time t are related through the heating rate α by the equation

$$T = \alpha t + T_o \tag{6.7}$$

where T_o is an initial temperature. Differentiating Equation 6.7 gives

$$dT = \alpha\,dt \tag{6.8}$$

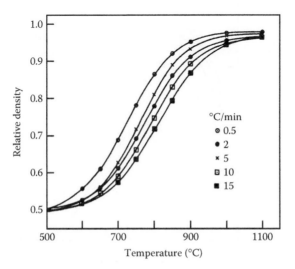

FIGURE 6.19 Relative density versus temperature for ZnO powder compacts with the same initial density (0.50 ± 0.01), sintered at constant heating rates of 0.5 to 15°C/min.

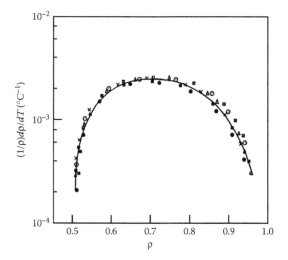

FIGURE 6.20 Change in the densification strain per unit change in temperature as a function of relative density, determined from the data of Figure 6.19. The data for the different heating rates fall on the same master curve.

According to Equation 6.8, the densification rate $(1/\rho)d\rho/dt$ can be obtained from the data of Figure 6.20 by multiplying by α. As Figure 6.21 shows, the densification rate increases with increasing heating rate. Higher heating rates also lead to a finer grain size (Figure 6.22). As an indication of the change in grain size, scanning electron micrographs of the compacts sintered to 1100°C show that the grain size is reduced by a factor of ~2 when the heating rate is increased from 0.5°C/min to 5°C/min (Figure 6.23).

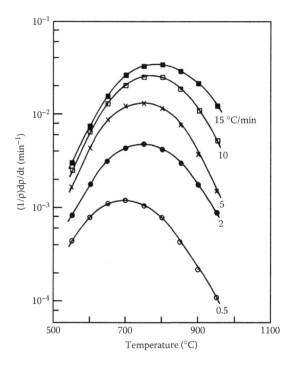

FIGURE 6.21 Densification rate versus temperature determined from the data of Figure 6.19. Above ~700°C, the rate is approximately proportional to the heating rate.

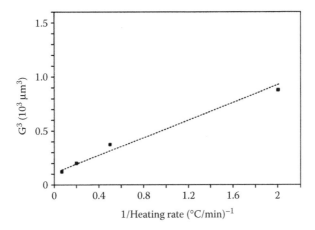

FIGURE 6.22 Cube of the final grain size versus the inverse of the heating rate for ZnO.

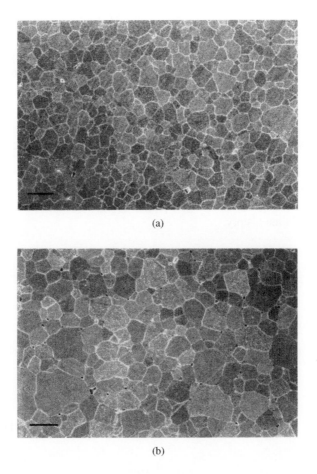

FIGURE 6.23 Scanning electron micrographs of polished and etched surfaces of ZnO powder compacts sintered to 1100°C at constant heating rates of (a) 5°C/min and (b) 0.5°C/min. (Bar = 10 μm.)

A representation of constant heating rate data should take into account the simultaneous occurrence of densification and coarsening. As a first approximation, we can modify the theoretical equations for isothermal sintering to account for the effect of changing temperature on the densification and coarsening processes. Following Equation 2.126, the linear densification rate, equal to one-third the volumetric densification rate $(1/\rho)d\rho/dt$, can be written

$$\dot{\varepsilon}_\rho = \frac{H_1 D(T)\phi^{(m+1)/2}\Sigma}{G^m(T,t)kT} \tag{6.9}$$

where $\dot{\varepsilon}_\rho$ is now a function of both temperature and time, H_1 is a constant, the diffusion coefficient D is now a function of temperature, ϕ is the stress intensity factor, Σ is the sintering stress, $G(T,t)$ represents a coarsening function that depends on temperature and time, k is the Boltzmann constant, and m is an integer that depends on the diffusion mechanism ($m = 2$ for lattice diffusion and $m = 3$ for grain boundary diffusion). The diffusion coefficient can be written

$$D(T) = D_o \exp\left(-\frac{Q_d}{RT}\right) \tag{6.10}$$

where D_o is a constant, R is the gas constant, and Q_d is the activation energy for the densification process. The coarsening function may be expected to have the same form as the grain growth equation (Equation 3.37) and for isothermal sintering at a given T can be written as

$$G^m(t) = G_o^m + A\exp\left(-\frac{Q_c}{RT}\right)t \tag{6.11}$$

where G_o is the initial grain size, A is a constant, and Q_c is the activation energy for the coarsening process. For constant heating rate sintering, Equation 6.11 can be modified to give the grain size at any temperature and time history:

$$G^m(T,t) = G_o^m + A\int_{t_o}^{t}\exp\left(-\frac{Q_c}{RT}\right)dt \tag{6.12}$$

Substituting for dt from Equation 6.8 gives

$$G^m(T,t) = G_o^m + \frac{A}{\alpha}\int_{T_o}^{T}\exp\left(-\frac{Q_c}{RT}\right)dT \tag{6.13}$$

Since the integral is a function of T only, when $G^m \gg G_o^m$, Equation 6.13 can be written in the form

$$G^m \approx \frac{1}{\alpha}F_1(T) \tag{6.14}$$

where $F_1(T)$ is a function of temperature only. At any temperature, G^m is proportional to $1/\alpha$, and when $m = 3$, the grain size data in Figure 6.22 satisfy Equation 6.14.

The densification rate follows from Equation 6.9 for the known relationship for $G(T,t)$. When $G^m \gg G_o^m$, Equation 6.14 can be used, and the densification rate becomes

$$\dot{\varepsilon}_\rho \approx \alpha F_2(T) \tag{6.15}$$

where $F_2(T)$ is a function of temperature only. The densification rate at any temperature is predicted to be proportional to the heating rate α, which is consistent with the data shown in Figure 6.21 for $T > {\sim}700°C$. To find the density, Equation 6.15 must be integrated, giving

$$\frac{1}{3}\int_{\rho_o}^{\rho} \frac{d\rho}{\rho} \approx \alpha \int_{t_o}^{t} F_2(T)dt \tag{6.16}$$

Substituting for dt from Equation 6.8 gives

$$\frac{1}{3}\ln\left(\frac{\rho}{\rho_o}\right) \approx \int_{T_o}^{T} F_2(T)\,dT \tag{6.17}$$

According to this equation, for compacts with the same green density ρ_o, the density at any temperature is independent of the heating rate and is a function of the temperature only. This result shows good agreement with the density data in Figure 6.19.

6.3.3.10 Master Sintering Curve

Experimentally, it is found that Σ is approximately constant and that ϕ is a function of the density only (see Chapter 2). For a given powder consolidated to a given green microstructure, if it is assumed that the grain size is dependent on the sintered density only, then, with slight rearrangement, Equation 6.9 becomes

$$\frac{k}{H_1 D_o \Sigma}\int_{\rho_o}^{\rho} \frac{G^m(\rho)}{3\rho\phi^{(m+1)/2}(\rho)}\,d\rho = \int_{t_o}^{t} \frac{1}{T}\exp\left(-\frac{Q_d}{RT}\right)dt \tag{6.18}$$

The integral on the left-hand side of Equation 6.18 contains density-dependent terms only and can be taken to quantify the effects of the microstructure on densification, whereas the right-hand side depends only on Q and the temperature–time schedule. Putting

$$F(\rho) = \frac{k}{H_1 D_o \Sigma}\int_{\rho_o}^{\rho} \frac{G^m(\rho)}{3\rho\phi^{(m+1)/2}(\rho)}\,d\rho \tag{6.19}$$

and

$$\Theta(T,t) = \int_{t_o}^{t} \frac{1}{T}\exp\left(-\frac{Q_d}{RT}\right)dt \tag{6.20}$$

then Equation 6.18 can be written

$$F(\rho) = \Theta(T,t) \tag{6.21}$$

The relationship between ρ and $F(\rho)$ is referred to as the *master sintering curve*.

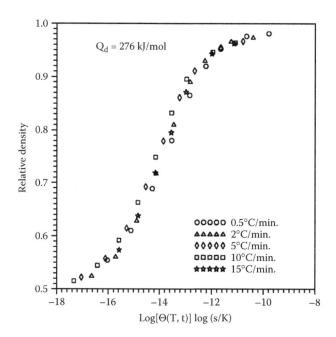

FIGURE 6.24 Master sintering curve for ZnO constructed from the data in Figure 6.19.

One way to construct a master sintering curve is to determine the grain size–density trajectory and the relationship for $\phi(\rho)$ and then to integrate the combined functions in Equation 6.19. Another way is to make use of the equality in Equation 6.21. This is the approach taken by Su and Johnson [48], who used a different formulation of the sintering model equation, corresponding to Equation 6.9. The master sintering curve can be constructed from a series of sintering experiments for different times or temperatures. Each experiment gives the measured value of ρ, and the corresponding value of $\Theta(T,t)$ may be found by performing the integration shown in Equation 6.20 for the temperature–time schedule if Q is known or can be approximated. The most economical method of generating the master sintering curve is to perform constant heating rate sintering experiments at four or five heating rates using a dilatometer. Figure 6.24 shows the master sintering curve determined from the data for ZnO shown in Figure 6.19.

Once the master sintering curve has been determined for a particular powder system, it can be used to predict the sintering behavior of similar powder compacts under arbitrary temperature–time schedules. In this way, it can serve a useful function for designing sintering schedules to reach a required final density. However, the limitations of the procedure should be understood if it is to be used effectively in practice. The procedure can be applied only to compacts of the same starting powder that have been consolidated to the same green density by the same forming method. It is also assumed that the grain size and geometrical parameters depend on the density only and that one diffusion mechanism dominates in the sintering process.

6.3.3.11 Multistage Sintering

Multistage sintering, involving the introduction of extended temperature plateaus or more complex temperature–time schedules, is often used in practice to achieve specific chemical or microstructural features. An example is the *two-peak sintering* technique used for the ion-conducting ceramic sodium beta-alumina, in which the sintering schedule consists of two separate peak temperatures [49]. Optimum results for the microstructure and strength were obtained in this case by reducing the isothermal sintering temperature by ~150°C after sintering at the first peak temperature of ~1500°C.

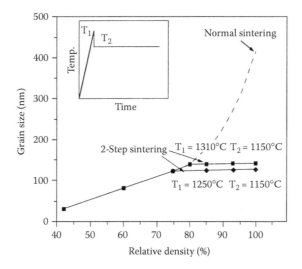

FIGURE 6.25 Grain size of Y_2O_3 in two-step sintering. (Heating schedule shown in inset.). The grain size remains almost constant in the second sintering step, despite density improvement to 100%. (From Chen, I.W., and Wang, XH., Sintering dense nanocrystalline ceramics without final-stage grain growth, *Nature*, 404, 168, 2000. With permission.)

A modification of the two-peak sintering technique has been shown to result in dramatically reduced grain growth during the final stage of sintering for a few ceramics formed from submicron powders [50–52], leading to the production of ceramics with high density and fine grain size. The technique is particularly useful for the production of nanocrystalline ceramics, where the final grain size must be <50–100 nm in order to achieve the desired enhancement in properties. The method uses a heating schedule consisting of two steps: the compact is first heated to a higher temperature (T_1) to achieve a relative density greater than ~0.75 and then cooled down and held at a lower temperature (T_2) until it is almost fully dense. Figure 6.25 compares the grain size G versus relative density ρ trajectory for undoped Y_2O_3 powder compacts using this type of two-step sintering and for conventional (normal) sintering in which the compact was heated at 10°C/min and held at 1550°C for 1 h. The microstructure of fully dense Y_2O_3 doped with 1 at% Mg is shown in Figure 6.26.

FIGURE 6.26 Microstructure of fully dense Y_2O_3 doped with 1 at% Mg. The material was sintered to 1080°C in step 1, reaching a relative density of 76%, then for 20 h at 1000°C in step 2. (Courtesy of I.-W. Chen.)

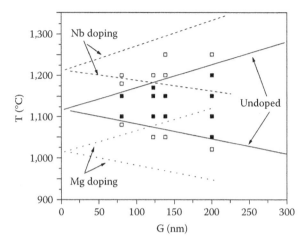

FIGURE 6.27 Kinetic window for reaching full density without grain growth. Solid symbols are temperature and grain size conditions of successful second-step sintering runs for Y_2O_3. Open symbols above the upper boundary show conditions for which grain growth occurred in the second step, and open symbols below the lower boundary show conditions that did not produce full density in the second step. Similar boundaries are shown for Y_2O_3 doped with 1 at% Nb and 1 at% Mg. (From Chen, I.W., and Wang, X.H., Sintering dense nanocrystalline ceramics without final-stage grain growth, *Nature,* 404, 168, 2000. With permission.)

This material was sintered to 1080°C ($\rho = 0.76$) and then held for 20 h at 1000°C. The feasibility of using this two-step method for achieving densification without grain growth in the final stage relies on the suppression of grain boundary migration while keeping the densifying mechanism active. For nanoscale powders, the dominant densifying mechanism is expected to be grain boundary diffusion. The conditions for which this is possible in the second-step sintering of undoped Y_2O_3 are shown in Figure 6.27, where the filled symbols indicate the conditions where full density was achieved without grain growth. Above this region, grain growth was observed, whereas below this region, full density was not achieved. The data suggest that grain boundary migration in Y_2O_3 may involve a mechanism with a higher activation energy than grain boundary diffusion, so it is suppressed at the lower sintering temperature used in the second step. The use of dopants provides an additional parameter for optimizing the conditions for achieving full density without grain growth in the final stage, as shown in Figure 6.27 for Mg- and Nb-doped Y_2O_3.

As described in Chapter 5, the use of another type of two-step sintering technique resulted in a reduction in the average grain size and an improvement in the microstructural homogeneity of ZnO, MgO, and Al_2O_3 [53–55]. In this method, the first heating stage consists of an extended hold at a temperature below the onset of shrinkage (50 h at 800°C for Al_2O_3), where the occurrence of a limited amount of coarsening, presumably by surface diffusion or vapor transport, has the effect of improving the homogeneity of the porous powder compact. In the second stage, sintering of the homogenized microstructure at normal temperatures results in a refinement of the microstructure. Excessive coarsening in the first stage should be avoided, and this is commonly achieved through trial-and-error experimentation to determine the optimum conditions of time and hold temperature.

6.3.3.12 Fast Firing

Fast firing is an example of heating cycle control where the powder compact is heated rapidly and subjected to a short sintering time at high temperature (e.g., 10 min at 1800°C–1900°C for Al_2O_3). The process has been shown to be effective for achieving high density and fine grain size in a few ceramics (e.g., Al_2O_3 and $BaTiO_3$), but, because of the rapid heating, the process is suitable only for the production of small or thin-walled articles. The powder compact is pushed through a short (5–10 cm) hot zone in a tube furnace, such as that shown in Figure 6.1, at constant speeds ranging from ~0.25 cm/min to ~20 cm/min. The temperature of the hot zone and the speed at which the sample is pushed through the lower temperature region to the hot zone determine the rate of heating of the sample.

The concept underlying fast firing, put forward by Brook [56], was outlined in Chapter 3. The most favorable situation is one where the activation energy for densification is greater than that for coarsening, in which case the ratio of the densification rate to the coarsening rate is larger at higher temperatures. The faster the sample is heated through the lower-temperature region, where the ratio of densification rate to coarsening rate is unfavorable, the better the expected result. The argument, therefore, is for the use of rapid heating and short sintering times at high temperature.

To predict which powder systems may benefit from the use of fast firing, it is necessary to know the controlling mechanism for the processes of densification and coarsening and, in addition, to have reliable data for the activation energies for the appropriate diffusion coefficients. In most cases, this information is unavailable or incomplete. The best way to determine the effectiveness of the fast firing technique is, therefore, to try it. However, some data can be found for a few systems, and it is worth considering the effectiveness of fast firing in these systems.

For Al_2O_3, the densification mechanism for fine powders at moderate to high temperatures is that of lattice diffusion of the cation with an activation energy of 580 kJ/mol. Surface diffusion is believed to be the dominant coarsening mechanism, and activation energies derived from sintering studies are relatively low (230–280 kJ/mol). On this basis, fast firing should work for Al_2O_3. In practice, fast firing is indeed found to be effective for the production of Al_2O_3 with high density and fine-grained microstructure, as the data in Figure 3.56 show [57].

For MgO, activation energies for surface diffusion (coarsening mechanism) range from 360 to 450 kJ/mol, and those for cation lattice diffusion (densification mechanism) range from 150 to 500 kJ/mol. The activation energies for lattice diffusion determined from sintering studies cover a narrower range (250–500 kJ/mol) but are still not precise enough to avoid an overlap with the surface diffusion values. As the grain size increases, densification by oxygen diffusion in the grain boundary is believed to become significant. The activation energy for the oxygen diffusion mechanism is in the range of 250–300 kJ/mol, which is significantly lower than that for surface diffusion. For densification controlled by the grain boundary diffusion of oxygen, firing to a higher temperature will clearly lead to a reduction in the ratio of the densification rate to the coarsening rate (an effect opposite to the one desired). Although the information for the rate-controlling mechanism and activation energies is fairly imprecise, experiments show that fast firing is ineffective for MgO. This finding is consistent with the general argument.

Fast firing has been used successfully to produce dense $BaTiO_3$ ceramics with a fine-grained microstructure [58]. The fast-fired samples contained smaller grains at a given density compared with those prepared by conventional sintering (Figure 6.28). For this system, information on the mechanisms of densification and coarsening and the activation energies is incomplete.

6.3.3.13 Rate-Controlled Sintering

In rate-controlled sintering, the densification rate is coupled with the temperature control of the sintering furnace in such a way as to keep the shrinkage rate (or densification rate) at a constant value or below a given value [59,60]. The result is a fairly complicated temperature schedule that at times approaches the multi-staged sintering schedules. Although the underlying mechanisms of rate-controlled sintering are not fully understood, beneficial effects on microstructure have been reported for a few ceramics, including cases where chemical reactions occur during densification [61,62]. Manipulation of the temperature schedule is necessarily limited by the thermal impedance of the sintering furnace and by the size of the ceramic sample.

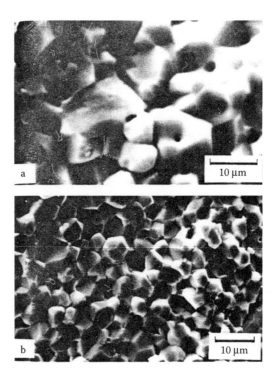

FIGURE 6.28 Scanning electron micrographs of the fractured surfaces of $BaTiO_3$ materials containing 2 mol% excess BaO (a) conventionally sintered at 1370°C for 26 h and (b) fast-fired at 1470°C for 5 min.

6.3.4 SINTERING ATMOSPHERE

The sintering atmosphere can have several important effects on densification and microstructural development, and, in many instances, it can have a decisive effect on the ability to reach high density with controlled grain size. Atmosphere is not directly considered in the sintering models, but the theories can provide a basis for understanding the effects arising from atmospheric conditions. The important effects are associated with both physical phenomena, as in the case of the atmospheric gas trapped in isolated pores in the final stage of sintering, and chemical phenomena, particularly when the volatility, ionic oxidation state, and defect chemistry of the system can be modified.

6.3.4.1 Gases in Pores

In the final stage of sintering, as the pores pinch off and become isolated, gas from the sintering atmosphere (or from volatile species in the solid) becomes trapped in the pores. At the point when the pores become isolated, the pressure of the trapped gas is equal to the pressure of the sintering atmosphere. Further sintering is influenced by the solubility of the gas atoms in the solid, as has been demonstrated by Coble [63] in the sintering of Al_2O_3. Coble found that that MgO-doped Al_2O_3 can be sintered to theoretical density in vacuum or in an atmosphere of H_2 or O_2, which diffuse out to the surface of the solid, but not in He, Ar, or N_2 (or, therefore, air), which have a limited solubility in Al_2O_3.

Information on the diffusivity of gases in ceramics is not readily available, so it is difficult to predict the effects of a given gas trapped in the pores. However, interstitial diffusion is expected to be more favorable for smaller molecules (e.g., H_2 and He) than for larger ones (e.g., Ar and N_2). Diffusion of larger molecules may occur by a vacancy mechanism, but this will depend on the

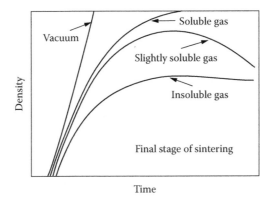

FIGURE 6.29 Schematic of the effect of atmospheric gas on final-stage densification, for vacuum, an inert gas that is insoluble in the solid phase, and two soluble gases (high and low solubility).

stoichiometry of the solid. Diffusion along the disordered grain boundaries can provide an easier path for smaller as well as larger molecules. Whenever it can be used, vacuum sintering (at least to the onset of pore closure) eliminates the problem of trapped gases.

When the gas has a high solubility in the solid, densification is unaffected by the gas trapped in the pores because rapid diffusion through the lattice or along the grain boundaries can occur during shrinkage (Figure 6.29). For an insoluble gas, as shrinkage of the isolated pores takes place, the gas is compressed and its pressure increases. Shrinkage stops when the gas pressure becomes equal to the compressive pressure of the driving force for sintering. Assuming for simplicity that the pores are spherical, the limiting density is reached when the gas pressure p in the pores is given by

$$p = 2\gamma_{sv}/r \qquad (6.22)$$

where γ_{sv} is the specific energy of the solid–vapor interface, and r is the limiting pore size when shrinkage stops. For an idealized microstructure in which the solid phase is a continuum and the pores have the same radius r, the limiting porosity P_f can easily be estimated. Applying the gas law $p_1 V_1 = p_2 V_2$, where p is the pressure and V is the volume, to the initial situation (when the pores just become isolated) and to the limiting situation (when shrinkage stops) gives

$$p_o N \left(\frac{4}{3} \pi r_o^3 \right) = p N \left(\frac{4}{3} \pi r^3 \right) \qquad (6.23)$$

where N is the number of pores per unit volume, assumed to be constant during the sintering; p_o is the pressure of the sintering atmosphere; and r_o is the radius of the pores when they become isolated. Substituting for p from Equation 6.22 gives

$$p_o r_o^3 = 2\gamma_{sv} r^2 \qquad (6.24)$$

The porosity P_o when the pores become isolated and the limiting porosity P_f are related by

$$\frac{P_o}{P_f} = \frac{r_o^3}{r^3} \qquad (6.25)$$

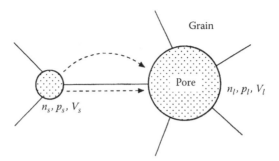

FIGURE 6.30 Sketch showing the parameters for two neighboring pores filled with a slightly soluble gas. A net diffusion of gas occurs from the small pore to the large pore.

Substituting for r from Equation 6.24, after some rearrangement, Equation 6.25 gives

$$P_f = P_o \left(\frac{p_o r_o}{2\gamma_{sv}} \right)^{3/2}$$

(6.26)

According to this equation, the limiting porosity is controlled essentially by the pressure of the sintering atmosphere and by the radius of the pores when they become pinched off. In practice, sintering in a vacuum (where applicable) and homogeneous packing of fine particles (giving fine pores) can improve the final density.

For a slightly soluble gas trapped in the pores, the diffusivity of the gas in the solid becomes low enough that it cannot escape to the surface in the time scale for sintering, but it is possible that diffusion of the gas between neighboring pores can occur. In this case, the final-stage sintering can be idealized as follows: After the pores become isolated, densification will become progressively slower because of the increasing gas pressure in the shrinking pores. Eventually a limiting density will be reached at which densification almost stops because the gas pressure in the pores is equal to the driving force for sintering. Any further changes during sintering will be controlled by the exchange of gases between neighboring pores. In general, there will be a distribution of pore sizes in the solid. Considering a small pore near to a large pore (Figure 6.30), initially, when the limiting density is reached, the pressure p_l in the large pore is given by

$$p_l = \frac{n_l kT}{V_l} = \frac{2\gamma_{sv}}{r_l}$$

(6.27)

where n_l is the number of molecules of gas in the large pore of radius r_l and volume V_l. Because of the higher gas pressure in the small pore, there will be a net diffusion of gas molecules from the small pore to the large pore. If n_s molecules of gas are transferred to the large pore, the new gas pressure in the large pore is

$$p_l' = \frac{(n_l + n_s)kT}{V_l} > \frac{2\gamma_{sv}}{r_l}$$

(6.28)

Since the gas pressure becomes greater than the driving force for sintering, the large pore must expand. The expansion of the large pore is greater than the shrinkage of the small pore, so the overall result is an increase in porosity. For this system, the densification behavior goes through a maximum, as sketched in Figure 6.30. This decrease in density after the maximum value is referred to as swelling (or bloating).

For polycrystalline solids, additional effects must be considered. The location of the pores will influence the gas diffusion path. When the pores are located within the grains, diffusion will depend on the gas solubility in the crystal lattice. For the more common case in which the pores are located at the grain boundaries, diffusion of the gas through the disordered grain boundary region provides an important additional path.

When microstructural coarsening occurs with a slightly soluble or insoluble gas trapped in the pores, pore coalescence can also lead to swelling. Consider two spherical pores of radius r_1, each containing n_1 molecules of an insoluble (or slightly soluble) gas. At the limiting situation when shrinkage stops, the gas pressure in each pore is given by

$$p_1 = \frac{2\gamma_{sv}}{r_1} = \frac{n_1 kT}{V_1} \tag{6.29}$$

The volume of the pore V_1 is equal to $(4/3)\pi r_1^3$, and substituting into Equation 6.29 gives

$$r_1 = \left(\frac{3n_1 kT}{8\pi\gamma_{sv}} \right)^{1/2} \tag{6.30}$$

When the two pores coalesce as a result of grain growth to form a single pore with a radius r_2, this pore contains $2n_1$ molecules of gas, and the limiting condition when it stops shrinking is given by

$$r_2 = \left(\frac{6n_1 kT}{8\pi\gamma_{sv}} \right)^{1/2} \tag{6.31}$$

Equation 6.30 and Equation 6.31 show that $r_2 = \sqrt{2}\, r_1$, so the volume V_2 is $2\sqrt{2}\, V_1$, which is greater than $2V_1$, the original volume of the two pores. This means that after coalescence, the pore will expand, leading to swelling and a decrease in density (Figure 6.31). Swelling is also often observed

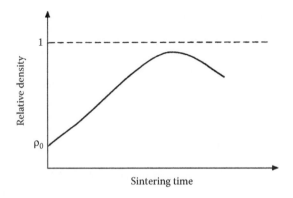

FIGURE 6.31 Schematic diagram showing the densification of a powder compact with a slightly soluble or insoluble gas in the pores. The density goes through a maximum with time, as a result of coarsening with gas trapped in the pores.

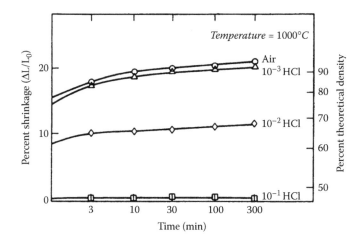

FIGURE 6.32 Shrinkage versus time for Fe_2O_3 powder compacts sintered at 1000°C in different partial pressures of HCl. (From Readey, D.W., Vapor transport and sintering, *Ceramic Trans.*, 7, 86, 1990. With permission.)

when relatively dense ceramics, previously hot pressed in a carbon die (or furnace), are annealed in an oxidizing furnace [64]. Carbon and sulfur impurities react with O_2 gas diffusing along the grain boundaries from the atmosphere to form insoluble gases such as CO, CO_2, and SO_2, and the pressures generated by these gases are often high enough to produce voids in the structure.

6.3.4.2 Vapor Transport

The heating schedule, as described earlier, provides an important process parameter for controlling the relative rates of densification and coarsening during sintering. Vapor transport, which can be varied by changing the composition or partial pressure of the gas in the sintering atmosphere, provides another important process parameter for microstructural control. An example is the work of Readey and coworkers [65–68], who studied the sintering of several oxides in reducing atmospheres. The gas in the atmosphere reacts with the powder to produce gaseous species that enhance vapor transport. An example is the sintering of ferric oxide, Fe_2O_3, in an atmosphere of HCl gas. The reaction can be described as follows:

$$Fe_2O_3(s) + 6HCl(g) \rightarrow Fe_2Cl_6(g) + 3H_2O(g) \tag{6.32}$$

The gaseous Fe_2Cl_6 produced in the reaction leads to enhanced vapor transport. Figure 6.32 shows the data for the shrinkage as a function of time at 1000°C for various partial pressures of HCl. Densification decreases as the partial pressure of the HCl, and, hence, the rate of vapor transport increases. For HCl pressures in excess of 10^{-2} MPa (0.1 atm), shrinkage is almost completely inhibited. As Figure 6.33 shows, the decrease in densification is caused by coarsening of the microstructure.

6.3.4.3 Water Vapor in the Sintering Atmosphere

Several studies have shown that water vapor in the atmosphere can also lead to a modification of the densification-to-coarsening ratio. For Al_2O_3, it was found that, when compared to sintering in dry H_2, sintering in H_2 containing water vapor enhanced the densification rate but reduced the coarsening rate during final-stage densification [69]. The overall result is an enhancement of the densification-to-coarsening ratio in H_2 containing water vapor, so the grain size versus density trajectory is displaced to higher density for a given grain size.

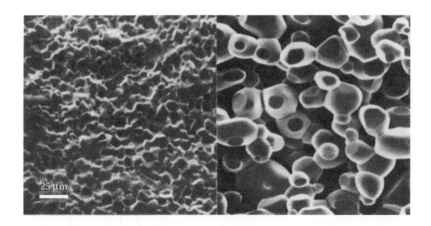

FIGURE 6.33 Scanning electron micrographs of fractured surfaces of Fe_2O_3 powder compacts sintered for 5 h at 1200°C in air (left) and 10% HCl (right). (From Readey, D.W., Vapor transport and sintering, *Ceramic Trans.*, 7, 86, 1990. With permission.)

In the case of MgO, the rates of densification and coarsening are both found to be enhanced by water vapor in the sintering atmosphere. Figure 6.34 shows the densification data for MgO powder compacts at various temperatures in static dry air and in flowing water vapor [70]. Comparison of the data at any given temperature shows faster densification in water vapor. The trajectories for grain size versus relative density for undoped MgO obtained in the same experiments

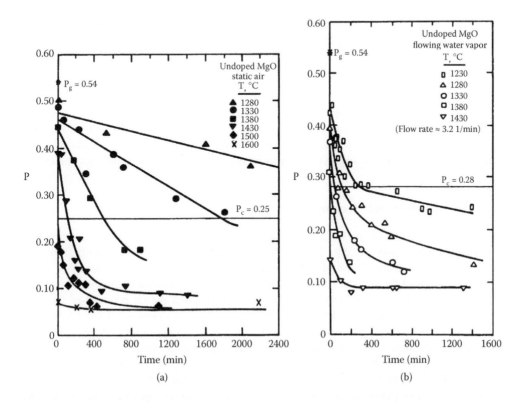

FIGURE 6.34 Porosity versus time for MgO powder compacts sintered at different temperatures in (a) static air and (b) flowing water vapor.

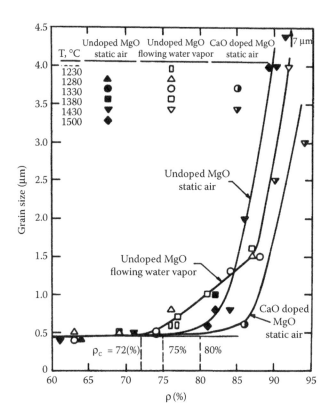

FIGURE 6.35 Grain size versus relative density for MgO powder compacts sintered in air and in flowing water vapor and for CaO-doped MgO powder compacts in static air.

(Figure 6.35) show that water vapor causes a slight enhancement in the grain size at any density during most of the intermediate stage ($\rho \approx 0.72$–0.85) but a reduction at higher density. The increased coarsening rate in MgO appears to be caused by enhanced surface diffusion [71], but the mechanism for the enhanced densification is unclear. Changes in the diffusivity and the interfacial energies (surface and grain boundary energies) appear to be the main factors responsible for the enhanced densification.

6.3.4.4 Volatilization and Decomposition

For some powder compositions, such as sodium β-alumina and lead-based electroceramics, evaporation of volatile components (e.g., Na and Pb) can occur during sintering, making it difficult to control the chemical composition and the microstructure of the sintered material. A common solution is to surround the powder compact with a coarse powder having the same composition as the compact, which leads to the establishment of an equilibrium partial pressure of the volatile component, reducing the tendency for evaporation from the powder compact. Figure 6.36 shows a schematic illustration of the system used for the sintering of sodium β-alumina in a laboratory-scale experiment. The ceramic is a good ionic conductor at temperatures even below 300°C and is used as a solid electrolyte in sodium–sulfur batteries that have electrodes of molten sulfur and sodium polysulfides. The sample, surrounded by a coarse powder with the same composition, is encapsulated in an alumina or platinum tube. The alumina tube is much cheaper but, due to the reaction with the powder, it can be used only a few times before falling apart. For extended use, platinum may have an advantage in overall cost.

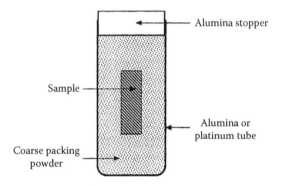

FIGURE 6.36 Laboratory-scale apparatus for the sintering of sodium β-alumina.

Evaporation of PbO in lead-based electroceramics can be represented as [72,73]

$$PbO(s) \rightleftharpoons PbO(g) \rightleftharpoons Pb(g) + 1/2 O_2 \qquad (6.33)$$

Lead is poisonous, so the lead loss must be controlled and the evaporated lead must be contained. In practice, this is achieved by surrounding the sample with lead-based powder compositions [74], such as a mixture of PbO and $PbZrO_3$ for lead-lanthanum-zirconium titanate (PLZT), to provide a positive vapor pressure in a closed Al_2O_3 crucible (Figure 6.37). With the controlled atmosphere apparatus, PLZT can be sintered to full density (Figure 6.38), yielding materials with a high degree of transparency.

For many systems, sintering must be carried out at temperatures where the decomposition of the powder becomes important. If the rate of decomposition is high, the porosity generated by the weight loss limits the attainment of high density. Generally, the weight loss during sintering should be kept below a few percent if high density is to be achieved. Silicon nitride (Si_3N_4) is a good example of a system in which the effects of decomposition are important. It shows significant

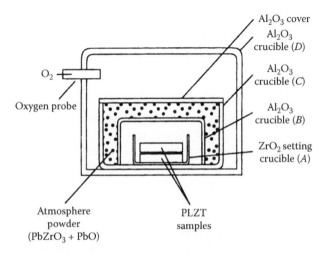

FIGURE 6.37 Apparatus for the sintering of lead-lanthanum-zirconium titanate (PLZT) in controlled atmosphere.

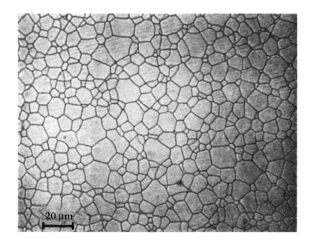

FIGURE 6.38 Scanning electron micrograph of PLZT microstructure produced by sintering in controlled atmosphere.

decomposition at the high temperatures (1700°C–1900°C) required for densification. The decomposition reaction can be written as

$$Si_3N_4 \rightleftharpoons 3Si(g) + 2N_2(g) \qquad (6.34)$$

The vapor pressure of the N_2 gas generated by the decomposition is ~0.1 atm at 1700°C and increases to ~1 atm at 1875°C, so significant weight loss can occur if the decomposition is not controlled. One solution is to surround the sample with a powder of the same composition in a closed graphite crucible and sinter in N_2 gas at normal atmospheric pressure. A better method involves raising the N_2 gas pressure in the sintering atmosphere to values of 10–20 atm or higher, so that the reaction in Equation 6.34 is driven to the left [75]. Nitrogen gas has a fairly high solubility in Si_3N_4, so the unfavorable sintering effects produced by gases trapped in pores are avoided even for such high gas pressures.

6.3.4.5 Oxidation Number

The sintering atmosphere influences the oxidation state of certain cations, particularly those of the transition elements (e.g., Cr), and control of the oxidation state has been shown to have a significant effect on the densification of chromium-containing oxides [76–79]. Figure 6.39 shows the porosity of three chromites as a function of oxygen partial pressure in the sintering atmosphere after sintering for 1 h at temperatures in the range of 1600°C–1740°C. The chromites show little densification if the oxygen partial pressure is greater than ~10^{-10} atm, but relative densities of 99% or greater are obtained by sintering in an oxygen partial pressure of ~10^{-12} atm when the Cr ion is stabilized in its trivalent state (as Cr_2O_3). At oxygen partial pressures greater than ~10^{-12} atm, Cr_2O_3 becomes unstable and vaporizes as CrO_2 or CrO_3, and the high vapor pressure enhances evaporation/condensation processes, leading to neck growth and coarsening but little densification.

The *valence* of an atom is equal to the number of bonds it forms in the most satisfactory formulation of the substance. For example, the valence of the oxygen atom is 2. On the other hand, *oxidation number* (or oxidation state), based on an ionic view of the substance, is the charge on an atom in the most plausible ionic formulation of the substance. Often the valence and the magnitude of the oxidation number are the same, as for example in MgO, in which the oxidation

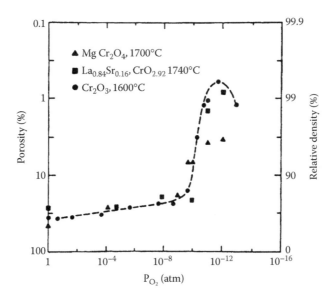

FIGURE 6.39 Final porosity versus oxygen partial pressure in the atmosphere for sintered chromites. (Courtesy of H.U. Anderson.)

state of the oxygen atom is −2. However, many atoms form stable compounds where the oxidation state is different from the valence. In the case of Cr, the principal oxidation numbers of the chromium atom are +6 (as, for example, in CrO_3), +4 (CrO_2), +3 (Cr_2O_3), +2 (CrO), and 0 (Cr metal). The oxidation number of the Cr atom changes readily from +6 in an atmosphere with a high oxygen partial pressure to +4, +3, +2, and finally 0 as the oxygen partial pressure decreases. Each oxidation state, except for the +3 state, corresponds to an oxide with a fairly high vapor pressure, so in the case of Cr_2O_3, sintering must be carried out in an atmosphere in which Cr atoms are maintained in the +3 oxidation state if high density is to be achieved.

The conditions for thermal stability of a substance can be calculated from standard thermochemical data. The results are most usefully plotted on an *Ellingham diagram*. Figure 6.40 shows an example of the calculations for the chromium oxide system [77]. The standard free energy of formation (per mole of oxygen) for three oxides is plotted versus temperature. Also shown (on the right) are the oxygen partial pressures (in atmospheres) in equilibrium with the reactions, along with the partial pressures of the gas mixtures (H_2–water vapor and CO–CO_2) required to produce the desired oxygen partial pressure. For example, Figure 6.40 indicates that at 1600°C (the vertical line), the oxygen partial pressure must not be below $10^{-11.7}$ (or 2×10^{-12}) atm if reduction of Cr_2O_3 to Cr is to be avoided (the lowest diagonal line). The maximum density in Figure 6.39 (99.4% of the theoretical value) is achieved when the oxygen partial pressure is $\sim 2 \times 10^{-12}$ atm, a value that corresponds to the equilibrium partial pressure of oxygen between Cr_2O_3 and Cr.

As shown in Figure 6.39, the oxidation state control used for the sintering of Cr_2O_3 can also be applied to other chromites. However, the dramatic effects of oxidation state control observed for the chromites have not been repeated for other ceramic systems, so the approach appears to have limited applicability. The use of TiO_2 as a dopant [80] allows the sintering of Cr_2O_3 to be carried out at a significantly higher oxygen partial pressure (10^{-1} atm) and at a lower temperature (1300°C). The mechanism by which TiO_2 operates to modify the sintering is not clear, but it has been suggested that the Ti^{4+} ions replace the Cr ions that are in an oxidation state greater than 3, suppressing the formation of the volatile oxides CrO_2 and CrO_3. Another suggestion is that the addition of TiO_2 to Cr_2O_3 leads to the formation of chromium vacancies that enhance the sintering rate.

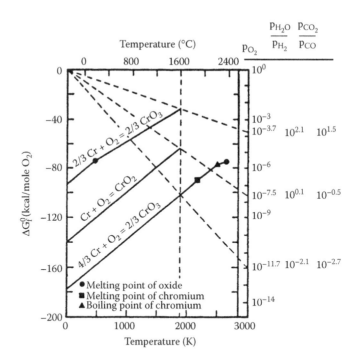

FIGURE 6.40 Standard free energy of formation of chromium oxides as a function of temperature. (Courtesy P.D. Ownby.)

Atmosphere control is also particularly important in the sintering of ferrites for magnetic applications, such as manganese–zinc ferrites and nickel–zinc ferrites, both of which have the spinel crystal structure [81]. In addition to limiting the evaporation of the volatile Zn component, the atmospheric conditions must also be controlled to give the right concentration of ferrous iron in the sintered material for achieving the required magnetic properties. The general formula of the stoichiometric compounds can be written $M_{1-x}Zn_xFe_2O_4$, where M represents Ni or Mn. Commercial compositions are often non-stoichiometric and must be carefully formulated to produce the desired properties. The nickel–zinc ferrites, with the general formula $Ni_{1-x}Zn_xFe_{2-y}O_4$, are formulated with a deficiency of iron (with y in the range $0 < y < 0.025$) to keep the sensitivity high and the magnetic loss low. The oxidation number of +2 is the most stable for both Ni and Zn. The +3 state for Fe is stable only if there is a slight deficiency of iron (or excess of oxygen). Nickel–zinc ferrites can, therefore, be sintered under a wide variety of oxidizing atmospheres (e.g., air or oxygen).

The manganese–zinc system provides an additional degree of complexity in the sintering of ferrites with the spinel structure. The spinel phase is stable only over a certain range of atmosphere and temperature conditions. These ferrites, with the general formula $Mn_{1-x}Zn_xFe_{2+y}O_4$, are formulated with an excess of iron ($0.05 < y < 0.20$). The concentration of Fe in the +2 oxidation state (i.e., "ferrous iron") is critical to the achievement of the desired properties of low magnetic loss and a maximum in the magnetic permeability. To produce the desired concentration of ferrous iron ($Fe^{2+}/Fe^{3+} \approx 0.05/2.00$), a two-stage sintering schedule is employed (Figure 6.41). The sample is first sintered at a temperature in the range of 1250°C–1400°C in an atmosphere of high oxygen partial pressure (0.3–1 atm) to minimize the evaporation of zinc. In this stage, the densification and grain growth are essentially completed. In the second stage, the sample is annealed at a lower temperature (1050°C–1200°C) in an atmosphere with low oxygen partial pressure (50–100 ppm) to establish the desired concentration of ferrous iron. Figure 6.42 shows the compositional path followed in the sintering schedule. During the first stage of the sintering schedule, the ferrite has

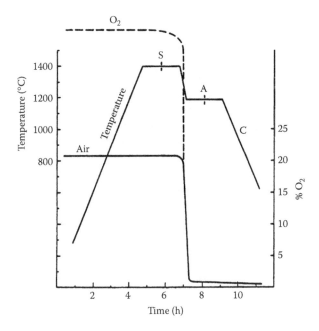

FIGURE 6.41 Schematic sintering cycle for manganese zinc ferrites.

a composition corresponding to S. In the second stage, equilibrium is quickly reached because of the high annealing temperature. The ferrite attains a different composition (corresponding to A) that depends on the temperature and atmosphere of annealing. Further cooling along C represents additional compositional changes if equilibrium is reached. In practice, with the cooling rates used and the decreasing temperature, the composition does not change significantly on cooling.

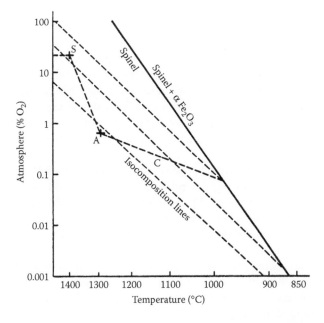

FIGURE 6.42 Schematic representation of the stoichiometry path followed during the sintering of manganese zinc ferrites.

6.3.4.6   Defect Chemistry and Stoichiometry

The sintering behavior of the highly stoichiometric oxides (e.g., Al_2O_3 and ZrO_2) shows little dependence on the oxygen partial pressure over a wide range of values. For other oxides, particularly the transition metal oxides, the oxygen partial pressure can have a significant influence on the concentration and type of lattice defects and, hence, the stoichiometry of the compound. The sintering of these oxides would therefore be expected to depend significantly on the oxygen partial pressure in the atmosphere. Oxygen ions have a significantly larger ionic radius than most cations and are often less mobile, so the diffusion of oxygen ions may often control the sintering rate. Thus, an atmosphere that produces an increase in the oxygen vacancy concentration may often enhance sintering. In practice, it is difficult to separate the effects of defect chemistry and stoichiometry from those of the nondensifying mechanisms (e.g., vapor transport and surface diffusion) brought about by the change in the atmosphere.

Reijnen [82] found that the sintering of $NiFe_2O_4$ and $NiAl_2O_4$ is severely retarded for powder compositions with an excess of Fe or Al and explained the behavior in terms of a reduction in the oxygen vacancy concentration. The densification and grain growth behavior of Fe_3O_4 as a function of oxygen partial pressure was studied by Yan [83] at three different temperatures. The type and concentration of the lattice defects are expected to vary over the wide range of oxygen partial pressure used in the experiments, but the sintered density is found to be insensitive to the atmosphere (Figure 6.43a). On the other hand, the oxygen partial pressure has a significant effect on the grain growth (Figure 6.43b). High oxygen partial pressure leads to small grain size, whereas the grain size is much larger at low oxygen partial pressure. Yan interpreted the data in terms of a mechanism in which oxygen lattice diffusion controlled the pore mobility. Since oxygen lattice diffusion is faster at lower oxygen partial pressure, an increase in the pore mobility leads to faster grain growth if the boundary mobility is controlled by the pore motion.

6.3.5   Production of Controlled Sintering Atmospheres

Atmospheres with a variety of compositions and oxygen partial pressures may be required for sintering, and it may also be necessary to control the rate of gas flow through the furnace. In most cases the required gas composition is available commercially (e.g., O_2, air, N_2, Ar, He, and H_2). Normally, Ar or He is used to provide an inert atmosphere, but the commercial gas contains a small amount of oxygen (10–100 ppm), so it may need to be purified by passing through an oxygen getter prior to entering the furnace. Nitrogen is a common atmosphere for the sintering of nitrides (e.g., Si_3N_4), but some care must be exercised in its use at very high temperatures. It reacts with refractory metal windings and heat shields to form nitrides, shortening the life of the furnace, and it also reacts with graphite to produce poisonous gases. Hydrogen is fairly explosive and must be handled carefully. However, inert gases such as He and Ar containing less than 5% H_2 are not explosive and can be handled safely.

Atmospheres with high oxygen partial pressure can be achieved with O_2 or air, whereas mixtures of O_2 and an inert gas (He, Ar, or N_2) can provide atmospheres with an oxygen partial pressure down to $\sim 10^{-2}$ atm. Lower oxygen partial pressures (below $\sim 10^{-4}$ atm) are commonly obtained by flowing mixtures of gases such as H_2–water vapor (H_2O) or CO–CO_2. Care must be exercised in the handling of both gas mixtures because H_2 is explosive and CO is poisonous, so it must be vented properly. Mixing of CO–CO_2 gas mixtures is performed prior to admission of the gas into the furnace (e.g., by passing the gases in the correct proportions through a column containing glass beads), and the total gas flow through the furnace must be adjusted to a fixed rate. For control of the oxygen partial pressure with H_2–H_2O mixtures, H_2 is bubbled through a gas washer containing distilled water at a fixed temperature.

The equilibrium oxygen partial pressure of a mixture at any given temperature is calculated by using standard thermodynamic equations and appropriate thermochemical data [84]. For example,

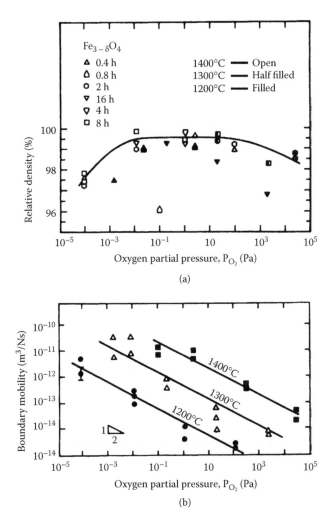

FIGURE 6.43 Data for (a) the relative density and (b) the boundary mobility as a function of the oxygen partial pressure in the atmosphere during the sintering of Fe_3O_4 powder compacts under the conditions of time and temperature indicated.

if an oxygen partial pressure of $10^{-11.7}$ atm at 1600°C is required (as indicated in Figure 6.40), this can be obtained by establishing the equilibrium

$$CO + \frac{1}{2}O_2 \rightarrow CO_2 \tag{6.35}$$

The reaction described in Equation 6.35 arises from the following two reactions:

$$CO(g) \rightarrow \frac{1}{2}O_2(g) + C(s), \quad \Delta G^\circ = 111,700 + 87.65T \quad J \tag{6.36}$$

$$C(s) + O_2(g) \rightarrow CO_2(g), \quad \Delta G^\circ = -394,100 - 0.84T \quad J \tag{6.37}$$

where G^o is the standard free energy change for the reaction. For the reaction described by Equation 6.35, we obtain

$$\Delta G^o = -282,400 + 86.81T \quad J \tag{6.38}$$

Putting $\Delta G^o = -RT \ln K_p$, where K_p is the equilibrium constant for the reaction, and $T = 1873$ K, gives $\ln K_p = 7.70$. The equilibrium constant is also given by

$$K_p = \frac{p_{CO_2}}{p_{CO}(p_{O_2})^{1/2}} \tag{6.39}$$

Substituting $p_{O_2} = 10^{-11.7}$ atm and $\ln K_p = 7.70$ into Equation 6.39 gives

$$\frac{p_{CO_2}}{p_{CO}} = 10^{-2.5} \tag{6.40}$$

The value in Equation 6.40 is very close to that indicated in Figure 6.40. For the same p_{O_2} of $10^{-11.7}$ atm at 1600°C, using the reaction

$$H_2 + \frac{1}{2}O_2 \rightarrow H_2O \tag{6.41}$$

and

$$\Delta G^o = -239,500 + 8.14T \ln T - 9.25T \quad J \tag{6.42}$$

it can be shown that

$$\frac{p_{H_2O}}{p_{H_2}} = 10^{-1.9} \tag{6.43}$$

Thus, if in the H_2–H_2O mixture, $p_{H_2} = 1$ atm, then p_{H_2O} must equal 0.013 atm, which is the saturated vapor pressure of water at 11.1°C. The required gas mixture can be produced by bubbling H_2 gas through pure water at 11.1°C, saturating the hydrogen with water vapor. When the reaction equilibrium is established at 1600°C, the oxygen partial pressure in the furnace will be $10^{-11.7}$ atm.

6.4 MICROWAVE SINTERING

Since about 1970, we have seen a growing interest in the use of microwaves for heating and sintering ceramics [86]. Microwave heating is fundamentally different from conventional heating, in which electrical resistance furnaces are typically used (Figure 6.44). In microwave heating, heat is generated internally by interaction of microwaves with the atoms, ions, and molecules of the material. Heating rates in excess of 1000°C/min can be achieved, and significantly enhanced densification rates have been reported. However, the control of microwave sintering can be difficult. The shape of the ceramic body affects the local heating rates, and the microwave frequency can play a significant role in the temperature gradients that develop in the body, so achieving sufficiently uniform heating is not straightforward.

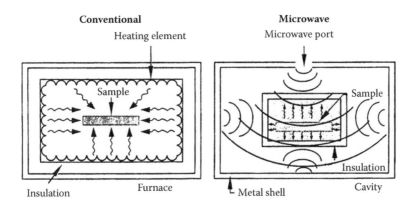

FIGURE 6.44 Heating patterns in conventional and microwave furnaces.

6.4.1 INTERACTION OF MICROWAVES WITH MATTER

Microwaves are electromagnetic waves with a frequency in the range of 0.3–300 GHz or, equivalently, with a wavelength in the range of 1 mm to 1 m, which is a wavelength range of the same order as the linear dimensions of most practical ceramics. In common with other electromagnetic waves, microwaves have electrical and magnetic field components, an amplitude and phase angle, and the ability to propagate (i.e., to transfer energy from one point to another). These properties govern the interaction of microwaves with materials and produce heating in some materials.

As sketched in Figure 6.45, depending on the electrical and magnetic properties of the material, microwaves can be transmitted, absorbed, or reflected. Microwaves penetrate metals only in a thin skin (on the order of 1 μm), so metals can be considered to be opaque to microwaves or to be good reflectors of microwaves. Most electrically insulating (or dielectric) ceramics, such as Al_2O_3, MgO, SiO_2, and glasses are transparent to microwaves at room temperature, but, when heated above a certain critical temperature T_c, they begin to absorb and couple more effectively

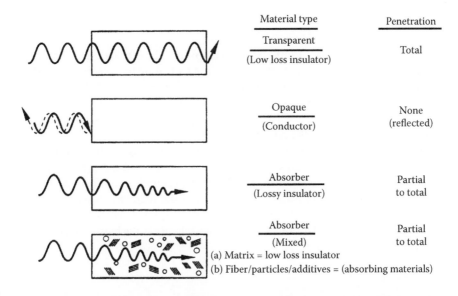

FIGURE 6.45 Schematic diagram illustrating the interaction of microwaves with materials.

with microwave radiation. Other ceramics, such as Fe_2O_3, Cr_2O_3, and SiC, absorb microwave radiation more efficiently at room temperature. The addition of a microwave-absorbing second phase to ceramics that are microwave-transparent can greatly enhance the interaction of the system with microwaves.

The degree of interaction between the microwave electric and magnetic field components with the dielectric or magnetic material determines the rate at which energy is dissipated in the material by various mechanisms. The properties of the material that are most important for the interaction are the permittivity ε for a dielectric material and the permeability μ for a magnetic material. Commonly, the relative permittivity ε_r (also called the dielectric constant), equal to $\varepsilon/\varepsilon_o$, and the relative permeability μ_r, equal to μ/μ_o, are used, where ε_o is the permittivity and μ_o the permeability of vacuum. In alternating fields, the behavior is best described with the help of complex quantities, which, in the case of the permittivity, is defined by

$$\varepsilon_r^* = \varepsilon_r' - j\varepsilon_r'' \qquad (6.44)$$

where ε_r^* is the complex relative permittivity, the prime and double prime represent the real and imaginary parts of the complex relative permittivity, and $j = \sqrt{-1}$. When microwaves penetrate the material, the electromagnetic field induces motion in the free and bound charges (e.g., electrons and ions) and in dipoles. The induced motion is resisted because it causes a departure from the natural equilibrium of the system, and this resistance, due to frictional, elastic, and inertial forces, leads to the dissipation of energy. As a result, the electric field associated with the microwave radiation is attenuated, and heating of the material occurs. The loss tangent, tan δ, defined by

$$\tan \delta = \varepsilon_r''/\varepsilon_r' \qquad (6.45)$$

is used to represent the losses arising from all mechanisms. The average power dissipated per unit volume of the material is given by

$$W = \pi E_o^2 f \varepsilon_o \varepsilon_r'' = \pi E_o^2 f \varepsilon_o \varepsilon_r' \tan \delta \qquad (6.46)$$

where E_o is the amplitude of the electric field, given by $E = E_o \exp(j\omega t)$, and f is the frequency, equal to $\omega/2\pi$. According to Equation 6.46, the power absorbed by the material depends on (1) the frequency and the square of the amplitude of the electric field and (2) the dielectric constant and the loss tangent of the material. In practice, these quantities are interdependent, so it is difficult to alter one without affecting the others. Nevertheless, Equation 6.46 shows, in a general way, the important parameters that control the power absorbed.

Since the electric field is attenuated as the microwaves propagate through the material, the depth of penetration of the microwaves into the material is an important parameter. If any linear dimension of the material is greater than the depth of penetration, uniform heating cannot be expected to occur. A few different parameters are used in the literature as measures of the depth of penetration. The *skin depth* D_s, commonly used for metals, gives the distance into the material at which the electric field falls to $1/e$ of its original value, where e is the base of the natural logarithm (equal to 2.718). It is given by

$$D_s = \frac{c}{\sqrt{2}\pi f \left(\varepsilon_r'\right)^{1/2} \tan \delta} \qquad (6.47)$$

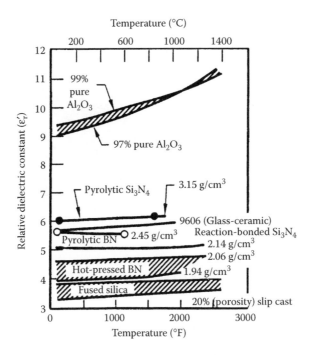

FIGURE 6.46 Dielectric constant (at a frequency of 8–10 GHz) versus temperature for some ceramics. (From Sutton, W.H., Microwave processing of ceramic materials, *Am. Ceram. Soc. Bulletin*, 68, 376, 1989. With permission.)

where c is the speed of light. For metals, D_s is on the order of 1 μm at microwave frequencies, whereas for insulating materials such as Al_2O_3, D_s is on the order of 1 m. For a given material, Equation 6.47 shows that the skin depth decreases with increasing frequency of the microwave radiation. Another version of the skin depth, often referred to as the *penetration depth*, gives the depth D_p at which the power is reduced to half of its value at the surface of the material. It is given by

$$D_p \approx \frac{c}{10 f (\varepsilon_r')^{1/2} \tan \delta} \tag{6.48}$$

Equation 6.46, Equation 6.47, and Equation 6.48 show that the most important parameters of a material that govern its interaction with microwave radiation are the dielectric constant ε_r' and the loss tangent, $\tan \delta$. During microwave heating, both ε_r' and $\tan \delta$ change with temperature, and knowledge of these changes is important for controlling microwave sintering. Figure 6.46 and Figure 6.47 show the variation in ε_r' and $\tan \delta$ for several ceramics during microwave heating at 8–10 GHz. Except for Al_2O_3, the ε_r' values do not change significantly, whereas the $\tan \delta$ values are far more significantly affected. For materials such as Al_2O_3, BN, SiO_2, and glass ceramics, $\tan \delta$ initially increases slowly with temperature until some critical temperature T_c is reached. Above T_c, the materials couple more effectively with the microwave radiation, and $\tan \delta$ increases more rapidly with temperature.

To overcome the problem of inefficient microwave heating at lower temperatures, the materials can be heated conventionally to T_c, when the microwaves couple more effectively with the sample to produce rapid heating and high sintering temperatures. Another approach involves partially coating the internal surfaces of a refractory cavity with a thin paste of SiC. When subjected to a microwave field, the microwaves penetrate the walls of the refractory cavity and interact with the highly absorbing SiC. The heat generated by the SiC heats the sample by radiation to T_c, at which point the microwaves couple more effectively with the sample.

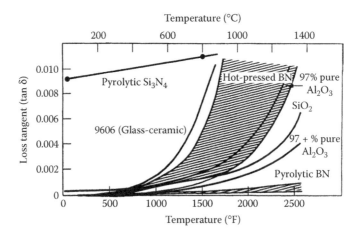

FIGURE 6.47 Dielectric loss tangent (at a frequency of 8–10 GHz) versus temperature for some ceramics. (From Sutton, W.H., Microwave processing of ceramic materials, *Am. Ceram. Soc. Bulletin*, 68, 376, 1989. With permission.)

6.4.2 MICROWAVE SINTERING TECHNIQUES

Microwave sintering of ceramics is quite straightforward. The ceramic body, usually contained in nonabsorbing or weakly absorbing insulation such as loose, nonconducting powder, is placed within a microwave cavity. It is possible to use a simple consumer microwave oven to achieve densification if the ceramic body is properly insulated [86], but achieving sufficiently uniform heating is difficult. Figure 6.48 shows the main components of laboratory-scale equipment for microwave sintering of ceramics [87]. The microwave generator is a key element in the system. Most generators are operated at a frequency of ~2.45 GHz, but a combination of frequencies ranging between ~2.45 GHz and 85 GHz has been proposed as providing more uniform heating [88]. Continuous microwave sintering has been reported, in which samples are passed through the microwave cavity [89].

During microwave heating, the absorption of electromagnetic energy raises the temperature of the entire sample, but heat loss from the surface causes the exterior to become cooler than the interior. Thus, in ceramics with poor thermal conductivity, large temperature gradients can develop

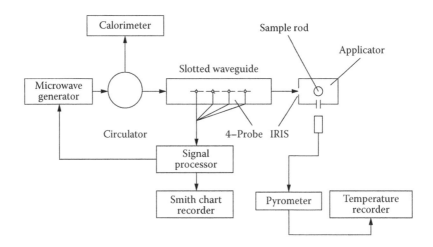

FIGURE 6.48 Schematic diagram of a microwave sintering system.

in the body. The shape of the ceramic body and the microwave frequency strongly influence the temperature gradients, and achieving sufficiently uniform heating of the body can be difficult. These thermal gradients, which are not as severe in conventional sintering, lead to undesirable density gradients during microwave sintering. At high heating rates commonly available with microwave sintering, these gradients lead to nonuniform properties and even cracking of the sample. Increased localized heating can also lead to thermal runaway, when the temperature of the sample increases rapidly, resulting in melting of the sample.

The energy absorbed during microwave sintering is proportional to the volume of the body, but the heat loss is proportional to the surface area, so the temperature gradient should depend on the surface-to-volume ratio. Ceramic components often have complex shapes, in which case the different regions, having different surface-to-volume ratios, would be subjected to different internal temperature gradients [90]. One approach for alleviating the temperature gradients involves insulating the sample with a nonabsorbing or weakly absorbing powder.

The microwave frequency also plays a significant role in the temperature gradients that can develop within the ceramic body [88]. At lower temperatures, when tan δ is small, and at lower microwave frequencies (2.45 GHz), the skin depth D_s is commonly larger than the sample dimensions (Equation 6.47), so a quasi-homogeneous energy absorption is expected inside the sample. At higher temperatures and frequencies, D_s can become smaller than the sample dimensions, so heating is concentrated at the surface. Thus, another approach to reducing the thermal gradients is to apply simultaneously two microwave sources operating at widely separated frequencies (e.g., 2.45 GHz and 30 GHz). By independently adjusting the power of each source, it is possible to control the rate and spatial distribution of the energy absorption within the sample.

Accurate measurement of the temperature in microwave sintering presents more difficulties than in conventional sintering. Because of interference from the microwave field, thermocouples cannot be relied upon to function properly [91]. The presence of thermocouples during microwave heating of low- and medium-loss ceramics (tan δ < 0.1) can locally distort the electromagnetic field and can lead to enhanced energy absorption, enhanced heat loss by conduction, and even thermal runaway. To avoid these difficulties as well as serious errors in the temperature measurement, a noncontact sensing system such as an optical pyrometer should be used whenever it is possible.

6.4.3 MICROWAVE SINTERING OF CERAMICS

Microwave heating has been used in the sintering of a variety of ceramics, and interest in the technique is growing. The most detailed studies of microwave sintering have been carried out with Al_2O_3, and the results allow a meaningful comparison of the sintering characteristics in microwave heating with those obtained for similar samples in conventional heating. The data show a significant enhancement of the sintering and grain growth rates when microwave heating is used, but the mechanisms responsible for this enhancement are unclear.

Figure 6.49 shows the data of Janney and Kimrey [92] for the densification of MgO-doped Al_2O_3 powder compacts during sintering by microwave heating (28 GHz) and by conventional heating. Similar green compacts (relative density of approximately 0.55) were used in the experiments, and they were heated at the same rate (50°C/min) to the required temperature, followed by isothermal sintering for 1 h. The conventional sintering data are in good agreement with those obtained by other researchers in similar experiments with the same powder. In contrast, the microwave-sintered compacts show significantly enhanced sintering rates. Shrinkage starts below ~900°C, and relative densities greater than 90% are achieved at temperatures as low as 950°C. As outlined earlier, accurate temperature measurement and control are commonly more difficult in microwave sintering. However, it appears unlikely that such a large reduction in the temperature (300°C–400°C) required to achieve a given density could be attributable to errors in temperature measurement. Activation energy data determined from the densification rates at different isothermal

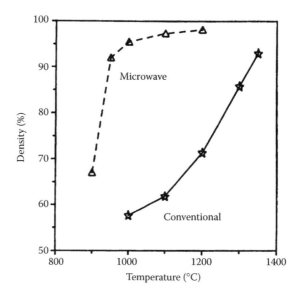

FIGURE 6.49 Relative density versus temperature for MgO-doped Al$_2$O$_3$ powder compacts during sintering by microwave heating and by conventional heating. (From Janney, M.A., and Kimrey, H.D., Diffusion-controlled processes in microwave-fired oxide ceramics, *Mater. Res. Soc. Symp. Proc.,* 189, 215, 1991. With permission.)

temperatures (Figure 6.50) show that the value for microwave sintering (170 kJ/mol) is considerably lower than that in conventional sintering (575 kJ/mol).

In a subsequent set of experiments [93], the grain growth kinetics of fully dense MgO-doped Al$_2$O$_3$ samples, fabricated by hot pressing, were measured during microwave heating (28 GHz) and conventional heating. The general evolution of the microstructure is roughly similar with both types

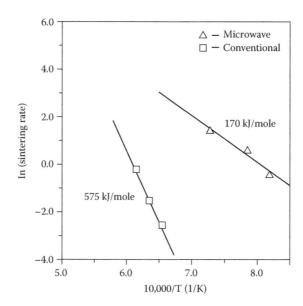

FIGURE 6.50 Arrhenius plot of the densification rate data (at a relative density of 0.80) for MgO-doped Al$_2$O$_3$ for sintering by microwave heating and by conventional heating. (From Janney, M.A., and Kimrey, H.D., Diffusion-controlled processes in microwave-fired oxide ceramics, *Mater. Res. Soc. Symp. Proc.,* 189, 215, 1991. With permission.)

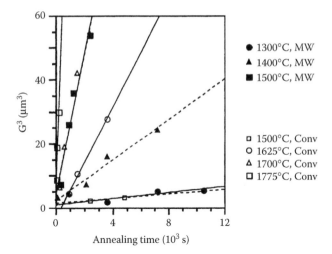

FIGURE 6.51 Grain growth kinetics of fully dense, hot pressed MgO-doped Al_2O_3 during annealing by microwave heating (*MW*) and by conventional heating (*Conv*). The data follow a cubic growth law in both cases, but grain growth is faster in microwave heating.

of heating, and the normal grain growth kinetics can be well described by a cubic law. However, as the data in Figure 6.51 show, the grain growth kinetics are considerably enhanced in microwave heating. The activation energy for grain growth in microwave heating (480 kJ/mol) is found to be ~20% lower than that in conventional heating (590 kJ/mol). These activation energies are expected to be appropriate to grain growth in Al_2O_3 controlled by fine second-phase $MgAl_2O_4$ particles in the grain boundary, because the MgO concentration used in the experiments (0.1 wt%) is significantly higher than the solid solubility limit for MgO in Al_2O_3 (~0.025 wt%).

Measurements of the rate of oxygen diffusion in sapphire (single-crystal Al_2O_3) with an ^{18}O tracer indicate that the diffusion coefficient for microwave heating is approximately twice the value for conventional heating [93]. Furthermore, the activation energy for diffusion in microwave heating (410 kJ/mol) is considerably lower than that for conventional heating (710 kJ/mol). This enhanced oxygen diffusion in sapphire indicates that neither free surfaces nor grain boundaries, as would be present in polycrystalline materials, are essential for the enhancement of diffusion.

6.4.4 Plasma Sintering

A variation of microwave sintering is sintering in a plasma generated in a microwave cavity or by a radio frequency induction-coupled method [94–98]. High heating rates (on the order of 100°C/min) and rapid sintering (linear shrinkage rates of up to 3% per second) can be achieved without specimen damage for small-diameter rods and tubes. Henrichsen et al. [98] achieved full densification for tubular specimens by translating the tube at a rate of 25 mm/min through an egg-shaped plasma with a height of ~50 mm. A schematic of the method is shown in Figure 6.52.

Significantly enhanced densification has been reported in plasma sintering for a variety of ceramics, but the mechanisms are not clear. In microwave-induced plasma sintering of Al_2O_3, Su and Johnson [97] found significantly enhanced densification rates when compared to conventional sintering under the same time–temperature schedule. These results indicate that, besides the rapid heating rate achievable in a plasma, an *athermal* effect, caused by the plasma itself, also contributes to the enhanced sintering. Su and Johnson attributed this athermal effect in Al_2O_3 to an increase in the concentration of Al interstitials. Diffusion of Al interstitials along grain boundaries is believed to be the dominant sintering mechanism, so the higher Al interstitial concentration contributes to the enhanced sintering in the plasma.

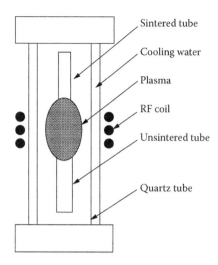

FIGURE 6.52 Schematic of plasma sintering apparatus.

6.4.5 PLASMA-ASSISTED SINTERING

Several attempts have been made at increasing the heating rates, giving rise to terms such as *superfast* or *ultrafast* sintering. One method, referred to as *spark plasma sintering*, involves passing a direct current (dc) pulse through a powder compact contained in a graphite die, under an applied pressure of 30–50 MPa (Figure 6.53). Specimen temperatures are difficult to assess in this method and are usually measured using an optical pyrometer focused on the graphite die wall. Both the die and the specimen are heated by the current pulse. Heating rates in excess of 600°C/min have been reported. The high heating rates are thought to be caused, in part, by spark discharges generated in the voids between the particles. Remarkably high densification rates may be achieved under such conditions, with minimal grain growth. This approach is particularly useful in producing dense ceramic bodies from nanoscale powders [99,100]. The specimen shapes that can be prepared by spark plasma sintering are limited to simple shapes, such as discs, that can be contained in the compression die.

A related method is a thermal explosive forming, in which a reactive mixture of components, for example, Ti and C, is heated up in a die and then ignited by passing an electrical current pulse. The process may perhaps be regarded as a self-propagating high-temperature synthesis under pressure. Formation of dense boride, nitride, and carbide composites has been reported [101].

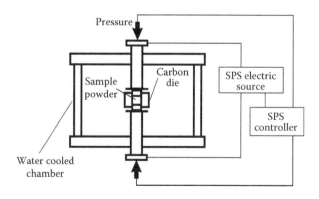

FIGURE 6.53 Schematic of spark plasma sintering (SPS) apparatus.

6.5 PRESSURE-ASSISTED SINTERING

Sintering with an externally applied pressure is referred to as *pressure-assisted sintering*, or simply as *pressure sintering*. There are three principal ways in which the pressure is applied to the sample, giving rise to the following methods:

Hot pressing, where the pressure is applied uniaxially to the powder in a die
Sinter forging, which is similar to hot pressing but without confining the sample in a die
Hot isostatic pressing, where the pressure is applied isostatically by means of a gas

Hot pressing is the most widely used of the three methods, whereas sinter forging sees little use in ceramics fabrication.

A key advantage of pressure-assisted sintering is the ability to enhance significantly the densification rate relative to the coarsening rate (see Chapter 2), guaranteeing, in most cases, the attainment of high density and fine grain size. The method is particularly effective for the highly covalent-bonded ceramics, such as SiC and Si_3N_4, and for ceramic matrix composites, which are commonly difficult to densify by conventional sintering. Since most properties normally improve with high density and fine grain size, superior properties are achieved. The enhancement of the densification rate also means that a lower sintering temperature or a shorter sintering time can be used in hot pressing than in conventional sintering. The significantly lower sintering temperature can be important for systems that contain volatile components or that suffer from decomposition at higher temperatures. In research and development studies, pressure-assisted sintering is well suited to the production of prototype materials for the investigation of microstructure–property relationships. Pressure-assisted sintering also provides an additional variable that is very useful for the study of the sintering mechanisms. It will be recalled that the kinetic data in conventional sintering are often difficult to interpret because of the simultaneous occurrence of multiple mechanisms. Pressure-assisted sintering significantly enhances the densification mechanisms, so the effect of the nondensifying mechanisms is reduced.

A disadvantage of pressure-assisted sintering is the high cost of production, so in industrial applications, the method is used only for the production of specialized, high-cost ceramics where high density must be guaranteed. Pressure-assisted sintering cannot easily be automated and, in the cases of hot pressing and sinter forging, the method is commonly limited to the fabrication of simple shapes. Although complex shapes can be produced by hot isostatic pressing, shape distortion can be a problem. Pressure-assisted sintering also has a size limitation. Large articles (e.g., > ~1 m) have been produced industrially, but the equipment becomes highly specialized and expensive.

6.5.1 HOT PRESSING

Figure 6.54 shows a schematic diagram of the main features of a laboratory-scale hot press [102]. Heat and pressure are applied to a sample, in the form of a powder or a green compact, in a die. Depending on the furnace, operating temperatures up to 2000°C or higher can be used, and typical operating pressures range from ~10 to ~75 MPa. Table 6.2 provides some information on the die materials that have been used in hot pressing. Graphite is the most common die material because it is inexpensive, is easily machined, and has excellent creep resistance at high temperatures. For pressures below ~40 MPa, standard graphite can be used, whereas for higher pressures, specialty graphite and more expensive refractory metal and ceramic dies can be used. Graphite oxidizes slowly below ~1200°C, so it can be exposed to an oxidizing atmosphere for short periods only. Above ~1200°C, it must be used in an inert or reducing atmosphere. A common problem is the reactivity of graphite toward other ceramics, which leads to a deterioration of the contact surfaces of the die or to sticking of the sample to the die wall. This problem can be alleviated by coating

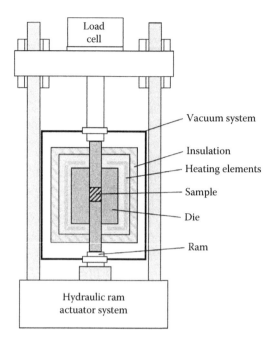

FIGURE 6.54 Schematic of the hot pressing process.

the contact surfaces of the die with boron nitride if the temperature remains below ~1350°C. Above this temperature, the boron nitride can react with graphite. Lining the graphite die with graphite foil is also useful for prolonging the life of the die.

The standard laboratory process leads to the production of simple shapes, such as discs. A variation allowing for some shape flexibility was developed by Lange and Terwilliger [103]. In this method, the green body is packed in coarse powder in the hot pressing die, developing a roughly isostatic pressure on the sample. More complex hot pressing systems have also been developed, such as a multibore machine for hot pressing nuclear reactor pellets [104].

TABLE 6.2
Die Materials for Hot Pressing

Material	Maximum Temperature (°C)	Maximum Pressure (MPa)	Comments
Nimonic alloys	1100	High	Creep and erosion problems
Mo (or Mo alloys)	1100	20	Oxidation unless protected
W	1500	25	Oxidation unless protected
Al_2O_3	1400	200	Expensive, brittle
BeO	1000	100	Expensive, brittle
SiC	1500	275	Expensive, brittle
TaC	1700	55	Expensive, brittle
WC, TiC	1400	70	Expensive, brittle
TiB_2	1200	100	Expensive, brittle
Graphite (standard)	2500	40	Severe oxidation above 1200°C
Graphite (special)	2500	100	Severe oxidation above 1200°C

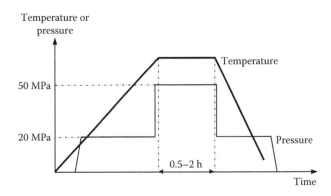

FIGURE 6.55 Schematic pressure–temperature schedule for hot pressing with a high-strength graphite die.

6.5.1.1 Hot Pressing Process Variables

The simplest procedure, often used in laboratory studies when kinetic data are required, is to apply the required pressure to the powder in the die and release the pressure prior to heating. The system is heated rapidly, and, on reaching the sintering temperature, the pressure is reapplied to the powder. After densification is completed, the pressure is released and the system is cooled to room temperature. More commonly in fabrication experiments, a moderate pressure (10–20 MPa) is applied to the system from the start, and, upon reaching the sintering temperature, the pressure is increased to the required value (40–50 MPa). The rate of densification can be deduced by following the ram displacement. Typically, the sintering temperature is chosen to achieve full densification in ~30 min to 2 h. Some guidance for selecting the appropriate hot pressing temperature may be obtained from pressure-sintering maps [105], but trial and error is more reliable. An applied pressure is often maintained during the cool-down period. A schematic of a pressure–temperature treatment is shown in Figure 6.55.

Because of the significant axial strain in hot pressing, texture may develop in the final microstructure. For powders that contain elongated particles, rotation of the particles under the action of the applied stress is an important mechanism for texture development [106]. Texturing may also develop for equiaxial particles, and this can be seen as follows: Since the die walls are fixed, the compaction of the powder during hot pressing occurs only in the direction of the applied pressure. A representative element of the powder (e.g., three grains) will undergo the same relative shape change as the overall compact (Figure 2.28). Thus, the grains become flatter in the direction of the applied pressure.

Hot pressing is less effective for very fine powders, particularly when they are well compacted prior to hot pressing, because the sintering stress for these powders can be greater than the pressure that the hot pressing die can tolerate. However, benefits may still be derived when the applied pressure enhances the rearrangement process or assists in the collapse of large pores. Insoluble gases trapped in residual pores can lead to swelling of the ceramic at elevated service temperatures, but this problem can be alleviated by hot pressing in a vacuum. Swelling can also occur in oxidizing atmospheres at elevated service temperatures if the ceramic is contaminated with carbonaceous impurities by hot pressing in graphite dies. For some powders, particularly those of highly covalent materials where solid-state diffusion is slow, it may still be difficult to achieve full densification by hot pressing. As in the conventional sintering route, solid solution or liquid-forming additives can be used in hot pressing to aid the densification process.

6.5.1.2 Analysis of Hot Pressing Data

Data for the densification kinetics during hot pressing can be determined from measurements of the change in the height of the sample as a function of time under known conditions of pressure, isothermal sintering temperature, and particle size of the powder. Commonly, the change in height is monitored continuously by measuring the ram displacement using a dial gage or a linear voltage displacement transducer. For a cylindrical die cavity with a constant cross-sectional area A_o, and assuming that the mass of the powder m_o is constant, the density of the sample at time t is given by $d = m_o/A_oL$, where L is the height of the sample. If d_f and L_f are the final density and height, respectively, of the hot pressed disc, then $d_f = m_o/A_oL_f$. Combining these two equations gives $\rho = \rho_f L_f/L$, where ρ is the relative density at time t and ρ_f is the final relative density of the hot pressed sample. Since $L = L_f + \Delta L$, where ΔL is the change in length of the sample, the relative density as a function of time can be determined from the data for ΔL as a function of time coupled with the measured values of L_f and ρ_f. In laboratory experiments, hot pressing data are often acquired at a given isothermal temperature for known values of the applied pressure. The most useful representation of the data is in terms of plots showing the relative density ρ versus log t, and the densification rate, defined as $(1/\rho)d\rho/dt$, versus ρ, for the applied pressures used [107].

As discussed in Chapter 2, the hot pressing models predict that the densification rate can be written in the general form

$$\frac{1}{\rho}\frac{d\rho}{dt} = \frac{AD\phi^{(m+n)/2}}{G^m kT}(p_a^n + \Sigma) \qquad (6.49)$$

where A is a constant, D is the diffusion coefficient for the rate-controlling species, G is the grain size, k is the Boltzmann constant, T is the absolute temperature, ϕ is the stress intensification factor (which is a function of density only), p_a is the applied stress, Σ is the sintering stress, and the exponents m and n depend on the mechanism of sintering (Table 2.7). According to Equation 6.49, a plot of the data for log (densification rate) versus log (applied pressure) gives the exponent n that can be used to determine the mechanism of densification. For the commonly used pressure range in hot pressing (20–50 MPa), data for most ceramics give $n \approx 1$, indicating that the densification process is diffusion controlled. This finding is not surprising in view of the impediments to dislocation motion caused by the strong bonding in most ceramics, along with the fine grain size, which favors diffusional mechanisms. Using starting powders with different particle sizes, data for the densification rate (at a given density and applied pressure) versus the grain size can be used to determine the exponent m, thus providing additional information on the densification mechanism. For diffusional mechanisms, because of the larger grain size exponent ($m = 3$) and the generally lower activation energy for grain boundary diffusion, finer powders and lower hot pressing temperatures favor grain boundary diffusion over lattice diffusion ($m = 2$).

6.5.1.3 Reactive Hot Pressing

Reactive hot pressing of a mixture of powders has been used successfully to produce composites [108,109]. Densification is more readily achieved in this case when compared to the reaction-sintering process described in Chapter 5, and this is due to the enhanced driving force for densification as well as the ability of the applied pressure to break down unfavorable microstructural features produced by the reaction, such as particle–particle bridges that constrain sintering.

A related process is hydrothermal hot pressing, in which powders are compacted under hydrothermal conditions. The powder is essentially mixed with a small amount of water and compacted in an autoclave at temperatures above the boiling point of water, typically between 100°C and 350°C. The design of the system should allow for fluid to leave the sample [110], and any residual water present in the sample after hydrothermal hot pressing is removed by drying. Ceramics such as hydroxyapatite [111] and amorphous titania [112] have been compacted in this way.

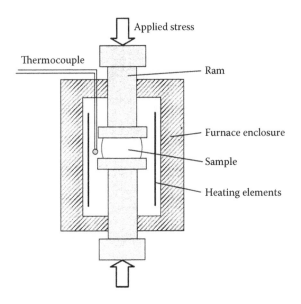

FIGURE 6.56 Schematic of the sinter forging process.

6.5.2 SINTER FORGING

Sinter forging, also referred to as hot forging, is similar to hot pressing but without the sample confined in a die (Figure 6.56). Typically, a green compact or a partially sintered compact is heated while a uniaxial stress is applied [113,114]. The applied uniaxial stress p_z gives rise to a hydrostatic component $p_z/3$ that provides an additional driving force for sintering, so Equation 6.49 can also be used to predict the densification rate in sinter forging if p_a is replaced by $p_z/3$. There is also a shear component of the applied uniaxial stress that causes creep of the sample. The high shear strains in sinter forging can serve to reduce processing flaws in the body by rearrangement of the grains during the initial and intermediate stage of sintering, producing a more homogeneous microstructure [115].

Uniaxial strains in the direction of the applied stress are significantly larger than in hot pressing, and these large strains have been used effectively to produce texturing or grain alignment in some ferroelectric ceramics to develop anisotropic dielectric properties [116–118]. For fine powders, use can be made of possible superplastic deformation modes in sinter forging to enhance texturing [119]. During the process, the strain rate has to be limited to avoid damaging the sample.

6.5.3 HOT ISOSTATIC PRESSING

In hot isostatic pressing [120,121], sometimes abbreviated HIP, the preconsolidated powder is first tightly enclosed in a glass or metal container (sometimes referred to as a can) that is sealed under a vacuum and then placed in a pressure vessel (Figure 6.57). Alternatively, the sample can be predensified to the closed porosity stage by conventional sintering, in which case the can is not needed in the subsequent HIP stage. The pressurizing gas is introduced by means of a compressor to achieve a given initial gas pressure (a few thousand psi), and the sample is heated to the sintering temperature. During the heat-up, the gas pressure increases further to the required value, and the can collapses around the sample, acting to transmit the isostatic pressure to the sample.

A schematic diagram of a pressure vessel containing a sample for HIP is shown in Figure 6.58. The pressure vessel is typically of the cold-wall design, in which the furnace is thermally insulated

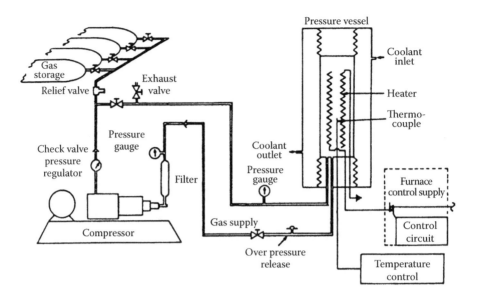

FIGURE 6.57 Simplified schematic diagram of a system used for hot isostatic pressing.

from the wall of the vessel and flowing water provides external cooling. It is penetrated, normally through the end closures, to supply power and control instrumentation to the furnace. Electrical resistance furnaces (e.g., graphite or Mo) and inert pressurizing gases (Ar or He) are commonly used. Depending on the furnace, HIP systems designed for use with inert gases can routinely be operated at temperatures up to ~2000°C and pressures up to ~200 MPa (30,000 psi). Systems for use with reactive gases (e.g., O_2 or N_2) are also available, but the temperature and pressure capability are significantly lower. The quality of the ceramics fabricated by HIP is perhaps the

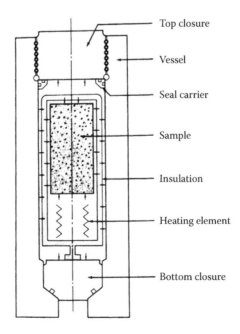

FIGURE 6.58 Schematic diagram of a pressure vessel with a sample for hot isostatic pressing.

highest obtainable by any pressure-sintering method, because externally heated dies cannot withstand the high pressures that can be applied in HIP.

6.5.3.1 Hot Isostatic Pressing: Process Variables

The densification of metal powders by HIP may sometimes involve filling the loose powder into a deformable metal container, but this method is unsuitable for most ceramic powders. Because of their fine size and low poured (or tap) density, considerable deformation and distortion of the material occurs during HIP. Most ceramic powders are first consolidated to form a shaped green article and, if required, lightly presintered to develop some strength for handling, prior to HIP. The shaped green article can be encapsulated in a glass or metal can, or it may be sintered to closed porosity, prior to HIP. Compacts with open porosity cannot be densified without the encapsulation, because the high pressure of the pressurizing gas in the pores resists the sintering stress. The materials commonly used for encapsulation consist of a thin-walled metal can (e.g., Mo or Ta) or a glass can. Following densification, the can is removed mechanically or chemically (by dissolution). The method by which the powder compact is sintered to closed porosity and then subjected to HIP is sometimes referred to as *sinter/HIP*. The main objective in sinter/HIP is to remove the residual porosity without significant coarsening, which can often be difficult to accomplish in conventional sintering.

The selection of the encapsulation route or the sinter/HIP route is often governed by the quality of the fabricated article and the cost. If significant coarsening occurs during the sintering step of the sinter/HIP route, the microstructure can be considerably coarser than that of a similar article produced by the encapsulation route. Dimensional control of the fabricated article may, however, be better in sinter/HIP because of the small shrinkage during the HIP stage. Reaction of the ceramic with the can often leads to degradation of the sample surface, so some surface machining may be necessary with the encapsulation route. The sinter/HIP route eliminates the cost of encapsulation, but the time required for the overall sintering plus HIP steps is longer than that for HIP of the encapsulated green article.

Because the green body can be shaped during forming or machined to the desired shape, dense bodies with complex shapes can be produced by HIP (Figure 6.59), which provides one of the most important advantages over hot pressing. However, inhomogeneous densification during HIP can lead to undesirable shape changes, such as distortion. Inhomogeneous densification may be caused by inadequate processing of the green article or by inadequate HIP practice, such as nonuniformity of the temperature and pressure over the dimensions of the sample. Temperature gradients are a common source of problems, particularly for large articles. As in conventional sintering, when the temperature of the powder compact is raised, heat diffuses inward from the surface.

FIGURE 6.59 Turbocharger impeller, formed by injection molding (left) and densified to full density by hot isostatic pressing (right).

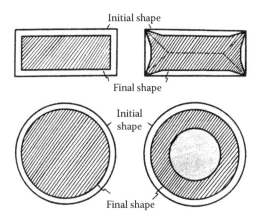

FIGURE 6.60 Consequences of inhomogeneous densification during hot isostatic pressing of an encapsulated cylindrical powder compact. (Top) Shape change and (bottom) density gradients with residual porosity and internal stress.

If heat flow through the compact is slow (as in most ceramic powders), a temperature gradient is set up. In most HIP operations, the sample is under a pressure during heat-up that causes the hotter surface layer to densify faster than the interior, giving rise to a dense skin. Heat conduction through the dense skin is faster than through the less dense interior, further adding to the temperature difference between the surface and the interior. Under certain circumstances, a densification front develops and propagates inward. This leads to large shape changes of the article, so the final geometry is no longer identical to that of the original article.

An analysis of a one-dimensional model for coupled heat flow and densification in a powder compact shows that the tendency toward shape distortion can be characterized by a dimensionless factor that measures the ratio of the densification rate to the heat-transfer rate [122]. When this ratio is less than 1, densification is nearly uniform, but when it is greater than 1, there is a tendency for a densification front to form and propagate inward from the surface. Extension of these ideas from one to three dimensions leads to new complications. As sketched in Figure 6.60 for a cylindrical shape, when densification is uniform, the cylinder (as expected) shrinks uniformly, but when nonuniform densification takes place, a dense shell forms around the cylinder. Pressure can be transmitted to the inner core only if the cylindrical shell creeps inward, a process that can be slow. A larger part of the pressure is carried by the shell, and the densification of the core all but ceases.

The problem of shape distortion caused by temperature gradients can be alleviated by allowing the sample to equilibrate at the sintering temperature before applying the full pressure, but this may be impractical in the simpler HIP equipment, or it may lead to a significant drop in the temperature if a large volume of gas is admitted quickly to the autoclave. The problem may also be reduced by using the sinter/HIP route.

6.6 CONCLUDING REMARKS

In this chapter, we considered the main methods for the densification of ceramic powder compacts by conventional sintering and by pressure-assisted sintering. The particle and green body characteristics have a significant effect on sintering, but even with proper powder preparation and consolidation procedures, successful fabrication still depends on the ability to control the microstructure through manipulation of the sintering process variables. In conventional sintering, the main process variables are the heating cycle (temperature–time schedule) and the sintering atmosphere. Proper manipulation of these two variables can provide considerable benefits for densification. Applied pressure provides an additional variable that is very effective for enhancing densification. The shrinkage during sintering can often be anisotropic, due to a pore shape anisotropy, particle alignment, or density variations in the green body, as well as to improper sintering conditions, such as temperature gradients in the furnace, friction between the sintering body and the substrate, or nonuniform application of the pressure during pressure-assisted sintering. Heating of the sample during sintering is commonly achieved with an electrical resistance furnace, but, in recent years, microwave sintering has attracted some interest. Experiments have shown considerable enhancement of sintering when microwave heating is used. Conventional sintering is the preferred method for densification because it is more economical than pressure-assisted sintering. When it is difficult or impossible to produce high density by conventional sintering, pressure-assisted sintering is used. The quality of the ceramic article produced by pressure sintering is often superior to that obtained in conventional sintering.

PROBLEMS

6.1 Derive the following elementary relations for the isotropic sintering of a powder compact:

(a) $\rho = \rho_o/(1 + \Delta L/L_o)^3$
(b) $\Delta\rho/\rho = -\Delta V/V_o$

where ρ is the relative density, L is the length, and V is the volume of the compact at any time; ρ_o, L_o, and V_o are the initial values; and $\Delta\rho$, ΔL, and ΔV represent the changes in the parameters.

6.2 For the two-dimensional geometrical models shown in Figure 6.13, determine the conditions under which shrinkage in the x-direction will exceed that in the y-direction for the cases when the dihedral angle is equal to (a) 180° and (b) 120°. State any assumptions that you make.

6.3 Consider the constant heating rate sintering of a ceramic powder compact. How would an increase in the heating rate from 1°C/min to 20°C/min be expected to influence the sintering behavior when:

(a) Coarsening is insignificant.
(b) Coarsening is significant.

6.4 Plot the limiting porosity in a powder compact as a function of the grain size (in the range of 0.1 μm to 100 μm) when sintering is carried out in an insoluble gas at atmospheric pressure. Assume that the pore size is one-third the grain size and that the specific surface energy of the solid–vapor interface is 1 J/m².

6.5 When sintered in an oxygen atmosphere at 1700°C for a few hours, an MgO-doped Al_2O_3 powder compact (starting particle size of approximately 0.2 μm) reaches theoretical density with an average grain size of ~10 μm. However, when sintered under similar conditions in air (1 atm), the compact reaches a limiting density of 95% of the theoretical. At the end of the intermediate stage, when the pores become pinched off, the average grain size and average pore size are found to be 2 μm and 0.5 μm, respectively. Estimate the average grain size and average pore size when the sample sintered in air reaches its limiting density, stating any assumptions that you make.

Further sintering of the sample in air leads to grain growth with pore coalescence, with the pores reaching an average size of 4 μm. Will the porosity be different? Explain. If the porosity is different, estimate how different.

6.6 When a CeO_2 powder (starting particle size of approximately 20 nm) is sintered at a constant heating rate of 5°C/min in air, the compact reaches a limiting density of ~95%

of the theoretical value at ~1300°C, after which the density decreases at higher temperatures (Figure 6.4). Discuss the factors that may be responsible for preventing the achievement of full density. What steps may be taken to achieve a density close to the theoretical value for this system?

6.7 In the sintering of Cr_2O_3 powder, the procedure adopted by a student calls for the powder to remain as Cr_2O_3 at all times above 800°C, never changing to Cr or CrO_2. The student decides to employ a flowing mixture of H_2 and water vapor for atmosphere control. Using the data in Figure 6.40 or otherwise, plot the limits of the H_2/H_2O ratio in the sintering atmosphere as a function of the temperature for the experiment.

6.8 Determine the skin depth for Al_2O_3 for microwave sintering at 1250°C when the microwave generator is operated at a frequency of (a) 2.5 GHz and (b) 85 GHz, assuming that the dielectric constant and tan δ values for Al_2O_3 at 1250°C are 10 and 0.005, respectively. Discuss qualitatively the expected temperature distribution in a ceramic body with a linear dimension of 30 cm during microwave sintering at each frequency. How can the temperature distribution in the body be made more uniform?

6.9 A student is given the task of sintering a fine-grained $BaTiO_3$ powder (average particle size of approximately 0.1 μm; purity greater than 99.9%) to a density greater than 98% of the theoretical value and with an average grain size not greater than 1 μm. Discuss how the student should attempt to accomplish the task.

6.10 Estimate the factor by which the driving force for densification is enhanced by hot pressing with an applied pressure of 40 MPa, relative to conventional sintering, when the particle size of the powder is (a) 20 μm and (b) 20 nm. Assume that the pore size is one-third the particle size and the specific surface energy of the solid–vapor interface is 1 J/m².

REFERENCES

1. Rahaman, M.N., *Ceramic Processing*, CRC Press, Taylor and Francis, Boca Raton, FL, 2006.
2. Underwood, E.E., *Quantitative Stereology*, Addison-Wesley, Reading, MA, 1970.
3. Fulrath, R.M., and Pask, J.A., Eds., *Ceramic Microstructures*, Wiley, New York, 1968.
4. Mendelson, M.I., Average grain size in polycrystalline ceramics, *J. Am. Ceram. Soc.*, 52, 443, 1969.
5. Wurst, J.C., and Nelson, J.A., Lineal intercept method for measuring grain size in two-phase polycrystalline ceramics, *J. Am. Ceram. Soc.*, 55, 109, 1972.
6. Zhou, Y.C., and Rahaman, M.N., Hydrothermal synthesis and sintering of ultrafine CeO_2 powders, *J. Mater. Res.*, 8, 1680, 1993.
7. Coble, R.L., Effects of particle size distribution in initial stage sintering, *J. Am. Ceram. Soc.*, 56, 461, 1973.
8. Onoda, G.Y., Jr. and Messing, G.L., Packing and sintering relations for binary powders, *Mater. Sci. Res.*, 11, 99, 1978.
9. Smith, J.P., and Messing, G.L., Sintering of bimodally distributed alumina powders, *J. Am. Ceram. Soc.*, 67, 238, 1984.
10. Yeh, T.S., and Sacks, M.D., Effect of green microstructure on sintering of alumina, *Ceramic Trans.*, 7, 309, 1990.
11. Yamaguchi, T., and Kosha, H., Sintering of acicular Fe_2O_3 particles, *J. Am. Ceram. Soc.*, 64, C84, 1981.
12. Edelson, L.H., and Glaeser, A.M., Role of particle substructure in the sintering of monosized titania, *J. Am. Ceram. Soc.*, 71, 225, 1988.
13. Slamovich, E.B., and Lange, F.F., Densification behavior of single-crystal and polycrystalline spherical particles of zirconia, *J. Am. Ceram. Soc.*, 73, 3368, 1990.
14. Yeh, T.-S., and Sacks, M.D., Effect of green microstructure on sintering of alumina, *Ceramic Trans.*, 7, 309, 1990.
15. Rhodes, W.W., Agglomerate and particle size effects on sintering of yttria-stabilized zirconia, *J. Am. Ceram. Soc.*, 64, 19, 1981.

16. Barringer, E.A., and Bowen, H.K., Formation, packing, and sintering of monodisperse TiO_2 powders, *J. Am. Ceram. Soc.*, 65, C199, 1982.

17. Barringer, E.A., and Bowen, H.K., Effect of particle packing on the sintered microstructure, *Appl. Phys. A*, 45, 271, 1988.

18. Liniger, E., and Raj, R., Packing and sintering of two-dimensional structures made from bimodal particle size distributions, *J. Am. Ceram. Soc.*, 70, 843, 1987.

19. Cesarano, J., III and Aksay, I.A., Processing of highly concentrated aqueous α-alumina suspensions stabilized with polyelectrolytes, *J. Am. Ceram. Soc.*, 71, 1062, 1988.

20. Rahaman, M.N., De Jonghe, L.C., and Chu, M.-Y., Effect of green density on densification and creep during sintering, *J. Am. Ceram. Soc.*, 74, 514, 1991.

21. Bruch, C.A., Sintering kinetics for the high density alumina process, *Am. Ceram. Soc. Bull.*, 41, 799, 1962.

22. Occhionero, M.A., and Halloran, J.W., The influence of green density on sintering, *Mater. Sci. Res.*, 16, 89, 1984.

23. Chen, P.-L., and Chen, I-W., Sintering of fine oxide powders: II, sintering mechanisms, *J. Am. Ceram. Soc.*, 80, 637, 1997.

24. Exner, H.E., and Giess, E.A., Anisotropic shrinkage of cordierite-type glass powder cylindrical compacts, *J. Mater. Res.*, 3, 122, 1988.

25. Rahaman, M.N., and De Jonghe, L.C., Sintering of spherical glass powder under a uniaxial stress, *J. Am. Ceram. Soc.*, 73, 707, 1990.

26. Olevsky, E.A., and Skorokod, V., Deformation aspects of anisotropic porous bodies sintering, *J. de Physique*, 4, 739, 1993.

27. Ch'ng, H.N., and Pan, J., Modeling microstructural evolution of porous polycrystalline materials and a numerical study of anisotropic sintering, *J. Comput. Phys.*, 204, 430, 2005.

28. Shui, A., Uchida, N., and Uematsu, K., Origin of shrinkage anisotropy during sintering for uniaxially pressed alumina compacts, *Powder Technol.*, 127, 9, 2002.

29. Shui, A., Kato, Z., Tanaka, S., Uchida, N., and Uematsu, K., Sintering deformation caused by particle orientation in uniaxially and isostatically pressed alumina compacts, *J. Europ. Ceram. Soc.*, 22, 311, 2002.

30. Krug, S., Evans, J.R.G., and ter Maat, J.H.H., Differential sintering in ceramic injection molding: particle orientation effects, *J. Europ. Ceram. Soc.*, 22, 173, 2002.

31. Raj, P.M., and Cannon, W.R., Anisotropic shrinkage in tapecast alumina: role of processing parameters and particle shape, *J. Am. Ceram. Soc.*, 82, 2619, 1999.

32. Patwardhan, J.S., and Cannon, W.R., Factors influencing anisotropic sintering in tape-cast alumina: effect of processing variables, *J. Am. Ceram. Soc.*, 89, 3019, 2006.

33. Raj, P.M., Odulena, A., and Cannon, W.R., Anisotropic shrinkage during sintering of particle-oriented systems – numerical simulations and experimental studies, *Acta Mater.*, 50, 2559, 2002.

34. Tikare, V., Braginsky, M., Olevsky, E., and Johnson, D.L., Numerical simulation of anisotropic shrinkage in a 2D compact of elongated particles, *J. Am. Ceram. Soc.*, 88, 59, 2005.

35. Stedman, S.J., Evans, J.R.G., Brook, R.J., and Hoffmann, M.J., Anisotropic sintering shrinkage in injection-molded composite ceramics, *J. Europ. Ceram. Soc.*, 11, 523, 1993.

36. Hoffmann, M.J., Nagel, A, Greil, P., and Petzow, G., Slip casting of SiC-whisker-reinforced Si_3N_4, *J. Am. Ceram. Soc.*, 72, 765, 1989.

37. Roeder, R.K., Trumble, K.P., and Bowman, K.J., Microstructure development in Al_2O_3 platelet-reinforced Ce-ZrO_2/Al_2O_3 composites, *J. Am. Ceram. Soc.*, 80, 27, 1977.

38. Ozer, I.O., Suvaci, E., Karademir, B., Missiaen, J.M., Carry, C.P., and Bouvard, D., Anisotropic sintering shrinkage in alumina ceramics containing oriented platelets, *J. Am. Ceram. Soc.*, 89, 1972, 2006.

39. Ling, H.C., and Yan, M.F., Second phase development in Sr-doped TiO_2, *J. Mater. Sci.*, 18, 2688, 1983.

40. Woolfrey, J.L., and Bannister, M.J., Nonisothermal techniques for studying initial-stage sintering, *J. Am. Ceram. Soc.*, 55, 390, 1972.

41. Bannister, M.J., Shape sensitivity of initial sintering equations, *J. Am. Ceram. Soc.*, 51, 548, 1968.

42. Cutler, I.B., Sintering of glass powders during constant rates of heating, *J. Am. Ceram. Soc.*, 52, 14, 1969.

43. Brinker, C.J., Scherer, G.W., and Roth, E.P., Sol → Gel → Glass: II. Physical and structural evolution during constant heating rate experiments, *J. Non-Cryst. Solids*, 72, 345, 1985.

44. Brinker, C.J., and Scherer, G.W., *Sol–Gel Science*, Academic Press, New York, 1990, Chap. 11.
45. Young, W.S., and Cutler, I.B., Initial sintering with constant rates of heating, *J. Am. Ceram. Soc.*, 53, 659, 1970.
46. Lange, F.F., Approach to reliable powder processing, *Ceramic Trans.*, 1, 1069, 1989.
47. Chu, M.-Y., Rahaman, M.N., De Jonghe, L.C., and Brook, R.J., Effect of heating rate on sintering and coarsening, *J. Am. Ceram. Soc.*, 74, 1217, 1991.
48. Su, H., and Johnson, D.L., Master sintering curve: a practical approach to sintering, *J. Am. Ceram. Soc.*, 79, 3211, 1996.
49. Duncan, J.H., and Bugden, W.G., Two-peak firing of β″alumina, *Proc. Brit. Ceram. Soc.*, 31, 221, 1981.
50. Chen, I.W., and Wang, X.H., Sintering dense nanocrystalline ceramics without final-stage grain growth, *Nature*, 404, 168, 2000.
51. Wang, X.H., Chen, P.L., and Chen, I.W., Two step sintering of ceramics with constant grain size, I. Y_2O_3, *J. Am. Ceram. Soc.*, 89, 431, 2006.
52. Wang, X.H., Deng, X.Y., Bai, H.L., Zhou, H., Qu, W.G., Li, L.T., and Chen, I.W., Two step sintering of ceramics with constant grain size, II. $BaTiO_3$ and Ni–Cu–Zn ferrite, *J. Am. Ceram. Soc.*, 89, 438, 2006.
53. Chu, M.-Y., De Jonghe, L.C., Lin, M.K.F., and Lin, F.J.T., Pre-coarsening to improve microstructure and sintering of powder compacts, *J. Am. Ceram. Soc.*, 74, 2902, 1991.
54. Lin, F.J.T., De Jonghe, L.C., and Rahaman, M.N., Microstructure refinement of sintered alumina by a two-step sintering technique, *J. Am. Ceram. Soc.*, 80, 2269, 1997.
55. Lin, F.J.T., De Jonghe, L.C., and Rahaman, M.N., Initial coarsening and microstructural evolution of fast-fired and MgO-doped alumina, *J. Am. Ceram. Soc.*, 80, 2891, 1997.
56. Brook, R.J., Fabrication principles for the production of ceramics with superior mechanical properties, *Proc. Br. Ceram. Soc.*, 32, 7, 1982.
57. Harmer, M.P., and Brook, R.J., Fast firing — microstructural benefits, *J. Br. Ceram. Soc.*, 80, 147, 1981.
58. Mostaghaci, H., and Brook, R.J., Production of dense and fine grain size $BaTiO_3$ by fast firing, *Trans. Br. Ceram. Soc.*, 82, 167, 1983.
59. Huckabee, M.L., Hare, T.M., and Palmour, H., III, Rate-controlled sintering as a processing method, in *Processing of Crystalline Ceramics*, Palmour, H., III, Davis, R.F., and Hare, R.T., Eds., Plenum Press, New York, 1978, 205.
60. Huckabee, M.L., Paisley, M.J., Russell, R.L., and Palmour, H., III, RCS — taking the mystery out of densification profiles, *J. Am. Ceram. Soc.*, 73, 82, 1994.
61. Agarwal, G., Speyer, R.F., and Hackenberger, W.S., Microstructural development of ZnO using a rate-controlled sintering dilatometer, *J. Mat. Res.*, 11, 671, 1996.
62. Ragulya, A.V., Rate-controlled synthesis and sintering of nanocrystalline barium titanate powder, *Nanostruct. Mater.*, 10, 349. 1998.
63. Coble, R.L., Sintering of alumina: effect of atmosphere, *J. Am. Ceram. Soc.*, 45, 123, 1962.
64. Bennison, S.J., and Harmer, M.P., Swelling of hot-pressed alumina, *J. Am. Ceram. Soc.*, 68, 591, 1985.
65. Readey, D.W., Vapor transport and sintering, *Ceramic Trans.*, 7, 86, 1990.
66. Quadir, T., and Readey, D.W., Microstructure development of zinc oxide in hydrogen, *J. Am. Ceram. Soc.*, 72, 297, 1989.
67. Lee, J., and Readey, D.W., Microstructure development in Fe_2O_3 in HCl vapor, in *Materials Science Research, Vol. 16, Sintering and Heterogeneous Catalysis*, Kuczynski, G.C., Miller, A.E., and Sargent, G.A., Eds., Plenum Press, New York, 1984, 145.
68. Readey, M.J., and Readey, D.W., Sintering TiO_2 in HCl atmospheres, *J. Am. Ceram. Soc.*, 70, C358, 1987.
69. Thompson, A.M., and Harmer, M.P., Influence of atmosphere on the final-stage sintering kinetics of ultra-high-purity alumina, *J. Am. Ceram. Soc.*, 76, 2248, 1993.
70. Wong, B., and Pask, J.A., Experimental analysis of sintering of MgO powder compacts, *J. Am. Ceram. Soc.*, 62, 141, 1979.
71. Anderson, P.J., and Morgan, P.L., Effects of water vapor on sintering of MgO, *Trans. Faraday Soc.*, 60, 930, 1964.
72. Northrop, D.A., Vaporization of lead zirconate–lead titanate materials, *J. Am. Ceram. Soc.*, 50, 441, 1967.

73. Northrop, D.A., Vaporization of lead zirconate–lead titanate materials: II, hot pressed compositions near theoretical density, *J. Am. Ceram. Soc.*, 51, 357, 1968.

74. Snow, G.S., Improvements in atmosphere sintering of transparent PLZT ceramics, *J. Am. Ceram. Soc.*, 56, 479, 1973.

75. Mitomo, M., Pressure sintering of Si_3N_4, *J. Mater. Sci.*, 11, 1103, 1976.

76. Ownby, P.D., and Jungquist, G.E., Final sintering of Cr_2O_3, *J. Am. Ceram. Soc.*, 55, 433, 1972.

77. Ownby, P.D., Oxidation state control of volatile species in sintering, *Mater. Sci. Res.*, 6, 431, 1973.

78. Anderson, H.U., Influence of oxygen activity on the sintering of $MgCr_2O_4$, *J. Amer. Ceram. Soc.*, 57, 34, 1974.

79. Anderson, H.U., Fabrication and property control of $LaCrO_3$-based oxides, *Mater. Sci. Res.*, 11, 469, 1978.

80. Callister, W.D., Johnson, M.L., Cutler, I.B., and Ure, R.W., Jr., Sintering chromium oxide with the aid of TiO_2, *J. Am. Ceram. Soc.*, 62, 208, 1979.

81. Reynolds, T., III, Firing, in *Treatise on Materials Science and Technology*, Vol. 9, Wang, F.F.Y., Ed., Academic Press, New York, 1976, 199.

82. Reijnen, P., Nonstoichiometry and sintering of ionic solids, in *Reactivity of Solids,* Mitchell, J.W., De Vries, R.C., Roberts, R.W., and Cannon, P., Eds., Wiley, New York, 1969, 99.

83. Yan, M.F., Grain growth in Fe_3O_4, *J. Amer. Ceram. Soc.*, 63, 443, 1980.

84. Gaskell, D.R., *Introduction to Metallurgical Thermodynamics*, 2nd ed., McGraw-Hill, New York, 1981.

85. Sutton, W.H., Microwave processing of ceramic materials, *Am. Ceram. Soc. Bulletin*, 68, 376, 1989.

86. Fang, Y., Agrawal, D.K., Roy, D., and Roy, R., Microwave sintering of hydroxyapatite ceramics, *J. Mater. Res.*, 9, 180, 1994.

87. Tian, Y.L., and Johnson, D.L., Ultrafine microstructure of Al_2O_3 produced by microwave sintering, *Ceramic Trans.*, 1, 925, 1988.

88. Birnboim, A., Gershon, D., Calame, J., Birman, A., Carmel, Y., Rodgers, J., Levush, B., Bykov, Y.V., Eremeev, A.G., Holoptsev, V.V., Semenov, V.E., Dadon, D., Martin, P.L., Rosen, M., and Hutcheon, R., Comparative study of microwave sintering of zinc oxide at 2.45, 30, and 83 GHz, *J. Amer. Ceram. Soc.*, 81, 1493, 1998.

89. Cheng, J., Agrawal, D., Roy, R., and Jayan, P.S., Continuous microwave sintering of alumina abrasive grits, *J. Mater. Processing Technol.*, 108, 26, 2000.

90. Birnboim, A., and Carmel, Y., Simulation of microwave sintering of ceramic bodies with complex geometry, *J. Am. Ceram. Soc.*, 82, 3024, 1999.

91. Pert, E., Carmel, Y., Birnboim, A., Olorunyolemi, T., Gershon, D., Calame, J., Lloyd, I.K., and Wilson, O.C., Jr., Temperature measurements during microwave processing: the significance of thermocouple effects, *J. Am. Ceram. Soc.*, 84, 1981, 2001.

92. Janney, M.A., and Kimrey, H.D., Diffusion-controlled processes in microwave-fired oxide ceramics, *Mater. Res. Soc. Symp. Proc.,* 189, 215, 1991.

93. Janney, M.A., Kimrey, H.D., Schmidt, M.A., and Kiggans, J.O., Grain growth in microwave-annealed alumina, *J. Am. Ceram. Soc.*, 74, 1675, 1991.

94. Johnson, D.L., and Rizzo, R.A., Plasma sintering of β″alumina, *Am. Ceram. Soc. Bulletin*, 59, 467, 1980.

95. Kim, J.S., and Johnson, D.L., Plasma sintering of alumina, *Am. Ceram. Soc. Bulletin*, 62, 620, 1983.

96. Kemer, E.L., and Johnson, D.L., Microwave plasma sintering of alumina, *Am. Ceram. Soc. Bulletin*, 64, 1132, 1985.

97. Su, H., and Johnson, D.L., Sintering alumina in microwave-induced oxygen plasma, *J. Am. Ceram. Soc.*, 79, 3199, 1996.

98. Henrichsen, M., Hwang, J., Dravid, V.P., and Johnson, D.L., Ultra-rapid phase conversion in beta″-alumina tubes, *J. Am. Ceram. Soc.*, 83, 2861, 2000.

99. Gao, L., Shen, Z., Miyamoto, H., and Nygren, M., Superfast densification of oxide/oxide ceramic composites, *J. Am. Ceram. Soc.*, 82, 1061, 1999.

100. Takeuchi, T., Tabuchi, M., and Kageyama, H., Preparation of dense $BaTiO_3$ ceramics with submicrometer grains by spark plasma sintering, *J. Am. Ceram.*, 82, 939, 1999.

101. Gutmanas, E., and Gotman, I., Dense high-temperature ceramics by thermal explosion under pressure, *J. Europ. Ceram. Soc.*, 19, 2381, 1999.

102. Vasilos, T., and Spriggs, R.M., Pressure sintering of ceramics, in *Progress in Ceramic Science*, Vol. 4, Burke, J.E., Ed., Pergamon Press, New York, 1966, 97.

103. Lange, F.F., and Terwilliger, G.R., The powder vehicle hot-pressing technique, *Bull. Am. Ceram. Soc.*, 52, 563, 1973.

104. Rigby, F., Development of hot pressing techniques at Springfields Nuclear Laboratories, *Proc. Brit. Ceram. Soc.*, 31, 249, 1981.

105. Wilkinson, D.S., and Ashby, M.F., The development of pressure sintering maps, *Mater. Sci. Res.*, 10, 473, 1975.

106. Kimura, T., Yoshimoto, T., Iida, N., Fujita, Y., and Yamaguchi, T., Mechanism of grain orientation during hot pressing of bismuth titanate, *J. Am. Ceram. Soc.*, 82, 85, 1989.

107. Rahaman, M.N., Riley, F.L., and Brook, R.J., Mechanism of densification during reaction hot-pressing in the system Si-Al-O-N, *J. Am. Ceram. Soc.*, 63, 648, 1980.

108. Zhang, G.J., Deng, Z.Y., Kondo, N., Yang, J.F., and Ohji, T., Reactive hot pressing of ZrB_2-SiC composites, *J. Amer. Ceram. Soc.*, 83, 2330, 2000.

109. Wen, G., Li, S.B., Zhang, B.S., and Guo, Z.X., Processing of in situ toughened B-W-C composites by reaction hot pressing of B_4C and WC, *Scripta Mater.*, 43, 853, 2000.

110. Yamasaki, N., Yanagisawa, K., Nishioka, M., and Kanahara, S., A hydrothermal hot-pressing method: apparatus and applications, *J. Mater. Sci. Lett.* 5: 355, 1986.

111. Hosoi, K., Hashida, T., Takahashi, H., Yamasaki, N., and Korenaga, T., New processing technique for hydroxyapatite ceramics by the hydrothermal hot-pressing method, *J. Amer. Ceram. Soc.*, 79, 2271, 1996.

112. Yanagisawa, K., Ioku, K., and Yamasaki, N., Formation of anatase porous ceramics by hydrothermal hot-pressing of amorphous titania spheres, *J. Amer. Ceram. Soc.*, 80, 1303, 1997.

113. He, Y.J., Winnubst, A.J.A., Verweij, A., and Burggraaf, A.J., Sinter-forging of zirconia toughened alumina, *J. Mater. Sci.*, 29, 6505, 1994.

114. Panda, P.C., Wang, J., and Raj, R., Sinter forging characteristics of fine-grained zirconia, *J. Am. Ceram. Soc.*, 71, C507, 1988.

115. Venkatachari, K.R., and Raj, R., Enhancement of strength through sinter forging, *J. Am. Ceram. Soc.*, 70, 514, 1987.

116. Takenaka, T., and Sakata, K., Grain orientation and electrical properties of hot-forged $Bi_4Ti_3O_{12}$, *Jpn. J. Appl. Phys.*, 19, 31, 1980.

117. Knickerbocker, J.U., and Payne, D.A., Orientation of ceramic microstructures by hot-forming methods, *Ferroelectrics*, 37, 733, 1981.

118. Patwardhan, J.S., and Rahaman, M.N., Compositional effects on densification and microstructural evolution of bismuth titanate, *J. Mater. Sci.*, 39, 133, 2004.

119. Kondo, N., Suzuki, Y., and Ohji, T., Superplastic sinter-forging of silicon nitride with anisotropic microstructure formation, *J. Amer. Ceram. Soc.*, 82, 1067, 1999.

120. Wills, R.R., Brockway, M.C., and McCoy, L.G., Hot isostatic pressing of ceramics, *Mater. Sci. Res.*, 17, 559, 1984.

121. Larker, H.T., Dense ceramic parts hot pressed to shape by HIP, *Mater. Sci. Res.*, 17, 571, 1984.

122. Li, W.B., Ashby, M.F., and Easterling, K.E., On densification and shape change during hot isostatic pressing, *Acta Metall.*, 35, 2831, 1987.

Appendix A

Physical Constants

Velocity of light, c	2.998×10^8 m s^{-1}
Permittivity of vacuum, ε_0	8.854×10^{-12} F m^{-1}
Permeability of vacuum, $\mu_0 = 1/\varepsilon_0 c^2$	1.257×10^{-6} H m^{-1}
Elementary charge, e	1.602×10^{-19} C
Planck constant, h	6.626×10^{-34} J s
Avogadro number, N_A	6.022×10^{23} mol^{-1}
Atomic mass unit, $m_u = 10^{-3}/N_A$	1.661×10^{-27} kg
Mass of electron, m_e	9.110×10^{-31} kg
Mass of proton, m_p	1.673×10^{-27} kg
Mass of neutron, m_n	1.675×10^{-27} kg
Faraday constant, $F = N_A e$	9.649×10^4 C mol^{-1}
Rydberg constant, R_∞	1.097×10^7 m^{-1}
Bohr magneton, μ_B	9.274×10^{-24} J T^{-1}
Gas constant, R	8.314 J K^{-1} mol^{-1}
Boltzmann constant, $k = R/N_A$	1.381×10^{-23} J K^{-1}
Gravitational constant, G	6.67×10^{-11} N m^2 kg^{-2}
Stefan-Boltzmann constant, σ	5.670×10^{-8} W m^{-2} K^{-4}
Standard volume of ideal gas, V_o	22.414×10^{-3} m^3 mol^{-1}
Acceleration due to gravity, g	9.780 m s^{-2}
(at sea level and zero degree latitude)	

Appendix B

SI Units—Names and Symbols

Quantity	Unit	Symbol	Relation to Other Units
Base units			
Length	meter	m	
Mass	kilogram	kg	
Time	second	s	
Electric current	ampere	A	
Temperature	kelvin	K	
Amount of substance	mole	mol	
Luminous intensity	candela	cd	
Supplementary units			
Plane angle	radian	rad	
Solid angle	steradian	sr	
Derived units with special names			
Frequency	hertz	Hz	s^{-1}
Temperature	degree Celsius	°C	$T(°C) = T(K) - 273.2$
Force	newton	N	$kg\ m\ s^{-2}$
Pressure and stress	pascal	Pa	$N\ m^{-2}$
Energy	joule	J	$N\ m$
Power	watt	W	$J\ s^{-1}$
Electric charge	coulomb	C	$A\ s$
Electric potential	volt	V	$J\ C^{-1}$
Electric resistance	ohm	Ω	$V\ A^{-1}$
Electric capacitance	farad	F	$C\ V^{-1}$
Magnetic flux	weber	Wb	$V\ s$
Magnetic flux density	tesla	T	$Wb\ m^{-2}$
Inductance	henry	H	$V\ s\ A^{-1}$
Luminous flux	lumen	lm	$cd\ sr$

Appendix C

Conversion of Units

Length	1 micron (μm) = 10^{-6} m
	1 ångström (Å) = 10^{-10} m
	1 inch (in.) = 25.4 mm
Mass	1 pound (lb) = 0.454 kg
Volume	1 liter (l) = 10^{-3} m^3
Density	1 gm cm^{-3} = 10^3 kg m^{-3}
Force	1 dyne = 10^{-5} N
Angle	$1° = 0.01745$ rad
Pressure and stress	1 lb in.$^{-2}$ (psi) = 6.89×10^3 Pa
	1 bar = 10^5 Pa
	1 atmosphere (atm) = 1.013×10^5 Pa
	1 torr = 1 mm Hg = 133.32 Pa
Energy	1 erg = 10^{-7} J
	1 calorie = 4.1868 J
	1 electron volt (eV) = 1.6022×10^{-19} J
Viscosity	
dynamic	1 poise = 0.1 Pa s
kinematic	1 stoke = 10^{-4} m^2 s^{-1}

Decimal Fractions and Multiples					
Fraction	Prefix	Symbol	Multiple	Prefix	Symbol
10^{-3}	milli	m	10^3	kilo	k
10^{-6}	micro	μ	10^6	mega	M
10^{-9}	nano	n	10^9	giga	G
10^{-12}	pico	p	10^{12}	tera	T
10^{-15}	femto	f	10^{15}	peta	P
10^{-18}	atto	a	10^{18}	exa	E

Appendix D

Ionic Crystal Radii (in units of 10^{-10} m)

Coordination Number = 6

Ag+	Al³⁺	As⁵⁺	Au+	B³⁺	Ba²⁺	Be²⁺	Bi³⁺	Bi⁵⁺	Br⁻	C⁴⁺	Ca²⁺
1.15	0.54	0.46	1.37	0.27	1.35	0.45	1.03	0.76	1.96	0.16	1.00

Cd²⁺	Ce⁴⁺	Cl⁻	Co²⁺	Co³⁺	Cr²⁺	Cr³⁺	Cr⁴⁺	Cs+	Cu+	Cu²⁺	Cu³⁺
0.95	0.87	1.81	0.75	0.55	0.80	0.62	0.55	1.67	0.77	0.73	0.54

Dy³⁺	Er³⁺	Eu³⁺	F⁻	Fe²⁺	Fe³⁺	Ga³⁺	Gd³⁺	Ge⁴⁺	Hf⁴⁺	Hg²⁺	Ho³⁺
0.91	0.89	0.95	1.33	0.78	0.65	0.62	0.94	0.53	0.71	1.02	0.90

I⁻	In³⁺	K+	La³⁺	Li+	Mg²⁺	Mn²⁺	Mn⁴⁺	Mo³⁺	Mo⁴⁺	Mo⁶⁺	N⁵⁺
2.20	0.80	1.38	1.03	0.76	0.72	0.83	0.53	0.69	0.65	0.59	0.13

Na+	Nb⁵⁺	Nd³⁺	Ni²⁺	Ni³⁺	O²⁻	OH⁻	P⁵⁺	Pb²⁺	Pb⁴⁺	Rb+	Ru⁴⁺
1.02	0.64	0.98	0.69	0.56	1.40	1.37	0.38	1.19	0.78	1.52	0.62

S²⁻	S⁶⁺	Sb³⁺	Sb⁵⁺	Sc³⁺	Se²⁻	Se⁶⁺	Si⁴⁺	Sm³⁺	Sn⁴⁺	Sr²⁺	Ta⁵⁺
1.84	0.29	0.76	0.60	0.75	1.98	0.42	0.40	0.96	0.69	1.18	0.64

Te²⁻	Te⁶⁺	Th⁴⁺	Ti²⁺	Ti³⁺	Ti⁴⁺	Tl+	Tl³⁺	U⁴⁺	U⁵⁺	U⁶⁺	V²⁺
2.21	0.56	0.94	0.86	0.67	0.61	1.50	0.89	0.89	0.77	0.73	0.79

V⁵⁺	W⁴⁺	W⁶⁺	Y³⁺	Yb³⁺	Zn²⁺	Zr⁴⁺
0.54	0.66	0.60	0.90	0.87	0.74	0.72

Coordination Number = 4

Ag+	Al³⁺	As⁵⁺	B³⁺	Be²⁺	C⁴⁺	Cd²⁺	Co²⁺	Cr⁴⁺	Cu+	Cu²⁺	F⁻
1.00	0.39	0.34	0.11	0.27	0.15	0.78	0.58	0.41	0.60	0.57	1.31

Fe²⁺	Fe³⁺	Ga³⁺	Ge⁴⁺	Hg²⁺	In³⁺	Li+	Mg²⁺	Mn²⁺	Mn⁴⁺	Na+	Nb⁵⁺
0.63	0.49	0.47	0.39	0.96	0.62	0.59	0.57	0.66	0.39	0.99	0.48

Ni²⁺	O²⁻	P⁵⁺	Pb²⁺	S⁶⁺	Se⁶⁺	Sn⁴⁺	Si⁴⁺	Ti⁴⁺	V⁵⁺	W⁶⁺	Zn²⁺
0.55	1.38	0.17	0.98	0.12	0.28	0.55	0.26	0.42	0.36	0.42	0.60

Coordination Number = 8

Bi³⁺	Ce⁴⁺	Ca²⁺	Ba²⁺	Dy³⁺	Gd³⁺	Hf⁴⁺	Ho³⁺	In³⁺	Na+	Nd³⁺	O²⁻
1.17	0.97	1.12	1.42	1.03	1.05	0.83	1.02	0.92	1.18	1.11	1.42

Pb²⁺	Rb+	Sr²⁺	Th⁴⁺	U⁴⁺	Y³⁺	Zr⁴⁺
1.29	1.61	1.26	1.05	1.00	1.02	0.84

Coordination Number = 12

Ba²⁺	Ca²⁺	La³⁺	Pb²⁺	Sr²⁺
1.61	1.34	1.36	1.49	1.44

Source: R.D. Shannon, *Acta Crystallographica*, A32, 751, 1976.

Appendix E

Density and Melting Point of Some Elements and Ceramics

Chemical Formula	Density (g/cm³)	Melting Point (°C)	Common Names
Ag	10.50	961	Silver
Al	2.70	660	Aluminum
AlN	3.255	2000^d	Aluminum nitride
Al_2O_3	3.986	2053	Alumina; corundum
$3Al_2O_3 \cdot 2SiO_2$	3.16	1850	Mullite
$Al_2O_3 \cdot TiO_2$	3.70	1200^d	Aluminum titanate
Au	19.30	1063	Gold
B_4C	2.50	2350	Boron carbide
BN	2.270	2500^s	Boron nitride (hexagonal)
	3.470	—	Boron nitride (cubic)
B_2O_3	2.55	450^d	Boric oxide
BaO	5.72	1972	Barium oxide (cubic)
$BaTiO_3$	6.02	1625	Barium titanate
BeO	3.01	2577	Beryllium oxide; bromelite
C	3.513	4400	Diamond
C	2.2	4480	Graphite
CaF_2	3.18	1418	Calcium fluoride; fluorite
CaO	3.34	2898	Calcium oxide; lime
$CaO \cdot Al_2O_3$	2.98	1605	Calcium aluminate
$CaO \cdot 2Al_2O_3$	2.91	1720	Calcium dialuminate
$CaO \cdot 6Al_2O_3$	3.69	1840	Calcium hexaluminate
$CaO \cdot Al_2O_3 \cdot 2SiO_2$	2.76	1551	Anorthite; lime feldspar
$CaO \cdot MgO \cdot 2SiO_2$	3.30	1391	Diopside
$CaO \cdot SiO_2$	2.92	1540	Calcium metasilicate; wollastonite
$CaO \cdot TiO_2$	3.98	1975	Calcium metatitanate; perovskite
CeO_2	7.65	2400	Cerium (IV) oxide
Ce_2O_3	6.20	2210	Cerium (III) oxide
Cr_2O_3	5.22	2435	Chromium oxide
Cu	8.96	1085	Copper
CuO	6.30–6.45	1326	Cupric oxide; tenorite
Cu_2O	6.00	1235	Cuprous oxide; cuprite
Fe	7.87	1536	Iron
FeO	5.70	1370	Ferrous oxide; wuestite
Fe_2O_3	5.25	1565	Ferric oxide; hematite
Fe_3O_4	5.17	1597	Ferroso-ferric oxide; magnetite
$FeTiO_3$	4.72	~1470	Ilmenite

Chemical Formula	Density (g/cm³)	Melting Point (°C)	Common Names
HfB$_2$	11.20	3250	Hafnium diboride
HfC	12.67	3890	Hafnium carbide
HfO$_2$	9.68	2845	Hafnia
K$_2$O	2.35	350^d	Potash
K$_2$O·Al$_2$O$_3$·4SiO$_2$	2.49	1686	Leucite
LaCrO$_3$	6.70	—	Lanthanum chromite
LaMnO$_3$	5.70	—	Lanthanum manganite
La$_2$O$_3$	6.51	2304	Lanthanum oxide; lanthana
LiF	2.64	848	Lithium fluoride
Li$_2$O	2.013	1570	Lithium oxide; lithia
MgO	3.581	2850	Magnesia; periclase
MgO·Al$_2$O$_3$	3.55	2135	Magnesium aluminate; spinel
MgO·SiO$_2$	3.19	1550^d	Magnesium metasilicate; clinostatite
2MgO·SiO$_2$	3.21	1900	Magnesium orthosilicate; forsterite
Mo	10.20	2617	Molybdenum
MoSi$_2$	6.24	2030	Molybdenum disilicide
NaCl	2.17	800	Sodium chloride
Na$_2$O	2.27	1132^d	Soda
Na$_2$O·Al$_2$O$_3$·2SiO$_2$	2.61	1526	Nephelite
Na$_2$O·Al$_2$O$_3$·4SiO$_2$	3.34	1000	Jadeite
Na$_2$O·Al$_2$O$_3$·6SiO$_2$	2.63	1100	Albite; soda feldspar
Na$_2$O·SiO$_2$	2.40	1088	Sodium metasilicate
Nb	8.57	2473	Niobium
Nb$_2$O$_5$	4.55	1500	Niobia
Ni	8.90	1453	Nickel
NiO	6.72	1955	Nickel oxide; bunsenite
PbO	9.35	—	Litharge (transforms to massicot)
	9.64	897	Massicot
PbO$_2$	9.64	300^d	Lead dioxide; plattnerite
Pb$_3$O$_4$	8.92	830	Red lead
PbS	7.60	1114	Lead sulfide; galena
PbTiO$_3$	7.9	1290	Lead titanate
Pt	21.45	1769	Platinum
Si	2.329	1414	Silicon
SiC	3.217	2700	Silicon carbide
Si$_3$N$_4$	3.184	1900^d	Silicon nitride
SiO$_2$	2.648	—	α-Quartz (transforms to β-quartz)
	2.533	—	β-Quartz (transforms to tridymite)
	2.265	—	Tridymite (transforms to cristobalite)
	2.334	1722	Cristobalite
	2.196	—	Silica (vitreous)
SrTiO$_3$	5.10	2080	Strontium titanate
Ta	16.65	2996	Tantalum
TaB$_2$	12.60	3100	Tantalum diboride
TaC	14.50	3880	Tantalum carbide
ThO$_2$	10.00	3250	Thoria
Ti	4.50	1668	Titanium
TiB$_2$	4.52	3200	Titanium diboride
TiC	4.91	3000	Titanium carbide
TiN	5.44	2950	Titanium nitride
TiO$_2$	4.23	1840	Titania; rutile
	3.89	—	Anatase (transforms to rutile)
	4.14	—	Brookite (transforms to rutile)

Chemical Formula	Density (g/cm³)	Melting Point (°C)	Common Names
UO_2	10.97	2825	Uranium dioxide; urania
W	19.3	3410	Tungsten
WC	15.6	2785	Tungsten carbide
WSi_2	9.87	2165	Tungsten disilicide
$Y2O_3$	5.03	2438	Yttrium oxide; yttria
$3Y_2O_3 \cdot 5Al_2O_3$	4.55	1970	Yttrium aluminum garnet (YAG)
Zn	7.13	420	Zinc
ZnO	5.606	1975	Zinc oxide; zincite
$ZnO \cdot Al_2O_3$	4.50	—	Zinc aluminate
ZnS	4.09	1700	Zinc sulfide; wurtzite
	4.04	—	Sphalerite
ZrB_2	6.09	3245	Zirconium diboride
ZrC	6.73	3420	Zirconium carbide
ZrO_2	5.58	—	Zirconia (monoclinic)
	6.10	—	Zirconia (tertragonal)
	5.68–5.91	2680	Zirconia (cubic)
$ZrO_2 \cdot SiO_2$	4.65	2250	Zirconium orthosilicate; zircon

[d]decomposition; [s]sublimation

Index

A

Abnormal grain growth 105, 107, 127–133
Activated sintering 177, 220–221
Activation energy 22, 24
Activity 205
Additives
  grain growth control 145
  liquid-phase sintering 178
Agglomerate 234
Aluminum oxide
  activation energy 333
  defect reaction 9, 273–274
  dihedral angle 274
  fast firing 333
  grain boundary film 193
  grain growth
    solute drag 280
    Zener pinning 140
  hot isostatic pressing map 89
  hot pressing map 87
  liquid-phase sintering 178
  microwave sintering 353–355
  MgO-doped 272, 273
  single crystal conversion 131
  sintering
    Al_2O_3/ZrO_2 composite 255–256
    atmosphere 334
    effect of particle packing 234
    grain size/density trajectory 167, 169
    microstructural map 167
    role of MgO 280–282
    thin films 260–262
  solute segregation 276, 281
  TiO_2-doped 274
  two-step sintering 236
Ambipolar diffusion 37–41
Anisotropic grain growth 107, 131
Anisotropic shrinkage 315–320
Arrhenius relation 121
Atomic flux 36–37
Avrami equation 291

B

Backstress 232, 246
Barium titanate
  fast firing 333
  single crystal conversion 131
Bjerrum length 194
Boltzmann equation 9

Boundary (*see* Grain boundary)
Brouwer diagram 16–18

C

Capillary stress 179, 187–191
Chemical potential 1, 29–34, 111, 142
  of atoms and vacancies 31–34
  of mixture of gases 30
  of pure substance 30
Coalescence
  grain 202, 208
  pore 145
Coarsening (*see also* Ostwald ripening)
  definition 106
  in liquid-phase sintering 209
  in porous compacts 45, 105
Coated particles 76, 256
Coatings (*see* Films)
Coble creep 83
Composite particles (*see* Coated particles)
Composite sphere model 244–245
Computer simulation (*see* Numerical modeling)
Constant heating rate sintering 2, 323
Constitutive equation
  incompressible viscous solid 248
  linear elastic solid 247
  porous viscous solid 248
  porous sintering material 265–266
Constrained sintering
  adherent films 258-264
  effect of rigid inclusions 240–257
  composite sphere model 244–245
  multilayers 264–265
Contact angle 182
Conventional sintering (*see* Sintering)
Correlation factor 28
Cosintering (cofiring) 264
Creep
  Coble mechanism 83
  equation for porous solids 100
  equation for dense solids 82, 83
  mechanisms 81–84
  Nabarro-Herring mechanism 82
  viscosity 95
Crystallization
  effect on sintering 289–297
  kinetics 291–293
Curvature
  effect on chemical potential 32–34
  effect on sintering stress 92
  effect on solubility 187

T - #0071 - 160425 - C0 - 254/178/22 [24] - CB - 9780849372865 - Gloss Lamination